Emerging Transportation Safety and Operations: Practical Perspectives

Emerging Transportation Safety and Operations: Practical Perspectives

Guest Editors

Deogratias Eustace
Bhaven Naik
Heng Wei
Parth Bhavsar

Basel • Beijing • Wuhan • Barcelona • Belgrade • Novi Sad • Cluj • Manchester

Guest Editors

Deogratias Eustace	Bhaven Naik	Heng Wei
Civil and Environmental	Department of Civil and	Department of Civil &
Engineering and Engineering	Environmental Engineering	Architectural Engineering &
Mechanics	Ohio University	Construction Management
University of Dayton	Athens	University of Cincinnati
Dayton	United States	Cincinnati
United States		United States

Parth Bhavsar
Department of Civil and
Environmental Engineering
Kennesaw State University
Marietta
United States

Editorial Office
MDPI AG
Grosspeteranlage 5
4052 Basel, Switzerland

This is a reprint of the Special Issue, published open access by the journal *Vehicles* (ISSN 2624-8921), freely accessible at: www.mdpi.com/journal/vehicles/special_issues/VP5V2662T4.

For citation purposes, cite each article independently as indicated on the article page online and using the guide below:

Lastname, A.A.; Lastname, B.B. Article Title. *Journal Name* **Year**, *Volume Number*, Page Range.

ISBN 978-3-7258-3154-8 (Hbk)
ISBN 978-3-7258-3153-1 (PDF)
https://doi.org/10.3390/books978-3-7258-3153-1

© 2025 by the authors. Articles in this book are Open Access and distributed under the Creative Commons Attribution (CC BY) license. The book as a whole is distributed by MDPI under the terms and conditions of the Creative Commons Attribution-NonCommercial-NoDerivs (CC BY-NC-ND) license (https://creativecommons.org/licenses/by-nc-nd/4.0/).

Contents

About the Editors vii

Preface ix

Deogratias Eustace
Emerging Transportation Safety and Operations: Practical Perspectives
Reprinted from: *Vehicles* 2024, 6, 2251-2256, https://doi.org/10.3390/vehicles6040110 1

Ernst Tomasch, Heinz Hoschopf, Karin Ausserer and Jannik Rieß
Required Field of View of a Sensor for an Advanced Driving Assistance System to Prevent Heavy-Goods-Vehicle to Bicycle Accidents
Reprinted from: *Vehicles* 2024, 6, 1922-1941, https://doi.org/10.3390/vehicles6040094 7

Victor Xu and Sheng Xu
The CornerGuard: Seeing around Corners to Prevent Broadside Collisions
Reprinted from: *Vehicles* 2024, 6, 1468-1481, https://doi.org/10.3390/vehicles6030069 27

Farnaz Zahedieh and Chris Lee
Impacts of a Toll Information Sign and Toll Lane Configuration on Queue Length and Collision Risk at a Toll Plaza with a High Percentage of Heavy Vehicles
Reprinted from: *Vehicles* 2024, 6, 1249-1267, https://doi.org/10.3390/vehicles6030059 41

Fernando Viadero-Monasterio, Luciano Alonso-Rentería, Juan Pérez-Oria and Fernando Viadero-Rueda
Radar-Based Pedestrian and Vehicle Detection and Identification for Driving Assistance
Reprinted from: *Vehicles* 2024, 6, 1185-1199, https://doi.org/10.3390/vehicles6030056 60

Wei Sun, Lili Nurliynana Abdullah, Puteri Suhaiza Sulaiman and Fatimah Khalid
Meta-Feature-Based Traffic Accident Risk Prediction: A Novel Approach to Forecasting Severity and Incidence
Reprinted from: *Vehicles* 2024, 6, 728-746, https://doi.org/10.3390/vehicles6020034 75

Anna Granà, Salvatore Curto, Andrea Petralia and Tullio Giuffrè
Connected Automated and Human-Driven Vehicle Mixed Traffic in Urban Freeway Interchanges: Safety Analysis and Design Assumptions
Reprinted from: *Vehicles* 2024, 6, 693-710, https://doi.org/10.3390/vehicles6020032 94

Manuela Vieira, Manuel Augusto Vieira, Gonçalo Galvão, Paula Louro, Mário Véstias and Pedro Vieira
Enhancing Urban Intersection Efficiency: Utilizing Visible Light Communication and Learning-Driven Control for Improved Traffic Signal Performance
Reprinted from: *Vehicles* 2024, 6, 666-692, https://doi.org/10.3390/vehicles6020031 112

Konstantinos Gkyrtis and Alexandros Kokkalis
An Overview of the Efficiency of Roundabouts: Design Aspects and Contribution toward Safer Vehicle Movement
Reprinted from: *Vehicles* 2024, 6, 433-449, https://doi.org/10.3390/vehicles6010019 139

Raj Bridgelall
Driving Standardization in Infrastructure Monitoring: A Role for Connected Vehicles
Reprinted from: *Vehicles* 2023, 5, 1878-1891, https://doi.org/10.3390/vehicles5040101 156

Martin Skoglund, Fredrik Warg, Anders Thorsén and Mats Bergman
Enhancing Safety Assessment of Automated Driving Systems with Key Enabling Technology Assessment Templates
Reprinted from: *Vehicles* **2023**, *5*, 1818-1843, https://doi.org/10.3390/vehicles5040098 . . . . . . **170**

About the Editors

Deogratias Eustace

Dr. Deogratias Eustace is a distinguished transportation engineer currently serving as a Professor and Chair of the Department of Civil and Environmental Engineering and Engineering Mechanics at the University of Dayton in Dayton, Ohio, USA. His research interests span transportation safety; traffic operations; emerging mobility services; travel demand modeling; ITS applications; CAV/AV impacts on traffic safety; non-motorized transportation; and statistical applications in transportation engineering. Dr. Eustace is particularly passionate about addressing the challenges of road safety and integrating innovative technologies into transportation systems.

Currently, Dr. Eustace is working on a project that is analyzing freeway median barrier-related crash severity in Ohio. He has led several high-impact projects, including the evaluation of safety effectiveness and development of crash modification factors of median cable barriers (MCBs) in Ohio, a study that has generated major media interest throughout the nation. He has also developed several data-driven models for traffic crash prediction and crash severity to enhance the safety of various road users.

Throughout his career, Dr. Eustace has been deeply involved in advancing the scientific understanding of transportation safety, planning, and operations. His research findings have been disseminated through peer-reviewed journals, technical reports, and technical conferences.

Dr. Eustace remains committed to mentoring the next generation of engineers and researchers, fostering an interdisciplinary approach to solving complex transportation challenges. He is highly active in international transportation professional groups such as the Transportation Research Board (TRB), Road Safety and Simulation (RSS), etc.

Bhaven Naik

Dr. Bhaven Naik is an associate professor in the department of civil and environmental engineering at Ohio University. He is a certified professional engineer (Nebraska and Ohio), professional traffic operations engineer, and also a road safety professional. Over the past 18 years, Dr. Naik has been involved in high-impact transportation systems engineering research in the areas of highway safety and human factors, traffic simulation modeling, geometric design, traffic operations and signal timing optimization, intelligent transportation systems, and more recently, CV/AV technologies. He has expertise in statistical methods and has been involved with projects requiring rigorous statistical analysis. Additionally, Dr. Naik has experience operating/managing a private design–build firm that was responsible for fabricating and erecting large-scale steel buildings, residential homes, shopping facilities, etc.

Dr. Naik has experience with providing technical guidance to state agencies and graduate student research thesis' and dissertations related to work involving roundabouts, highway rail grade crossings, geometric design elements, etc. He is an effective communicator of research procedures, results, and applications to state departments of transportation, transportation research boards, and the general public. He has authored manuscripts, reports, proposals, and presented to clients in addition to overseeing technical analysis, scheduling, and data collection. His skills also include establishing rapport with staff at all levels, project management, rapid situation assessment, and strategic planning.

Heng Wei

Dr. Heng Wei is a Professor and Director of the Civil Engineering program and ART-EngineS Transportation Research Lab at the University of Cincinnati. Dr. Wei has extensive research expertise and industrial experience in cooperative intelligent transportation systems (C-ITS), connected and automated vehicle (CAV) impacts on transportation infrastructure and control system design and operation, artificial intelligence (AI)-based informatics and GIS in transportation, and smart mobility in travel demand and environmental analysis. He has secured more than 50 research grants/projects from Ohio DOT, the Federal Highway Administration (FHWA), the US Environmental Protection Agency, the National Science Foundation, USDOT University Transportation Centers, and the industrial sector. His research has resulted in over 240 peer-reviewed and refereed papers, and he has authored three books and four book chapters and edited 22 international conference proceedings. He is the lead editor and a key author for the guidance-type book published by ASCE, *Disruptive Emerging Transportation Technologies*. He has received 35 professional awards/honors for his outstanding research and education achievements, including best paper awards by TRB/UTCs/transportation journals. He has advised six Ph.D. and two M.S. students in winning a total of 21 top awards at national and regional student paper competitions.

Dr. Wei has been serving on NCHRP panels for the following projects: 1) NCHRP Project 08-189: Developing a Data Repository to Help with TSMO Strategies Evaluations (since 2024); 2) NCHRP Project 25-75: A Streamlined Risk-Based Approach for NEPA Air Quality Analyses, (since 2024); 3) NCHRP Project 03-152: Managing the Life Cycle of Software for Evolving Traffic Management Systems (since 2024); and 4) NCHRP Synthesis 20-05/Topic 54-06: Ancillary Asset Data Stewardship and Data Models (SN5406) (2022–2024). He also is the Chair of the ASCE T&DI Emerging Technologies Council (2022 – 2028).

Parth Bhavsar

Parth Bhavsar is an assistant professor in the Department of Civil and Environmental Engineering at Kennesaw State University. His research interests include intelligent transportation systems (ITSs), transportation data analytics, connected and automated vehicle technology (CVT), and alternative fuel vehicles (AFVs). Dr. Bhavsar received his Ph.D. in 2013 and his M.S. in 2006 from Clemson University, South Carolina. He also has experience in the private sector in developing transportation engineering and planning solutions, specifically traffic micro-simulation projects. He has published in peer-reviewed journals such as *Transportation Research Part C: Emerging Technology*, *Transportation Research Part D: Transport and the Environment*, and *Transportation Research Record* journal of the Transportation Research Board. Dr. Bhavsar has received funding (over USD 1.9 million as a PI) from various agencies such as the National Science Foundation, the New Jersey Department of Transportation (NJDOT), the New Jersey Department of Community Affairs, and the U.S. Department of Transportation.

Preface

Transportation systems are the backbone of modern society, facilitating economic growth, connectivity, and access to essential services. However, evolving challenges such as new technologies, changing mobility patterns, and environmental concerns necessitate innovative strategies for maintaining safety and efficiency. Emerging Transportation Safety and Operations: Practical Perspectives bridges the gap between theoretical advancements and real-world practices, offering actionable insights for professionals like policymakers, engineers, planners, researchers, and students.

The reprint focuses on the following four key themes:

- *Technological innovations* explores how technologies like autonomous vehicles, connected infrastructure, and data analytics transform safety and operations.

- *Human factors and behavior* investigates how human decision-making, driver behavior, and pedestrian interactions can be enhanced by technology to influence and improve transportation safety by preventing collisions.

- *Infrastructure innovations for safety and efficiency* examines the interplay between roadway design innovations such as innovative intersections and interchanges and traffic safety.

- *Equitable and safe infrastructure for all road users* emphasizes strategies to ensure the safety of vulnerable road users (VRUs) such as pedestrians and cyclists when sharing roads with motorists.

Each chapter combines theory with real-world case studies and best practices, providing a holistic approach to addressing contemporary transportation challenges. Contributions from diverse experts underscore the importance of adaptive and collaborative solutions in shaping resilient transportation systems. This book includes topics contributed by thirty-three authors from nine different countries. By tackling pressing issues and promoting innovative thinking, this book aims to foster dialogue and action that will help future transportation systems meet the demands of a rapidly changing world.

The guest editors would like to thank the editorial team of the *Vehicles* journal that successfully organized this Special Issue. Special thanks go to Ms. Maria Chen, the Section Managing Editor, who worked tirelessly to support the guest editors and who was available promptly all the time whenever was needed. The guest editors also wish to thank all authors whose valuable work was published in this issue and the reviewers for evaluating the manuscripts and providing helpful suggestions.

Deogratias Eustace, Bhaven Naik, Heng Wei, and Parth Bhavsar
Guest Editors

Editorial

Emerging Transportation Safety and Operations: Practical Perspectives

Deogratias Eustace

Department of Civil and Environmental Engineering and Engineering Mechanics, University of Dayton, Dayton, OH 45469, USA; deustace1@udayton.edu

1. Introduction

Improving transportation traffic safety and operations is a global priority, with efforts focusing on both technological advancements and strategic planning [1]. Generally, researchers and practitioners in the transportation field are continuously seeking and utilizing modern approaches and technologies that can be implemented to improve the safety and efficiency of transportation systems. Modern concepts focus on the application of new strategies, data-driven methods, and innovative technologies to address contemporary challenges in transportation safety and operations [2,3]. There are several key focus areas that can be highlighted here due to their popularity and worldwide usage.

Advanced Traffic Management Systems (ATMSs) is an area that has become a major focus worldwide [4–6]. These systems include connected and automated vehicles (CAVs), which involve the integration of connected and autonomous vehicle technology into traffic systems to reduce human error and improve traffic flow [7,8], as well as smart traffic signals, which involve adaptive traffic signals that adjust in real time based on traffic conditions to reduce congestion and improve safety [9]. These systems also include Intelligent Transportation Systems (ITSs), which include the use of technologies like cameras, sensors, and data analytics to monitor and manage traffic conditions [10,11].

Another important aspect that transportation professionals are leveraging to their advantage is data-driven decision making. This involves big data and analytics [11], leveraging data from various sources, such as GPS, mobile devices, and traffic sensors, to make informed decisions regarding traffic patterns, safety measures, and infrastructure improvements. This decision making also includes predictive analytics, which uses historical data to predict future traffic conditions, identify high-risk areas, and proactively implement safety measures [11,12].

Vision zero and the safe system approach are two concepts that are currently being advocated for and explored worldwide. Vision zero is a strategy aimed at eliminating all traffic fatalities and severe injuries, while promoting safe and healthy mobility for all [13,14]. It focuses on the design of roads, vehicles, and systems that prioritize human life and on minimizing the consequences of human error. The safe system approach emphasizes a holistic view of the transportation system, accounting for human vulnerabilities by creating multiple layers of protection (such as safer road designs, vehicle safety features, and responsible road user behavior) [15,16].

The use of emerging technologies in transportation safety is another practice that is increasingly becoming common, and it is expected to be very influential in the future. This Special issue mentions just a few of these technologies. First, it discusses artificial intelligence (AI) and machine learning (ML): AI algorithms are used to detect unsafe driving behaviors, analyze crash data, and predict potential collision points in real time [17,18]. Next, it discusses drones and aerial surveillance: drones are used for traffic monitoring, accident investigation, and emergency response coordination [19]. It also focuses on vehicle-to-everything (V2X) communication, which enables vehicles to communicate with other

Citation: Eustace, D. Emerging Transportation Safety and Operations: Practical Perspectives. *Vehicles* **2024**, *6*, 2251–2256. https://doi.org/10.3390/vehicles6040110

Received: 3 December 2024
Accepted: 18 December 2024
Published: 23 December 2024

Copyright: © 2024 by the author. Licensee MDPI, Basel, Switzerland. This article is an open access article distributed under the terms and conditions of the Creative Commons Attribution (CC BY) license (https://creativecommons.org/licenses/by/4.0/).

vehicles, traffic signals, road infrastructure, and even pedestrians to enhance situational awareness and reduce crash risks [20,21].

Roadway infrastructure enhancement is another area being pursued to enhance safety and traffic operations. Some of the initiatives include dynamic road markings and signages that utilize LED technology for dynamic lane markings and road signs that change in response to traffic conditions or weather [22]; cable barriers and roadside safety features that involve improved designs for barriers, guardrails, and crash attenuators to enhance safety on highways and reduce the severity of collisions [23,24]; road diets [25] that involve the reconfiguration of roadways to reduce the number of lanes, lower traffic speeds, and create safer environments for pedestrians and cyclists.

Several sustainable and resilient transportation solutions are introduced. Examples include multimodal transportation systems [26] that encourage the integration of various modes of transportation (buses, bicycles, walking, etc.) to reduce reliance on single-occupancy vehicles, promote safety and reduce traffic congestion; complete street design [27], which involves designing roadways to accommodate all users, including pedestrians, cyclists, and public transit riders, not just motor vehicles; and resilient infra-structure planning [28] by ensuring transportation systems can withstand and recover quickly from extreme weather events and other disruptions.

This Special Issue aimed to share and provide a platform for sharing forward-thinking approach that combines technology, policy, data, and human behavior to create a safer, more efficient, and more inclusive transportation system.

2. Overview of Published Articles

The call for papers for this Special Issue received a good response: we received fifteen (15) submissions, and ten (10) of them were published. Table 1 summarizes an overview of the contributions to this Special Issue.

The article by Gkyrtis and Kokkalis [29] presents a brief overview of the contribution of roundabouts to road safety and the interactions between the safety and design elements of roundabouts. This article provides a discussion about current challenges in and the prospects of roundabouts. In addition, the article provides findings from the environmental assessment of roundabouts; these reveal insights into their expected use and performance with regard to the presence of autonomous vehicles which are anticipated to be the main vehicle types in the foreseeable future. It also provides an overview on the role and importance of simulation studies in the improvement of the design and operation of roundabouts in research on safer vehicle movements.

In the article by Skoglund et al. [30], the authors present an argument on the safety requirements of Automated Driving Systems (ADSs). The article's goal is to enhance the effectiveness of the assessment performed by a homologation service provider by using assessment templates based on refined requirement attributes that are linked to the operational design domain (ODD) and the use of Key Enabling Technologies (KETs). These include communication, positioning, and cybersecurity in the implementation of ADSs. This article contributes to the body of knowledge by (1) outlining a method for de-riving assessment templates for use in future ADS assessments; (2) demonstrating the method by analyzing three KETs with respect to such assessment templates; and (3) demonstrating the use of assessment templates on a use case, an unmanned (remotely assisted) truck in a limited ODD.

The article by Bridgelall [31] contributes to the urgent need for the efficient condition monitoring of road and rail infrastructure. The author argues that traditional methods are both costly and inadequate and thus advocates for the employment of vehicles with integrated sensors and cloud computing capabilities in order to provide a cost-effective, sustainable solution for comprehensive infrastructure monitoring. This article advocates for international standardization by providing compelling evidence that encompasses trends in transportation, economics, and patent landscapes by highlighting the advantages of such standards. It shows that by integrating data from diverse sources, agencies can optimize maintenance triggers and allocate funds more strategically, thus preserving vital transportation networks.

Table 1. Overview of the contributions to this Special Issue.

Authors	First Author Affiliation	Country	Focus
1. Gkyrtis, K.; Kokkalis, A.	Department of Civil Engineering, Democritus University of Thrace	Greece	Road design, safety and operation
2. Skoglund, M.; Warg, F.; Thorsén, A.; Bergman, M.	RISE—Research Institutes of Sweden	Sweden	Assess of automated driving systems
3. Bridgelall, R.	Transportation, Logistics, and Finance, College of Business, North Dakota State University	USA	Infrastructure monitoring and transportation economics
4. Vieira, M.; Vieira, M.A.; Galvão, G.; Louro, P.; Véstias, M.; Vieira, P.	DEETC-ISEL/IPL, R. Conselheiro Emídio Navarro	Portugal	Traffic management and intersection control
5. Granà, A.; Curto, S.; Petralia, A.; Giuffrè, T.	Department of Engineering, University of Palermo	Italy	Interchange design, surrogate safety measures, and traffic simulation
6. Sun, W.; Abdullah, L.N.; Sulaiman, P.S.; Khalid, F.	Computer Vision, Faculty of Computer Science and Information, Universiti Putra Malaysia	Malaysia	Traffic crash risk prediction and traffic safety management
7. Viadero-Monasterio, F.; Alonso-Rentería, L.; Pérez-Oria, J.; Viadero-Rueda, F.	Mechanical Engineering Department, Advanced Vehicle Dynamics and Mechatronic Systems, Universidad Carlos III de Madrid	Spain	Vehicle safety, intelligent vehicles
8. Zahedieh, F.; Lee, C.	Department of Civil and Environmental Engineering, University of Windsor	Canada	Surrogate safety measures and traffic simulation
9. Xu, V.; Xu, S.	Texas Academy of Mathematics and Science (TAMS), University of North Texas	USA	Radar detection, broadside collision avoidance
10. Tomasch, E.; Hoschopf, H.; Ausserer, K.; Rieß, J.	Vehicle Safety Institute, Graz University of Technology, Inffeldgasse 13/6, 8010 Graz,	Austria	Heavy vehicle–cyclist collision avoidance

An article by Vieira et al. [32] proposes an approach to enhance the efficiency of urban intersections by integrating Visible Light Communication (VLC) into a multi-intersection traffic control system. This article aims at introducing a procedure that can reduce waiting times for vehicles and pedestrians, that can improve overall traffic safety, and that can accommodate diverse traffic movements during multiple-signal phases. The proposed system utilizes VLC to facilitate communication among interconnected vehicles and infrastructure. The proposed system successfully reduces both waiting and travel times. Their study emphasizes the possibility of applying reinforcement learning in everyday traffic scenarios, showcasing the potential for the dynamic identification of control actions and improved traffic management.

In [33] of the Special Issue, Granà et al. investigate connected automated vehicle (CAV)–human driver interactions and estimate the potential conflicts by using traffic microsimulation and surrogate safety assessment measures. The article highlights how CAV presence can diminish conflicts, employing surrogate safety measures and real-world mixed traffic data, and assesses the safety and performance of freeway interchange configurations in Italy and the US across diverse urban contexts. The authors propose tools for optimizing urban layouts to minimize conflicts in mixed traffic environments, and their results show that adding auxiliary lanes enhances safety, particularly for CAVs and rear-end collisions. They conclude that when CAVs follow human-driven vehicles in near-identical conditions, more conflicts arise, emphasizing the complexity of CAV integration and the need for careful safety measures and roadway design considerations.

In their article, Sun et al. [34] developed an innovative traffic crash risk prediction model called StackTrafficRiskPrediction, which intends to improve the accuracy of predicting the severity of traffic crashes. The model combines multidimensional data analysis including environmental factors, human factors, roadway characteristics, and accident-related meta-features. In the model comparison, the StackTrafficRiskPrediction model achieved an accuracy of 0.9613, 0.9069, and 0.7508 in predicting fatal, serious, and minor crashes, respectively, showing that it significantly outperforms the traditional logistic regression model. The authors conducted an experiment that analyzed the severity of traffic crashes under different road, light and weather conditions involving drivers of different age groups and with different levels of driving experience. The results show that drivers between 31 and 50 years of age with 2 to 5 years of driving experience are more likely to be involved in serious crashes.

In their article, Viadero-Monasterio et al. [35] argue that although the advanced driver assistance systems that were introduced significantly reduced motor vehicle crashes by providing crucial support for high-speed driving and alerting drivers to imminent dangers, these systems still depend on the driver's ability to respond to warnings effectively. In this article, the authors developed a neural network model for the automatic detection and classification of objects in front of a vehicle, including pedestrians and other vehicles, using radar technology, in order to overcome this limitation. The proposed neural network model achieved a high accuracy rate, correctly identifying approximately 91% of objects in the test scenarios. The results demonstrate that this model can be used to inform drivers of potential hazards or to initiate autonomous braking and steering maneuvers to prevent collisions.

Zahedieh and Lee [36] evaluate the impacts of a toll information sign with different toll lane configurations on the queue length and collision risk at a toll plaza with an estimated high percentage of heavy vehicles (HVs). The toll information sign displays information about different toll payment methods for cars and HVs upstream of the toll booth. The authors used a traffic simulation model to assess the impacts of the toll plaza utilizing the Gordie Howe International Bridge under construction at the Windsor–Detroit international border crossing as a case study. The results show that the toll information sign upstream of the toll plaza and converting the toll lanes with multiple toll payment methods to electronic toll collection (ETC)-only lanes reduces queue length and collision risk but at the same time increases the number of HV-only lanes, because a higher percentage of HVs increases the lane change collision risk.

In order to reduce broadside collisions, Xu and Xu [37] tested CornerGuard, a prototype system they developed that senses objects around a corner to alert a car driver of an impending collision with a pedestrian or automobile that is not in the line of sight (LOS). CornerGuard leverages a microwave-transceiving radar sensor mounted on the car and a curved radio wave reflector installed at the corner to sense objects around the corner and detect a broadside collision threat. Field testing demonstrated that CornerGuard can effectively and consistently detect threats within a consistent range without blind spots under broad weather conditions.

In an article by Tomasch et al. [38], the authors use crash simulations to assess a warning and autonomously intervening assistance system that could prevent heavy vehicle trucks from crashes with cyclists that is capable of blind spot detection (BSD). BSD is supposed to overcome challenges caused by local sight obstructions such as fences, hedges, or inattentive cyclists. The assessment results showed that the BSD system could prevent 26.3–65.8% of crashes involving heavy vehicles, such as trucks, and cyclists.

Acknowledgments: This Special Issue was successfully organized with the support of the Editorial Team of the *Vehicles* journal. The Guest Editors also wish to thank all authors whose valuable work was published in this Special Issue and the reviewers for evaluating the manuscripts and providing helpful suggestions.

Conflicts of Interest: The authors declare no conflicts of interest.

References

1. World Health Organization. Regional Approach to the Decade of Action for Road Safety 2021–2030. 2022. Available online: https://www.who.int/teams/social-determinants-of-health/safety-and-mobility/decade-of-action-for-road-safety-2021-2030 (accessed on 10 September 2024).
2. Elassy, M.; Al-Hattab, M.; Takruri, M.; Badawi, S. Intelligent transportation systems for sustainable smart cities. *Transp. Eng.* **2024**, *16*, 100252. [CrossRef]
3. Wang, Y.; Zeng, Z. (Eds.) Overview of data-driven solutions. In *Data-Driven Solutions to Transportation Problems*; Elsevier: Amsterdam, The Netherlands, 2019; pp. 1–10.
4. Lakshminarasimhan, M. Advanced traffic management system using Internet of Things. In Proceedings of the 20th International Conference on Advanced Communication Technology (ICACT), Pyeongchang, Republic of Korea, 6–9 March 2016; pp. 1–9.
5. Gilmore, J.F.; Elibiary, K.J. *AI in Advanced Traffic Management Systems*; Technical Report WS-93-04; Association for the Advancement of Artificial Intelligence (AAAI): Washington, DC, USA, 1993.
6. Shaon MR, R.; Li, X.; Wu, Y.-J.; Ramos, S. Quantitative Evaluation of Advanced Traffic Management Systems using Analytic Hierarchy Process. *Transp. Res. Rec.* **2021**, *2675*, 610–621. [CrossRef]
7. Elliott, D.; Keen, W.; Miao, L. Recent advances in connected and automated vehicles. *J. Traffic Transp. Eng.* **2019**, *6*, 109–131. [CrossRef]
8. Talebpour, A.; Mahmassani, H.S. Influence of connected and autonomous vehicles on traffic flow stability and throughput. *Transp. Res. Part C Emerg. Technol.* **2016**, *71*, 143–163. [CrossRef]
9. Arifin, A.S.; Zulkifli, F.Y. Recent development of smart traffic lights. *IAES Int. J. Artif. Intell.* **2021**, *10*, 224–233.
10. Alam, M.; Ferreira, J.; Fonseca, J. Introduction to intelligent transportation systems. In *Intelligent Transportation Systems: Dependable Vehicular Communications for Improved Road Safety*; Springer: Berlin/Heidelberg, Germany, 2016; pp. 1–17.
11. Zhu, L.; Yu, F.R.; Wang, Y.; Ning, B.; Tang, T. Big data analytics in intelligent transportation systems: A survey. *IEEE Trans. Intell. Transp. Syst.* **2018**, *20*, 383–398. [CrossRef]
12. Torre-Bastida, A.I.; Del Ser, J.; Laña, I.; Ilardia, M.; Bilbao, M.N.; Campos-Cordobés, S. Big Data for transportation and mobility: Recent advances, trends and challenges. *IET Intell. Transp. Syst.* **2018**, *12*, 742–755. [CrossRef]
13. Kim, E.; Muennig, P.; Rosen, Z. Vision zero: A toolkit for road safety in the modern era. *Inj. Epidemiol.* **2017**, *4*, 1. [CrossRef]
14. Law, J.; Petric, A.T. Monitoring day and dark traffic collisions in Toronto neighbourhoods with implications for injury reduction and Vision Zero initiatives: A spatial analysis approach. *Accid. Anal. Prev.* **2024**, *207*, 107728. [CrossRef]
15. Naumann, R.B.; Sandt, L.; Kumfer, W.; LaJeunesse, S.; Heiny, S.; Lich, K.H. Systems thinking in the context of road safety: Can systems tools help us realize a true "safe systems" approach? *Curr. Epidemiol. Rep.* **2020**, *7*, 343–351. [CrossRef]
16. Job, R.S.; Truong, J.; Sakashita, C. The ultimate Safe System: Redefining the Safe System approach for road safety. *Sustainability* **2022**, *14*, 2978. [CrossRef]
17. Halim, Z.; Kalsoom, R.; Bashir, S.; Abbas, G. Artificial intelligence techniques for driving safety and vehicle crash prediction. *Artif. Intell. Rev.* **2016**, *46*, 351–387. [CrossRef]
18. Olugbade, S.; Ojo, S.; Imoize, A.L.; Isabona, J.; Alaba, M.O. A review of artificial intelligence and machine learning for incident detectors in road transport systems. *Math. Comput. Appl.* **2022**, *27*, 77. [CrossRef]
19. Hildmann, H.; Kovacs, E. Using unmanned aerial vehicles (UAVs) as mobile sensing platforms (MSPs) for disaster response, civil security and public safety. *Drones* **2019**, *3*, 59. [CrossRef]
20. Hasan, M.; Mohan, S.; Shimizu, T.; Lu, H. Securing vehicle-to-everything (V2X) communication platforms. *IEEE Trans. Intell. Veh.* **2020**, *5*, 693–713. [CrossRef]
21. Ouaissa, M.; Ouaissa, M.; Houmer, M.; El Hamdani, S.; Boulouard, Z. A secure vehicle to everything (v2x) communication model for intelligent transportation system. In *Computational Intelligence in Recent Communication Networks*; Ouaissa, M., Boulouard, Z., Ouaissa, M., Guermah, B., Eds.; Springer International Publishing: Cham, Switzerland, 2022; pp. 83–102. [CrossRef]
22. Choi, W.; Sung, H.; Chong, K. Impact of Illuminated Road Signs on Driver's Perception. *Sustainability* **2023**, *15*, 12582. [CrossRef]
23. Zou, Y.; Tarko, A.P.; Chen, E.; Romero, M.A. Effectiveness of cable barriers, guardrails, and concrete barrier walls in reducing the risk of injury. *Accid. Anal. Prev.* **2014**, *72*, 55–65. [CrossRef]
24. Jiga, G.; Stamin, Ş.; Popovici, D.; Dinu, G. Study of shock attenuation for impacted safety barriers. *Procedia Eng.* **2014**, *69*, 1191–1200. [CrossRef]
25. Huang, H.F.; Stewart, J.R.; Zegeer, C.V. Evaluation of lane reduction "road diet" measures on crashes and injuries. *Transp. Res. Rec.* **2002**, *1784*, 80–90. [CrossRef]
26. Kumar, P.P.; Parida, M.; Swami, M. Performance evaluation of multimodal transportation systems. *Procedia-Soc. Behav. Sci.* **2013**, *104*, 795–804. [CrossRef]

27. Hui, N.; Saxe, S.; Roorda, M.; Hess, P.; Miller, E.J. Measuring the completeness of complete streets. *Transp. Rev.* **2018**, *38*, 73–95. [CrossRef]
28. Seager, T.P.; Clark, S.S.; Eisenberg, D.A.; Thomas, J.E.; Hinrichs, M.M.; Kofron, R.; Jensen, C.N.; McBurnett, L.R.; Snell, M.; Alderson, D.L. Redesigning resilient infrastructure research. In *Resilience and Risk: Methods and Application in Environment, Cyber and Social Domains*; Springer: Dordrecht, The Netherlands, 2017; pp. 81–119.
29. Gkyrtis, K.; Kokkalis, A. An Overview of the Efficiency of Roundabouts: Design Aspects and Contribution toward Safer Vehicle Movement. *Vehicles* **2024**, *6*, 433–449. [CrossRef]
30. Skoglund, M.; Warg, F.; Thorsén, A.; Bergman, M. Enhancing Safety Assessment of Automated Driving Systems with Key Enabling Technology Assessment Templates. *Vehicles* **2023**, *5*, 1818–1843. [CrossRef]
31. Bridgelall, R. Driving Standardization in Infrastructure Monitoring: A Role for Connected Vehicles. *Vehicles* **2023**, *5*, 1878–1891. [CrossRef]
32. Vieira, M.; Vieira, M.A.; Galvão, G.; Louro, P.; Véstias, M.; Vieira, P. Enhancing Urban Intersection Efficiency: Utilizing Visible Light Communication and Learning-Driven Control for Improved Traffic Signal Performance. *Vehicles* **2024**, *6*, 666–692. [CrossRef]
33. Granà, A.; Curto, S.; Petralia, A.; Giuffrè, T. Connected Automated and Human-Driven Vehicle Mixed Traffic in Urban Freeway Interchanges: Safety Analysis and Design Assumptions. *Vehicles* **2024**, *6*, 693–710. [CrossRef]
34. Sun, W.; Abdullah, L.N.; Sulaiman, P.S.; Khalid, F. Meta-Feature-Based Traffic Accident Risk Prediction: A Novel Approach to Forecasting Severity and Incidence. *Vehicles* **2024**, *6*, 728–746. [CrossRef]
35. Viadero-Monasterio, F.; Alonso-Rentería, L.; Pérez-Oria, J.; Viadero-Rueda, F. Radar-Based Pedestrian and Vehicle Detection and Identification for Driving Assistance. *Vehicles* **2024**, *6*, 1185–1199. [CrossRef]
36. Zahedieh, F.; Lee, C. Impacts of a Toll Information Sign and Toll Lane Configuration on Queue Length and Collision Risk at a Toll Plaza with a High Percentage of Heavy Vehicles. *Vehicles* **2024**, *6*, 1249–1267. [CrossRef]
37. Xu, V.; Xu, S. The CornerGuard: Seeing around Corners to Prevent Broadside Collisions. *Vehicles* **2024**, *6*, 1468–1481. [CrossRef]
38. Tomasch, E.; Hoschopf, H.; Ausserer, K.; Rieß, J. Required Field of View of a Sensor for an Advanced Driving Assistance System to Prevent Heavy-Goods-Vehicle to Bicycle Accidents. *Vehicles* **2024**, *6*, 1922–1941. [CrossRef]

Disclaimer/Publisher's Note: The statements, opinions and data contained in all publications are solely those of the individual author(s) and contributor(s) and not of MDPI and/or the editor(s). MDPI and/or the editor(s) disclaim responsibility for any injury to people or property resulting from any ideas, methods, instructions or products referred to in the content.

Article

Required Field of View of a Sensor for an Advanced Driving Assistance System to Prevent Heavy-Goods-Vehicle to Bicycle Accidents

Ernst Tomasch [1,*], Heinz Hoschopf [1], Karin Ausserer [2] and Jannik Rieß [2,3]

1. Vehicle Safety Institute, Graz University of Technology, Inffeldgasse 13/6, 8010 Graz, Austria; hoschopf@tugraz.at
2. Factum-Apptec Ventures GmbH, Slamastraße 43, 1230 Vienna, Austria; karin.ausserer@factum.at (K.A.); jannik.riess@austriatech.at (J.R.)
3. Austria Tech—Gesellschaft des Bundes für Technologiepolitische Maßnahmen GmbH, Raimundgasse 1/6, 1020 Vienna, Austria
* Correspondence: ernst.tomasch@tugraz.at

Abstract: Accidents involving cyclists and trucks are among the most severe road accidents. In 2021, 199 cyclists were killed in accidents involving a truck in the EU. The main accident situation is a truck turning right and a cyclist going straight ahead. A large proportion of these accidents are caused by the inadequate visibility in an HGV (Heavy Goods Vehicle). The blind spot, in particular, is a significant contributor to these accidents. A BSD (Blind Spot Detection) system is expected to significantly reduce these accidents. There are only a few studies that estimate the potential of assistance systems, and these studies include a combined assessment of cyclists and pedestrians. In the present study, accident simulations are used to assess a warning and an autonomously intervening assistance system that could prevent truck to cyclist accidents. The main challenges are local sight obstructions such as fences, hedges, etc., rule violations by cyclists, and the complexity of correctly predicting the cyclist's intentions, i.e., detecting the trajectory. Taking these accident circumstances into consideration, a BSD system could prevent between 26.3% and 65.8% of accidents involving HGVs and cyclists.

Keywords: Blind Spot Detection; heavy goods vehicle; truck; autonomous brake; right turn accidents

1. Introduction

Although the number of traffic fatalities in the EU is decreasing, approximately 20,000 people are still killed on the roads in the EU every year [1]. Approximately 1900 cyclists (9% of all road fatalities) die every year, and, compared to other road users, the number of fatally injured cyclists has remained almost constant over the past years. Passenger cars are the most common type of opponent involved in fatal accidents with cyclists in the EU [1]. Accidents involving HGVs, however, are more severe because of the high mass and the likelihood of being run over [2–7]. In total, 199 (11%) of the fatally injured cyclists are victims in accidents with heavy goods vehicles [1]. In some countries, the number of cyclist killed in HGV collisions is up to 30% [8,9]. Wang and Wei [10] reported that in Taiwan, 75% of vulnerable road users were killed in HGV accidents. It has been shown that cyclists are at greater risk of accidents simply because of the presence of HGVs [11] and that HGV-bicycle accidents tend to have more severe consequences for the cyclists involved than any other type of accident [12]; in addition, trucks are more frequent in fatal bicycle accidents [3]. Studies on fatal cycling accidents in London have shown that HGVs were the most common vehicle category in accidents involving cyclist fatalities [13]. Kim et al. [12] associate the involvement of a truck in a crash with a significant increase in the likelihood of fatal injuries to cyclists in the US. Lee and Abdel-Aty [14] found, in an analysis of data from Florida (US), that the larger size of the truck correlated with an increased likelihood of serious injury to pedestrians at intersections. Adminaite et al. [15] describe accidents between trucks and vulnerable road users as particularly problematic and point out that the main reason for these collisions is the limited field of vision of truck drivers, so that vulnerable road users are particularly susceptible to being in the blind spot and being overlooked by truck drivers. Different studies reported the major cause of truck versus bicyclist crashes is the inadequate visibility condition when bicyclists are in the vehicle's blind spot [2,5,16,17]. The most frequent accident scenario is therefore an HGV turning right and the cyclist going straight ahead and getting hit by the front or right side of the vehicle [2]. Pokorny et al. [8] reported that 12% of collisions between trucks and cyclists were a direct result of the blind spot. These accidents were, on average, more severe than other types of accidents between trucks and cyclists. In other studies, almost 20% of accidents between HGVs and cyclists are reported within this scenario [2,18–20]. In the Netherlands, 41% of accidents between HGVs and cyclists are blind spot accidents [21].

Why do truck drivers not see vulnerable road users in their blind spots, even though trucks are equipped with numerous mirrors? Talbot et al. [22] mention three possible causes. First, the drivers are looking in the right direction, but they fail to see the cyclists. As a second cause, they mention the need to pay attention to other road users due to the volume of traffic, which was also identified by Summala et al. [23]. Such an accident situation is highly dynamic and thus, third, the drivers look at the blind spot, but not at the time when the cyclists would be visible in the mirrors. Due to the large number of mirrors, drivers also need considerable time to check all mirrors, which can sometimes take up to four seconds [24]. The correct adjustment of mirrors is defined in Directive 2003/97/EC [25]. The mirrors are adjusted when the truck and the other participant are stationary. In many cases, however, traffic scenarios are dynamic situations, i.e., the participants are moving relative to each other, and therefore do not reflect several situations [22].

There are several ways to reduce the number of cyclists being fatally injured. The risk can be reduced by an increase in the risk awareness of all parties involved, i.e., vehicle drivers as well as cyclists (e.g., the blind spot problem, [5]). Infrastructure measures (e.g., separate signal phases [26,27]) can also be implemented. Advance driver assistance systems (ADAS) can also have a positive impact on the avoidance of accidents with cyclists (e.g., blind spot monitoring [28]). The expectation is that ADAS will be highly effective in terms of accident prevention. That is why the European Commission has decided that, from 2022 onwards, new vehicles will only be registered with systems designed to detect and warn vulnerable road users [29].

There has not yet been sufficient research into the extent to which these systems could influence cycling accidents. The objective of the study is to investigate the minimum longitudinal and lateral view of an ADAS in preventing heavy goods vehicle versus bicycle accidents.

2. Literature on the System Effectiveness

New truck models introduced to the market with a gross vehicle weight of more than 3.5 tonnes must be equipped with a BSD system from 2022, and generally all newly registered trucks (also with a gross weight of more than 3.5 tonnes) from 2024 [30]. Wilmink et al. [31] estimate that a BSD system could prevent approximately 39 fatalities and approximately 1900 injuries to vulnerable road users in Europe every year. A study by the Insurance Institute for Highway Safety (IIHS) quantified the potential of a BSD system at 79 fatalities and 39,000 injuries per year (Insurance Institute for Highway Safety, 2010), cited in [10]. According to Kingsley [32], 5.9% of accidents involving trucks could be avoided with a BSD system. The extent to which vulnerable road users are also affected was not specified in the study. According to Kühn et al. [33], a turning assistant would result in a potential accident avoidance rate of 42.8% between trucks and cyclists or pedestrians. In terms of injury severity, a turning assistant could prevent 31.4% of fatalities, 43.5% of serious injuries, and 42.1% of minor injuries. According to Wang and Wei [10], a BSD system would have a potential of 24% for accidents involving pedestrians, 10% for cyclists, and 11% for accidents involving motorcycles. In the truck-bicycle accidents identified by Hoedemaeker et al. [34], which were associated with the blind spot, approx. 71% could potentially be avoided with a BSD system. Silla et al. [21] estimated the potential at approx. 78%. In a before-and-after study, Tomasch and Smit [35] estimated the potential of an aftermarket BSD system at up to one third, assuming that accidents are reduced to the same extent as the warning messages after activation of an aftermarket assistant. In a prospective simulation study of accident data, an average accident avoidance potential of 15% was identified for such a system [36].

The ADAC (Allgemeiner Deutscher Automobil-Club e.V.) examined nine different aftermarket turning assistants in real-life test conditions [37]. None of the systems were rated "very good" (grade range between 1.0 and 1.5). Two of the systems were rated "good" (grade range 1.6 to 2.5), two systems were in the range between 2.6 and 3.5 (grade "satisfactory"), and one system was rated 4.4 ("sufficient", grade range 3.6 to 4.5). Four of the systems examined failed the tests and were rated "poor" (grade range worse than 4.6). Good systems did not produce any false positives and were able to detect cyclists even at a greater distance from the truck. Furthermore, the communication between the system and the truck drivers was described as "easy". Moreover, such systems are characterized by the fact that vulnerable road users and static objects can be distinguished. Vulnerable road users are detected in advance at different speeds, distances, and in different test scenarios and drivers are warned. Inadequate turning assistants produced a high number of false positives and only had a small field of view, causing drivers to not be warned in time. Some systems only recognize cyclists when they are overtaking the truck, but not when the cyclists are riding next to the truck or the truck is overtaking the cyclists. In one of the systems tested, the warning only worked if the turn signal was also on. The best system is also the most expensive assistant, with the top three systems being the most expensive. For the systems evaluated, a higher price correlates with the overall rating. Cheap systems can therefore only inadequately meet the complex requirements of road traffic.

3. Materials

The accidents used in this study are based on the road accident database CEDATU (Central Database for In-Depth Accident Analysis) [38]. The data collection is entirely retrospective. It is based on court accident data. These data are collected by the police and contain general information about the accident, such as the road users involved, age, vehicle data, etc. The police take pictures of the accident scene and prepare a sketch of the accident scene. They take pictures of the vehicles and interview the road users and witnesses involved in the accident. Injury data are collected by the hospital and are included in the court data.

Unfortunately, access to data is not granted for every court case of interest, which leads to a bias in CEDATU compared to Austrian national statistics. The aim is to have a dataset in CEDATU that is fully equivalent to the national statistics, but this will take time, as only approximately 200 to 300 cases can be investigated per year due to limited human resources.

Out of the approximately 4750 road accidents in CEDATU, 38 accidents of HGVs with cyclists were available for the study. Most of them are accident scenarios in which the HGV was turning right and the cyclist was going straight ahead (accident type number 312, Figure 1). This corresponds very well with the national statistics Austria, in which the right-turning HGV is also the most frequent type of accident involving a cyclist. Furthermore, crossing accidents (accident type number 511), accidents at entrances (accident type number 948), and accidents in which the HGV is turning right or left (accident type number 622 and 611) are of importance. Overtaking accidents (accident type number 112) and lane change accidents (accident type number 121 and 123) are the second most important in the national statistics but are underrepresented in the CEDATU sample. Approximately three quarters of the accidents in the sample took place in an urban area, which corresponds very well with the figures in the national statistics (Table 1). The sample of CEDATU accidents covers approximately 60% of the accidents in the national statistics.

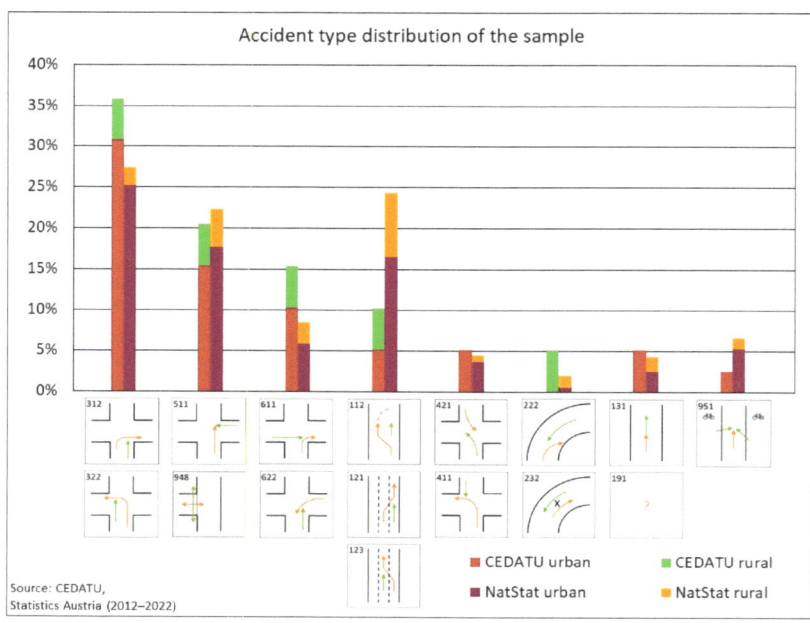

Figure 1. Accident type distribution of the HGV vs. cycle collisions analyzed. A description of the accident types is given in Appendix A.

Table 1. Accident type distribution of HGV vs. cycle collisions in the sample analyzed and the national statistics between 2012 and 2022.

Accident Type	CEDATU			National Statistics			CEDATU			National Statistics		
	Urban	Rural	Total	Urban	Rural	Total	Urban	Rural	Total	Urban	Rural	Total
312, 322	12	2	14	128	11	139	30.8%	5.1%	35.9%	25.2%	2.2%	27.4%
511, 948	5	2	8	90	23	113	12.8%	5.1%	20.5%	17.8%	4.5%	22.3%
611, 622	4	2	6	30	13	43	10.3%	5.1%	15.4%	5.9%	2.6%	8.5%
112, 121, 123	2	2	4	84	39	123	5.1%	5.1%	10.3%	16.6%	7.7%	24.3%
411, 421	2	0	2	19	4	23	5.1%	0.0%	5.1%	3.7%	0.8%	4.5%
222, 232	0	2	2	3	7	10	0.0%	5.1%	5.1%	0.6%	1.4%	2.0%
131, 191	2	0	2	13	9	22	5.1%	0.0%	5.1%	2.6%	1.8%	4.3%
951	1	0	1	27	7	34	2.6%	0.0%	2.6%	5.3%	1.4%	6.7%
Total	28	10	38	394	113	507	71.8%	25.6%	100.0%	77.7%	22.3%	100.0%

4. Method

The methodology used is referred to as a counterfactual simulation [39–41]. Correspondingly driving situations are evaluated by means of a before-and-after analysis using a "what-if simulation" approach [41]. Another expression for this method is prospective safety performance assessment of pre-crash technology by virtual simulation [42,43] and is currently being developed as an ISO standard [44]. The methodology has already been applied in several studies e.g., [28,42,45,46]. In this method, an accident scenario is simulated twice. The present study investigates real accidents. In the first simulation run, these real accidents are reconstructed and referred to as the baseline. In the second step, these reconstructed accidents are simulated again but the vehicles are virtually equipped with an ADAS (the "what-if simulation"). This simulation is referred to as the treatment. Within the treatment simulation, the minimum longitudinal and lateral view of an ADAS that is able to completely avoid a collision is evaluated.

4.1. Accident Reconstruction (Baseline)

Accident reconstruction is used to analyze traffic accidents in detail. Accidents are divided into a pre-collision, collision, and post-collision phases and certain parameters, such as initial speed, reaction time, braking deceleration, collision speed, collision angle, etc., are calculated. All factors that have a significant influence on the accident are taken into consideration (e.g., road conditions, weather conditions, speed limit on the road or speed limit, road width, etc.). The purpose is to calculate all phases of the accident sequence in terms of space and time [47–49]. Accident reconstruction is carried out with the simulation software PC-Crash [50], which is used by accident experts and in accident research. PC-Crash has been validated in various studies [51–53].

The reconstruction methods used are described in Burg and Moser [49] and Hugemann [54]. A key parameter is the collision speed of the truck and bicycle. This is a function of the final positions of the parties involved the position of the collision. A multi-body simulation in PC-Crash is used to calculate the collision speed [53]. As the mass ratio between the pedestrian and the truck is so high, the change in collision speed during the impact of the truck can be neglected. The initial speed of the vehicle is calculated from the course of the road, road conditions, the skid marks on the road (if any), the tachograph or EDR data, and witness reports. On the basis of these data, a time-speed-acceleration history and the trajectory followed in the pre-crash phase are calculated. The pre-crash phase is reconstructed for up to five seconds, as proposed in Schubert et al. [55]. The time-speed-acceleration history and the trajectory followed are used as the baseline for the treatment simulations.

A symptomatic accident is given in Figure 2. The HGV intends to turn right at the intersection and the cyclist is going straight ahead. The pictures below reflect the situation from the truck driver's perspective at different times. Seven seconds before the collision, the truck is approaching the junction. The cyclist is approaching the junction on the right-hand side on a cycle path. At this point, the cyclist is not visible on the right-hand side, nor is he visible in the right-hand side mirrors. Six seconds before the collision, the light turns green. At the same time, the cyclist becomes visible to the driver. Now it can become very complex for the truck driver. As soon as the traffic light turns green, the vehicles start to accelerate. The truck driver must now pay attention to the vehicles in front. At the time when he wants to turn right, he must also keep an eye on the oncoming traffic, as turning right sometimes requires steering slightly to the left. This is about two seconds before the collision. The cyclist is now invisible again, obstructed by the right-hand parts of the truck cab. He is also not visible in the side mirrors. The truck driver has two options as to how the cyclist could proceed. Either the cyclist turns right and follows the cycle lane, or he goes straight ahead and crosses the junction. Obviously, the truck driver should have been more cautious and stopped when in doubt.

The accident is now reconstructed to such an extent that a sufficiently long time history of the speed and acceleration before the collision is available. Figure 3 shows the relative trajectory of a cyclist in relation to the HGV as an example of an accident with a truck turning right and a cyclist going straight ahead. The collision position was on the right side of the truck. The relative movement looks quite strange. This is because the truck first has to steer slightly to left in order to be able to turn right at the junction. Due to the different relative speeds at certain points in time, the cyclist is in front of the truck and then falls behind when the truck is traveling faster than the cyclist.

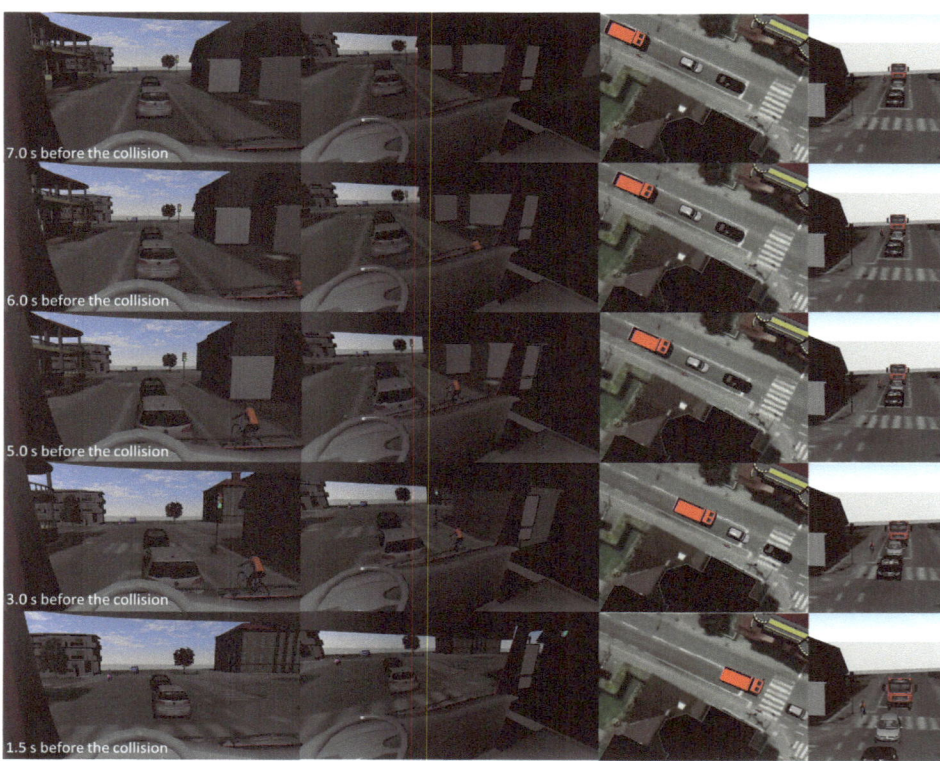

Figure 2. Junction accident with right-turning HGV crossing the bicycle's path and at different time steps (far left: driver's view; left mid: driver's view when paying attention to the right side; right mid: top view; far right: scenario view from opposite direction).

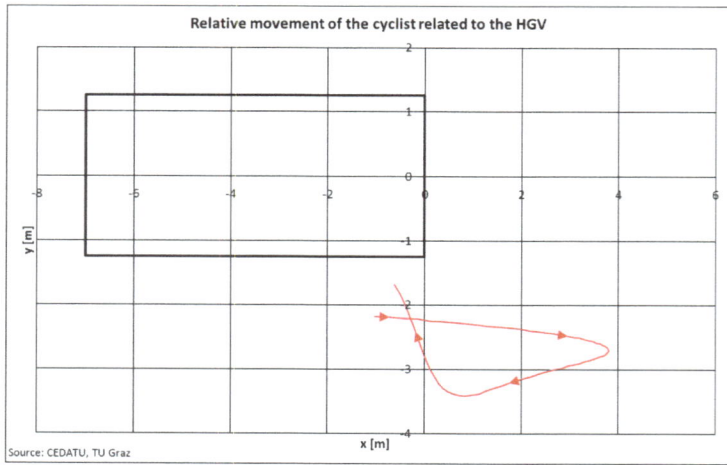

Figure 3. Relative trajectory of the cyclist in relation to the HGV. The rectangle represents a generic shape of the HGV. The red line with directional markings represents the cyclist's movement relative to the truck.

4.2. Treatment Simulation ("What-If Simulation")

After the accident had been fully reconstructed, the simulations were used to calculate at what point the truck still had the possibility to stop in time and avoid a collision. The treatment simulations have now been used to determine the necessary x and y distances required by an assistance system to detect cyclists. Figure 4 shows the required longitudinal and lateral distances of a generic assistance system that is able to detect a cyclist. At the time of the system response, the cyclist must be fully within the sensor's field of view for at least 150 ms [56,57]. The system response is either a warning to the driver or autonomous braking. After triggering the warning, a driver reaction time of 0.8 s [58–61] was taken into consideration. A comprehensive summary of driver reaction times can be found in Green [62]. A reaction time of 0.8 s should be considered sufficient for a driver who is not under the influence of substances (alcohol, medication, drugs) or is fatigued or distracted. An actuator time of 0.2 s was assumed for the reaction time of the autonomous system [56,57]. After the reaction time, the braking phase started. The build-up time to maximum deceleration was set at 0.5 s [49]. The maximum deceleration depends on road conditions [49], but all the accidents investigated were on dry roads. New trucks reliably achieve braking acceleration in a range between 7 and 8 m/s^2 [63]. Since most of the accidents in the accident data involved old and new vehicles and the condition of the tires is not known, the deceleration was limited to 5 m/s^2.

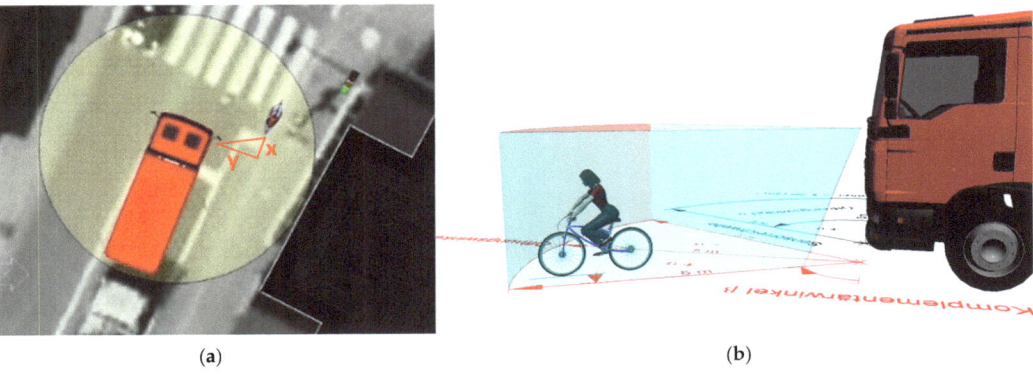

Figure 4. (a) Longitudinal and lateral distances for detection of cyclists and (b) vertical view of an assistance system that fully detects the cyclist.

Based on the assumptions made, a warning system would require a visibility range of approximately one meter in the longitudinal direction and 3.5 m in the lateral direction (Figure 5). An autonomously intervening system would require approximately 0.3 m in the longitudinal direction and a lateral view of 3.3 m to avoid this collision. The collision takes place at the end of the bicycle's trajectory, which is in the front lateral area of the truck. The field of view of an assistance system is based on complete avoidance of a collision independent of visual obstructions caused by any objects (e.g., hedges, fences, etc.). The field of view therefore refers to an ideal system, and thus the ranges refer to theoretical configurations.

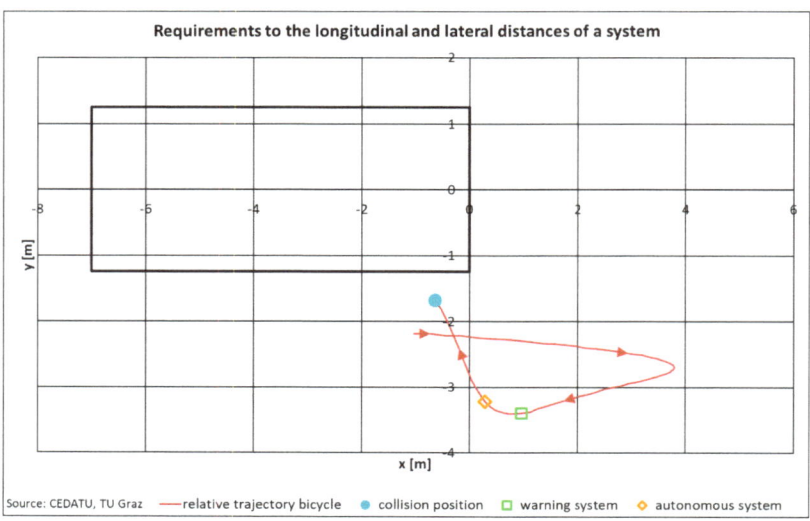

Figure 5. Theoretical required longitudinal and lateral distances for proper detection of a cyclist to avoid a collision. The rectangle represents a generic shape of the HGV. The red line with directional markings represents the cyclist's movement relative to the truck. The rectangle markers indicate the positions where a particular system needs to take action.

5. Results

All road accidents in the sample were reconstructed and the effectiveness of an assistance system evaluated. Different accident types require different sensor configurations. For accidents involving trucks turning to the right, lateral view is of particular importance. However, rearward-facing observation is also required for trucks. For an ideal warning system, a lateral view of approximately 10 m is required for this type of accident (Figure 6). A forward-facing view of approximately 2.5 m and a rearward-facing view of approximately 17.5 m is required. Somewhat different ranges are required for an ideal autonomous system (Table 2). For left-turning and crossing accidents, the requirements differ significantly from those for right-turning accidents. If the truck is on the priority road, a much longer view in the longitudinal and lateral direction is necessary. For these accidents, a longitudinal view of approximately 58 m and lateral view of approximately 21 m are required. When the truck is on the non-priority road, significantly closer ranges are required and the system will need to monitor the area directly surrounding the truck. Approximately 8 m would be necessary to the front and to the side. The longest ranges in the longitudinal direction are for accidents with oncoming traffic. This category also includes turning off accidents as these are also accidents with oncoming traffic. However, the relative speeds are considerably lower in this case. Only a few traffic accidents in the sample were available for an in-depth analysis of accidents when overtaking or changing lanes. The problem here is rather a falling cyclist due to the suction effect of the truck or the cyclist's instability during the overtaking maneuver. In another type of accident, some cyclists are very careless and cross the road without paying sufficient attention to traffic. For this type of accident, a system would require a huge longitudinal view and also have a sufficient range to the side in order to stop in time.

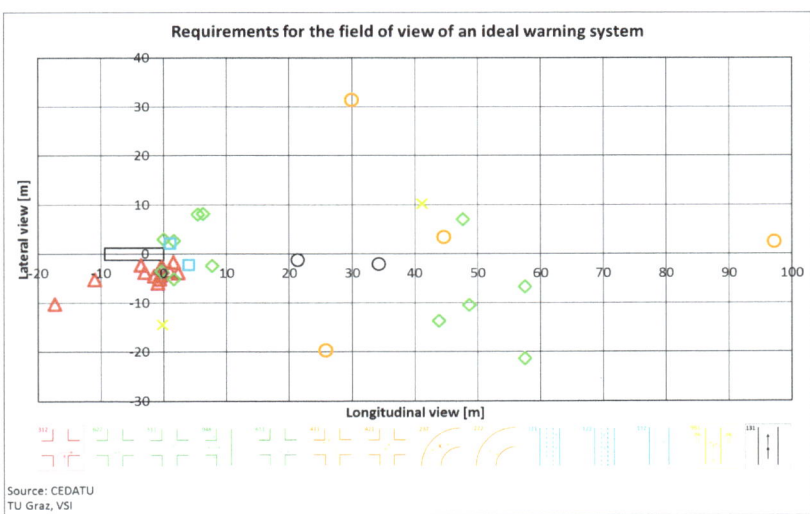

Figure 6. Theoretical requirements for the longitudinal and lateral field of view of an ideal warning system for the different accident types. A description of the accident types is given in Appendix A.

The requirements for a sensor vary depending on the location of the accident (Table 2). In urban areas, a warning system would need to have a forward-facing view of up to 97 m and a lateral view of up to 14 m. In rural areas, this would be a forward-facing view of up to 58 m and a lateral view of up to 31 m. With an autonomous system, the forward-facing view would be up to 76 m and the lateral view up to 13 m in urban areas. In rural areas, the forward-facing view should be up to 47 m and the lateral view up to 23 m. For both systems, the rearward-facing view would be up to 11 m in urban areas and up to 17 m in rural areas.

Table 2 summarises the requirements for a warning or ideal autonomous intervention system in terms of longitudinal and lateral field of view. The "Avoidance" column indicates whether the accident can be avoided or not, or whether further requirements are necessary, with information on the circumstances identified in the "Circumstances" column.

Without taking into consideration specific accident circumstances and further requirements to the system, 10 (26.3%) road accidents are potentially avoidable. Three cases could not be avoided due to sight obstructions, and in three cases the cyclist had violated the rules, i.e., ignored the priority of the truck. In eight accidents, a collision might be prevented if the cyclist's driving path was known in advance, i.e., it would have to be determined in advance that the cyclist would cross the truck's driving path.

In UN R 151 [64], test scenarios are defined for the assessment of blind spot assistance systems to avoid collisions with cyclists. In the test, a constant speed is defined for both the truck and the bicycle. Tests in which the truck is stationary and is moving-off and the bicycle is in the blind spot are described in UN R 159 [65]. In the UN R 159 test, the truck accelerates in a straight line and the cyclist moves parallel in the same direction. However, the two regulations do not take into consideration the starting to move and turning scenarios. For this reason, the two factors, speed and starting to move, were taken into account when assessing the effectiveness of the system. In six and three accidents, respectively, the speed was below 10 km/h or the truck was starting to move. These accidents can only be avoided if the system also works under these conditions. Finally, ten accidents cannot be prevented at all.

Table 2. Theoretical requirements for the longitudinal and lateral field of view of a warning system and an ideal autonomous intervening system for the different accident types and circumstances that influence avoidability. Specific accident types are aggregated (Gx). U (urban) and R (rural) indicate the accident location.

Case	Type	Accident Group	Site	Warning System x [m]	Warning System y [m]	Autonomous System x [m]	Autonomous System y [m]	Avoidance	Circumstances
#01	312	G1	U	−11.0	−5.2	−9.2	−5.4	yes	-
#02	312	G1	R	−17.3	−10.2	−16.7	−10.6	further requirements	1
#03	312	G1	U	1.5	−1.7	1.5	−1.6	further requirements	5
#04	312	G1	U	−0.7	−5.1	−0.2	−4.0	further requirements	5
#05	312	G1	R	−3.6	−2.3	−2.4	−2.1	further requirements	5
#06	312	G1	U	−0.6	−4.3	−0.7	−3.5	yes	-
#07	312	G1	U	−0.9	−5.9	−0.2	−5.4	yes	-
#08	312	G1	U	−1.5	−4.5	−1.6	−4.5	yes	-
#09	312	G1	U	1.0	−3.4	0.3	−3.2	yes	-
#10	312	G1	U	2.3	−3.9	0.8	−3.6	further requirements	2
#11	312	G1	U	−3.0	−3.8	−3.4	−3.8	yes	-
#12	312	G1	U	−0.2	−2.5	−0.5	−2.6	further requirements	6
#13	312	G1	U	−0.5	−2.8	−0.7	−2.9	yes	-
#14	322	G1	U	34.3	−2.1	25.7	−1.9	further requirements	2
#15	511	G2	R	57.5	−21.3	46.0	−15.8	no	2
#16	511	G2	U	1.6	2.7	1.8	1.9	further requirements	6
#17	511	G2	U	−0.3	−3.5	−0.1	−2.8	further requirements	5
#18	622	G2	U	0.0	3.0	0.0	3.0	further requirements	5
#19	948	G2	R	47.6	7.1	36.9	5.4	no	4
#20	948	G2	U	1.6	−5.1	1.4	−3.4	further requirements	6
#21	611	G2	R	7.7	−2.4	5.0	−2.0	yes	-
#22	622	G2	U	6.3	8.2	4.2	5.7	further requirements	5
#23	622	G2	U	43.8	−13.6	35.1	−9.6	no	1, 3
#24	622	G2	R	57.5	−6.7	46.8	−4.8	no	3
#25	622	G2	U	5.5	8.1	3.6	5.5	yes	-
#26	622	G2	U	48.6	−10.4	38.6	−8.6	further requirements	2
#27	222	G3	R	30.0	31.4	23.7	23.0	no	8
#28	232	G3	R	25.9	−19.8	20.5	−12.1	further requirements	1
#29	411	G3	U	44.7	3.4	33.5	2.9	no	3
#30	421	G3	U	97.3	2.4	75.7	1.8	further requirements	2
#31	112	G4	R	32.7	2.2	22.9	0.6	no	7
#32	121	G4	R	0.0	0.0	0.0	0.0	no	4
#33	121	G4	U	0.0	0.0	0.0	0.0	no	4
#34	123	G4	U	2.5	−2.2	1.7	−2.2	further requirements	2
#35	951	G5	U	41.1	10.3	31.1	8.1	further requirements	2
#36	951	G5	U	−0.2	−14.4	1.4	−13.3	further requirements	2
#37	131	G6	U	21.4	−1.4	15.4	−1.2	yes	-
#38	191	G6	U	0.0	0.0	0.0	0.0	no	4

[1] Sight obstructions. [2] Driving path of cyclist. [3] Rule violation of the cyclist. [4] No collision in real accident. [5] Speed below 10 km/h. [6] HGV starting to move. [7] Other circumstances.

6. Discussion

6.1. Accident Location

Accidents in urban areas differ from those on rural roads due to a variety of conditions. Although speed limits are much lower in urban areas, obstructions such as parked vehicles or objects on the roadside can make it much more difficult to see and detect the cyclist. The mean forward field of view in the urban area is 23.6 m (standard deviation: 3.3 m), which is significantly lower than in the rural area at 37 m (standard deviation: 9.7 m). This is related to the lower driving speed in urban areas and situations at junctions where the truck starts to move. However, a significantly longer forward-facing field of view was required in the urban area than in the rural area. This relatively long field of view was determined for accident type 421, where the road users are moving towards each other. In order to avoid this accident, it would be necessary to react at a point where it is not yet foreseeable that a road user will turn at the junction. At the moment when the intention of the road user becomes apparent, the collision cannot be avoided. It is therefore necessary to know the intention and the trajectory in advance. Obviously, it is very difficult to judge the driven trajectory in situations where road users are travelling in opposite directions and one road user is turning into the driven lane of the other road user. In these accident situations, the trajectory must be known and the intention to cross one's own lane must be known in advance. Otherwise, a collision cannot be avoided.

The necessary requirements for a sensor do not only depend on the location of the accident, but also on the type of accident. Accident analysis has clearly shown that in many cases of truck accidents, there is insufficient visibility of the close surroundings, so a sensor system that monitors the close proximity of the truck would make a significant contribution to accident prevention. This is particularly true of turning accidents involving cyclists travelling straight ahead and accidents at junctions where cyclists cross the road directly in front of the truck, usually when the truck is starting to move.

6.2. Effectiveness with an Infinite Sensor Range

With regard to all accidents investigated, an assistance system with an infinite sensor range would be able to prevent 26.3% of accidents involving trucks and cyclists. If all accidents classified as possibly avoidable on the basis of the existing test conditions of UN R 151 [64] and UN R 159 [65] were also covered by the assistance system, effectiveness would increase to 50.0%. If local visual obstructions could also be removed, effectiveness could be increased to 57.9%. Effectiveness could be increased to 65.8% as long as the cyclist complies with the traffic rules, e.g., does not violate the right of way.

Only a few studies in the literature refer to the avoidance potential of BSD systems in truck and bicycle accidents. Hoedemaeker et al. [34] estimate the potential of a BSD system at approximately 71%. Silla et al. [21] estimates the potential of a BSD system at approximately 78%. According to Wang and Wei [10], a BSD system would have a potential of 10%. Other studies includes pedestrians, too. Based on a natural driving study, Tomasch and Smit [35] estimate the potential of a blind spot assistant at up to a third. Kühn et al. [33] estimates the potential for accident avoidance between trucks and cyclists or pedestrians at 42.8%.

The evaluation of the above studies is based on a purely descriptive analysis of defined pre-crash scenarios, without considering accident reconstruction with relative movement of the road users and without distinguishing between different intervention strategies. It is only an assessment of how many accidents could potentially be avoided.

6.3. Effectiveness with Different Sensor Ranges

6.3.1. Ideal Conditions

The required field of view in the longitudinal and lateral directions for an ideal warning or ideal autonomously intervening system is based on full avoidance of a collision regardless of local sight obstructions, rule violations by cyclists, etc. An ideal warning system would therefore require a forward-facing view of approximately 98 m and rearward-facing view of approximately 18 m. Laterally, a system would need to be able to detect a cyclist within approximately 31 m to the left and 21 m to the right. These requirements take into consideration all bicycle accidents and not just accidents when turning right. An ideally autonomous system requires a forward-facing field of view of approximately 76 m and a rearward-facing view of approximately 17 m. To the side, the system would have to be able to detect a cyclist within a range of approximately 23 m to the left and approximately 16 m to the right. At the specified distances, the driver or the system would have to intervene in order to prevent a collision. However, this also indicates that the cyclist would have to be fully detected even before that point. Table 3 indicates different ranges of a sensor and the number of accidents that are potentially covered by this range. Without consideration of any specific accident circumstances a sensor range of 10 m would cover 50% of the accidents. Up to approximately 50 to 60 m, the proportion of accidents increases. With a range of more than 60 m, only a few more accidents will be covered by a warning system, and 50 m by an autonomous system. The accident analysis revealed that in some accidents, the cyclist fell and either had no contact with the truck or skidded against the truck. It is therefore not possible to maximize the potential to one hundred percent.

Table 3. Maximum (theoretical) effectiveness of an assistance system with different sensor ranges and proportion of accidents that could be potentially covered by the sensor without consideration of specific accident circumstances in relation to all 38 accidents investigated.

System		Sensor Range									
		10 m	20 m	30 m	40 m	50 m	60 m	70 m	80 m	90 m	100 m
Warning system	effectiveness	50.0%	57.9%	63.2%	71.1%	84.2%	89.5%	89.5%	89.5%	89.5%	92.1%
	number of accidents	19	22	24	27	32	34	34	34	34	35
Autonomous system	effectiveness	52.6%	60.5%	71.1%	84.2%	89.5%	89.5%	89.5%	92.1%	92.1%	92.1%
	number of accidents	20	23	27	32	34	34	34	35	35	35

6.3.2. Real Conditions

In addition to the accidents in which the cyclist fell without physical contact with the truck, there are other factors that influence whether an accident is completely avoidable, e.g., sight obstructions, rule violation cyclist (not yielding at red lights), cyclist's trajectory, etc.

Table 4 contains a summary of the accident circumstances with the required sensor ranges. If all the circumstances are taken into consideration, for instance, 21.1% (8 out of 38) of collisions can be avoided if the sensor range is 10 m. However, without the circumstances of the accident, the potential to avoid the accident would increase. The number of avoidable accidents in the "avoidance" category is taken as the reference. Any elimination of an accident circumstance would increase the potential of an assistance system. At a range of 10 m, however, there were obviously no sight obstructions, rule violations by the cyclist, or other circumstances that would have had an impact on the effectiveness. Nevertheless, if the cyclist's driving trajectory were known, the effectiveness of the assistance system could be increased from 8 potentially avoidable accidents to 10, i.e., 26.3%. For the circumstances speed or starting to move, the effectiveness could be increased to 36.8% (from 8 to 14) and 28.9% (from 8 to 11), respectively. If there are infrastructural obstructions to visibility such as bushes, hedges, fences, etc., even an assistance system cannot detect the cyclist. Other significant circumstances are rule violations by the cyclist. Taking all accident circumstances into consideration, a total of 19 accidents could be covered with a range of 10 m (Table 3). If the system only has to warn or intervene when there is no possibility of the cyclist stopping in time before crossing, it is simply not possible to avoid an accident. Otherwise, false positives would increase and reduce the acceptance of such systems.

Table 4. Number of accidents that could be potentially covered by a warning assistance system with different sensor ranges with consideration of specific accident circumstances in relation to all 38 accidents investigated.

Circumstances	Sensor Range									
	10 m	20 m	30 m	40 m	50 m	60 m	70 m	80 m	90 m	100 m
Avoidance	8	9	10	10	10	10	10	10	10	10
Sight obstructions	0	1	2	2	3	3	3	3	3	3
Cyclist's trajectory	2	3	3	4	6	7	7	7	7	8
Rule violation cyclist	0	0	0	0	2	3	3	3	3	3
Speed	6	6	6	6	6	6	6	6	6	6
Starting to move	3	3	3	3	3	3	3	3	3	3
Other	0	0	0	2	2	2	2	2	2	2
Total	19	22	24	27	32	34	34	34	34	35

However, with a greater range of the system and taking the mentioned circumstances into consideration, the effectiveness of the system could be significantly increased.

6.3.3. Accident Type Groups Under Real Conditions

A warning assistance system with a field of view of 360° and a range of 10 m would be able to cover 19 accidents; 11 of these accidents apply to trucks turning right and cyclists going straight ahead. Only six of these could be avoided without further limitations such as sight obstructions, information about the cyclist's driving trajectory, speed, etc. (Table 5). Group G2 refers to the left-turning and crossing accidents and is the second largest group of accidents that can be prevented by an assistance system with a range of 10 m. The two main circumstances of the accident are speed and moving-off, so that a total of seven accidents could be prevented if these two circumstances were taken into consideration.

Table 5. Number of accidents that could be potentially covered by a warning assistance system with a field of view of 360° and a range of 10 m and specific accident circumstances separated into different accident groups.

Circumstances	Accident Type Group						
	G1	G2	G3	G4	G5	G6	Total
Avoidance	6	2	0	0	0	0	8
Sight obstructions	0	0	0	0	0	0	0
Cyclist's trajectory	1	0	0	1	0	0	2
Rule violation cyclist	0	0	0	0	0	0	0
Speed	3	3	0	0	0	0	6
Starting to move	1	2	0	0	0	0	3
Other	0	0	0	0	0	0	0
Total	11	7	0	1	0	0	

The technologies currently available (camera, RADAR (Radio Detection And Ranging), LIDAR (Light Detection And Ranging), etc.) make it possible to monitor the vehicle's surroundings. These technologies are combined (=sensor fusion) to optimise detection and provide an accurate understanding of the environment. Most technologies have ranges that extend well beyond the immediate area around the vehicle [66]. However, the lateral visibility range of the systems tested by ADAC [37] was less than 6 m. One system was unable to detect cyclists at a distance of more than 2.5 m.

Meanwhile, the unit cost of sensors is decreasing and is now quite low, depending on the technology [67]. Although there are relatively cheap aftermarket systems available, they performed less well in the tests [37]. The best performing aftermarket turning safety assistance systems were associated with the highest costs.

The requirements for a turning assistant differ quite significantly with regard to the field of view for avoiding all the cyclist accidents investigated. To avoid these accidents, an ideal warning or an ideal autonomous system would have to be able to monitor at least approximately 6 m to the right, but also approximately 18 m to the rear. With these specifications, 13 accidents could be fully avoided, as long as the circumstances of the accident are taken into consideration. In three accidents, the collision speed would be less than the UN R 151 [64] test speed, and they are labeled as not explicitly avoidable or only avoidable if an assistance system also fulfilled this criterion. Another accident could only be prevented if the assistance system were able to detect the cyclist when starting to move and turning at the same time. In one accident, sight obstructions were present and in another accident, the trajectory of the cyclist would have to be known in order to avoid the accident.

6.3.4. Cyclist Behavior

In many cases, the behavior of the cyclist, i.e., prediction of the trajectory, is an essential factor to prevent accidents. Studies show that cyclist behaviour is very difficult to predict in advance [34,68,69] so that a collision can still be prevented, especially if the cyclist is travelling parallel to the vehicle and turns into the vehicle's path just before the collision. Predicting what a cyclist will do often depends on strong indicators, such as head movement [70], hand gesture to indicate change of direction [69], etc.

If the cyclist is coming from a clearly visible cycle path and is crossing the road, the cyclist could often stop in time before the collision, but the HGV would have to react much earlier to avoid a collision in case the cyclist does not stop. However, if the cyclist stops as required, unnecessary error warnings ("false-positives") are issued and the system is deemed unreliable with a poor acceptance [71], meaning that drivers ignore warnings [72] and stop using it [73]. False-positives are therefore a major challenge. Furthermore, rule violations, e.g., ignoring the priority of the HGV, running red lights, are a very common contributing factor in accidents [74–76] and are a considerable burden on the assistance system. If the driver of the HGV reacts when the cyclist starts crossing the road, the accident cannot be avoided. If the system reacts earlier, this can lead to many false positives and would have a huge impact on the acceptance of an assistance system. In real-world tests, a high number of false positives have already been identified, particularly when the system was unable to distinguish between stationary objects and vulnerable road users [37].

Cyclists sometimes fall only because the HGV has overtaken them. This can only be prevented if there is sufficient distance between the cyclist and the truck. The Austrian Road Traffic Regulations [77] specifies a minimum distance of 1.5 m in urban areas and 2 m on rural roads. However, this can sometimes be very difficult to monitor. An assistance system that continuously monitors the minimum distance could help.

7. Limitations

Only 38 road accidents were available for analysis in this study. Although a similar number of accidents were available for analysis in other studies [34], the number of cases analysed is not sufficient to draw general conclusions. Nevertheless, these different situations can contribute to an insight and to the definition of requirements for systems that may be able to prevent such accidents. However, the cases showed that the accident situations can be very complex and the requirements vary from accident to accident.

Overtaking accidents or lane change accidents are very biased in the sample, i.e., there are many more accidents in the national statistics than in the CEDATU sample.

In the overtaking accidents, the lateral distance between the truck and the cyclist was not examined. Accidents where there was no contact between the truck and the cyclist were therefore classified as unavoidable.

In the treatment simulation, an attentive driver was assumed for the warning system. This driver responded with a reaction time of 0.8 s. It was assumed that the drivers were not under the influence of substances, nor were they fatigued, distracted, psychologically stressed, or aggressive.

The market penetration rate has been assumed to be 100%.

8. Conclusions

Although there are a large number of mirrors on trucks, sight conditions while driving are no longer optimal. The reason for this is that, in accordance with European regulations, the mirrors are evaluated in a stationary condition [22]. In addition, drivers also need a certain amount of time to properly check the mirrors, which can take up to four seconds [24]. An assistance system is able to monitor the surrounding of the truck continuously. The most significant benefit would be the prevention of accidents in blind spots, i.e., HGV turning right and cyclist going straight ahead.

Nevertheless, it is not possible to prevent all accidents involving cyclists. The main problems are sight obstructions due to fences, bushes, hedges, etc., rule violations by the cyclists, and missing information on the cyclist's trajectory. For specific types of accident, in particular with oncoming traffic, the cyclist's intention to cross the truck's driving trajectory must be known in advance in order to prevent a collision. Otherwise, it is not possible to avoid a collision due to physical limits, e.g., road friction and maximum possible deceleration.

It is estimated that an assistance system could potentially prevent between 26.3% and 57.9% of accidents involving HGVs and cyclists, depending on whether the identified accident circumstances are taken into consideration.

The maximum theoretical effectiveness of a driver warning system with an infinite sensor range is 26.3% without further consideration of circumstances. If the system successfully meets the existing UN R 151 [64] and UN R 159 [65] test conditions, the effectiveness increases to 50%. If local sight obstructions are also removed, the effectiveness increases to 57.9%. One of the most difficult challenges is predicting the cyclist's riding behaviour, i.e., the trajectory. If the cyclist's trajectory can be very accurately predicted in advance, the effectiveness increases to 78.9%. This is rather unlikely, as the intention of the cyclist can only be recognised at a very late stage, which makes it extremely difficult to avoid a collision. At the least, the collision speed could be reduced.

It has been shown that several accidents occur in the immediate vicinity of the truck. A system with a range of 10 m could potentially prevent 50% of the accidents studied.

Author Contributions: Conceptualization, E.T.; methodology, E.T.; formal analysis, E.T. and H.H.; investigation, E.T. and H.H.; data curation, H.H.; writing—original draft preparation, E.T.; writing—review and editing, H.H., K.A. and J.R.; project administration, E.T.; funding acquisition, E.T., H.H. and K.A. All authors have read and agreed to the published version of the manuscript.

Funding: This research was funded by the Austrian Federal Ministry for Climate Action, Environment, Energy, Mobility, Innovation and Technology (BMK) by grants of the Austrian Road Safety Fund (VSF), grant number GZ. BMK-2021-0.773.146 and The APC was funded by TU Graz Open Access Publishing Fund. Open Access Funding by the Graz University of Technology.

Data Availability Statement: The original contributions presented in the study are included in the article, further inquiries can be directed to the corresponding author.

Conflicts of Interest: Author Karin Ausserer is employed by Faclum-Applec Ventures GmbH. Author Jannik Rieß was employed by the Faclum-applec ventures GmbH and now is employed by Austria Tech. The remaining authors declare that the research was conducted in the absence of any commercial or financial relationships that could be construed as a potential conflict of interest.

Appendix A

Table A1. Accident type classification in Austria.

Accident Type	Pictogram	Description
112		collision between two vehicles driving in the same direction after overtaking on the left side and returning to the original lane
121		collision between two vehicles driving in the same direction due to changing into the right lane
123		collision between two vehicles driving in the same direction due to changing into the left lane
131		rear-end collision into moving vehicle on a straight section

Table A1. *Cont.*

Accident Type	Pictogram	Description
191		
222		leaving the road to the right due to a vehicle proceeding in the opposite direction in a right-hand curve, without a collision
232		Lateral (sliding) collision between two vehicles proceeding in opposite direction in a curve
312		collision of a vehicle which is turning right at a junction with another vehicle which is passing by and moving straight
322		collision of a vehicle which is turning left at a junction with another vehicle which is overtaking or passing by and moving straight
411		collision of a vehicle which is turning left with another vehicle which is proceeding from the opposite direction and moving straight
421		lateral (sliding) collision between two vehicles turning left in opposite direction
511		collision at a junction of two vehicles proceeding at right angles to each other
611		collision at a junction of a vehicle which is turning right and a vehicle from left which is crossing straight
622		collision at a junction of a vehicle which is turning left with a vehicle from left which is moving straight

Table A1. Cont.

Accident Type	Pictogram	Description
948		collision by the entrance of a building or plot
951		Collision with a cyclist entering the lane

References

1. European Commission. *Collision Matrix: Road Traffic Fatalities in the EU in 2021*; European Commission: Brussles, Belgium, 2023.
2. Kockum, S.; Örtlund, R.; Ekfjorden, A.; Wells, P. *Volvo Trucks Safety Report 2017*; Volvo Trucks: Göteborg, Sweden, 2017.
3. Ackery, A.D.; McLellan, B.A.; Redelmeier, D.A. Bicyclist deaths and striking vehicles in the USA. *Inj. Prev. J. Int. Soc. Child Adolesc. Inj. Prev.* 2012, 18, 22–26. [CrossRef] [PubMed]
4. Bíl, M.; Bílová, M.; Dobiáš, M.; Andrášik, R. Circumstances and causes of fatal cycling crashes in the Czech Republic. *Traffic Inj. Prev.* 2016, 17, 394–399. [CrossRef] [PubMed]
5. Niewoehner, W.; Berg, A.F. Endangerment of Pedestrians and Bicyclists at Intersections by Right Turning Trucks. In Proceedings of the 19th International Technical Conference on the Enhanced Safety of Vehicles (ESV), Washington, DC, USA, 6–9 June 2005.
6. McCarthy, M.; Gilbert, K. Cyclist road deaths in London 1985–1992: Drivers, vehicles, manoeuvres and injuries. *Accid. Anal. Prev.* 1996, 28, 275–279. [CrossRef] [PubMed]
7. Edwards, A.; Barrow, A.; O'Connell, S.; Krishnamurthy, V.; Khatry, R.; Hylands, N.; McCarthy, M.; Helman, S.; Knight, I. *Analysis of Bus Collisions and Identification of Countermeasures*, 1st ed.; TRL: Wokingham, UK, 2018.
8. Pokorny, P.; Drescher, J.; Pitera, K.; Jonsson, T. Accidents between freight vehicles and bicycles, with a focus on urban areas. *Transp. Res. Procedia* 2017, 25, 999–1007. [CrossRef]
9. OECD; ITF. *Cycling, Health and Safety*; OECD: Paris, France, 2013.
10. Wang, M.-H.; Wei, C.-H. Potential Safety Benefit of the Blind Spot Detection System for Large Trucks on the Vulnerable Road Users in Taiwan. In Proceedings of the 2016 5th International Conference on Transportation and Traffic Engineering (ICTTE 2016), Lucerne, Switzerland, 6–10 July 2016; Volume 81, p. 2007.
11. Vandenbulcke, G.; Thomas, I.; Panis, L.I. Predicting cycling accident risk in Brussels: A spatial case-control approach. *Accid. Anal. Prev.* 2014, 62, 341–357. [CrossRef]
12. Kim, J.-K.; Kim, S.; Ulfarsson, G.F.; Porrello, L.A. Bicyclist injury severities in bicycle-motor vehicle accidents. *Accid. Anal. Prev.* 2007, 39, 238–251. [CrossRef]
13. Manson, J.; Cooper, S.; West, A.; Foster, E.; Cole, E.; Tai, N.R.M. Major trauma and urban cyclists: Physiological status and injury profile. *Emerg. Med. J. EMJ* 2013, 30, 32–37. [CrossRef]
14. Lee, C.; Abdel-Aty, M. Comprehensive analysis of vehicle-pedestrian crashes at intersections in Florida. *Accid. Anal. Prev.* 2005, 37, 775–786. [CrossRef]
15. Adminaite, D.; Allsop, R.; Jost, G. *Making Walking and Cycling on Europe's Roads Safer: PIN Flash Report 29*; European Transport Safety Council (ETSC): Brussels, Belgium, 2015.
16. Evgenikos, P.; Yannis, G.; Folla, K.; Bauer, R.; Machata, K.; Brandstaetter, C. Characteristics and Causes of Heavy Goods Vehicles and Buses Accidents in Europe. *Transp. Res. Procedia* 2016, 14, 2158–2167. [CrossRef]
17. Summerskill, S.; Marshall, R. *Understanding Direct and Indirect Driver Vision in Heavy Goods Vehicles—Summary Report*; Loughborough University: Loughborough, UK, 2016.
18. Alrutz, D. *Unfallrisiko und Regelakzeptanz von Fahrradfahrern: [Bericht zum Forschungsprojekt FE 82.262: Unfallrisiko, Konfliktpotenzial und Akzeptanz der Verkehrsregelungen von Fahrradfahrern]*; Bundesanstalt für Straßenwesen (BASt): Bergisch Gladbach, Germany, 2009.
19. Schindler, R.; Piccinini, G.B. Truck drivers' behavior in encounters with vulnerable road users at intersections: Results from a test-track experiment. *Accid. Anal. Prev.* 2021, 159, 106289. [CrossRef]
20. HVU. Ulykker Mellem Højresvingende Lastbiler og Ligeudkørende Cyklister: Rapport nr. 4. 2006. Available online: https://www.ft.dk/samling/20061/almdel/reu/bilag/107/318507.pdf (accessed on 23 February 2022).
21. Silla, A.; Leden, L.; Rämä, P.; Scholliers, J.; van Noort, M.; Bell, D. Can cyclist safety be improved with intelligent transport systems? *Accid. Anal. Prev.* 2017, 105, 134–145. [CrossRef] [PubMed]

22. Talbot, R.; Reed, S.; Barnes, J.; Thomas, P.; Christie, N. *Pedal Cyclist Fatalities in Pedal Cyclist Fatalities in London: Analysis of Police Collision Files (2007–2011)*; UCL; Loughborough University: Loughborough, UK, 2014.
23. Summala, H.; Pasanen, E.; Räsänen, M.; Sievänen, J. Bicycle accidents and drivers' visual search at left and right turns. *Accid. Anal. Prev.* **1996**, *28*, 147–153. [CrossRef] [PubMed]
24. Mole, C.D.; Wilkie, R.M. Looking forward to safer HGVs: The impact of mirrors on driver reaction times. *Accid. Anal. Prev.* **2017**, *107*, 173–185. [CrossRef]
25. *Type-Approval of Devices for Indirect Vision and of Vehicles Equipped with These Devices*; European Parliament and European Council: London, UK, 2003.
26. Richter, T.; Sachs, J. Turning accidents between cars and trucks and cyclists driving straight ahead. *Transp. Res. Procedia* **2017**, *25*, 1946–1954. [CrossRef]
27. Pokorny, P.; Pitera, K. Truck-bicycle safety: An overview of methods of study, risk factors and research needs. *Eur. Transp. Res. Rev.* **2019**, *11*, 29. [CrossRef]
28. Hoschopf, H.; Tomasch, E. Limitations and challenges of avoiding HGV-VRU accidents through advanced driver assistance systems. In Proceedings of the 8th International Conference on ESAR "Expert Symposium on Accident Research", Hanover, Germany, 19–20 April 2018.
29. European Commission. *Road Safety: Commission Welcomes Agreement on New EU Rules to Help Save Lives*; European Commission: Brussels, Belgium, 2019.
30. *Type-Approval Requirements for Motor Vehicles and Their Trailers, and Systems, Components and Separate Technical Units Intended for such Vehicles, as Regards Their General Safety and the Protection of Vehicle Occupants and Vulnerable Road Users. Regulation (EU)*; European Parliament and European Council: London, UK, 2019.
31. Wilmink, I.; Janssen, W.; Jonkers, E.; Malone, K.; van Noort, M.; Rämä, P.; Sihvola, N.; Kulmala, A.; Schirokoff, G.; Lind, T.; et al. *Deliverable D4: Impact Assessment of Intelligent Vehicle Safety Systems: Final Report and Integration of Results and Perspectives for Market Introduction of IVSS. Version 2.0*; SWOV: Hague, The Netherlands, 2008.
32. Kingsley, K.J. *Evaluating Crash Avoidance Countermeasures Using Data from FMCSA/NHTSA's Large Truck Crash Causation Study*; National Highway Traffic Safety Administration: Washington, DC, USA, 2009.
33. Kuehn, M.; Hummel, T.; Bende, J. Advanced Driver Assistance Systems for Trucks—Benefit Estimation from Real-Live Accidents. In Proceedings of the 22nd International Technical Conference on the Enhanced Safety of Vehicles (ESV), Washington, DC, USA, 13–16 June 2011; pp. 1–12.
34. Hoedemaeker, D.M.; Doumen, M.; de Goede, M.; Hogema, J.H.; Brouwer, R.F.T.; Wennemers, A.S. *Modelopzet Voor Dodehoek Detectie en Signalerings Systemen (DDSS)*; TNO: Soesterberg, The Netherlands; Stichting Wetenschappelijk Onderzoek Verkeersveiligheid (SWOV): Hague, The Netherlands, 2010.
35. Tomasch, E.; Smit, S. Naturalistic driving study on the impact of an aftermarket blind spot monitoring system on the driver's behaviour of heavy goods vehicles and buses on reducing conflicts with pedestrians and cyclists. *Accid. Anal. Prev.* **2023**, *192*, 107242. [CrossRef]
36. Smit, S.; Tomasch, E. *Rundum-Sicht im Straßenverkehr*; Bundesministerium für Verkehr, Innovation und Technologie: Vienna, Austria, 2020.
37. ADAC. Lkw-Abbiegeassistenten im Test: So Verhindern Sie Schwere Unfälle. Available online: www.adac.de/rund-ums-fahrzeug/tests/assistenzsysteme/lkw-abbiegeassistent/ (accessed on 4 February 2022).
38. Tomasch, E.; Steffan, H.; Darok, M. Retrospective Accident Investigation Using Information from Court. In *Transport Research Arena*; TRA: Ljubljana, Slovenia, 2008.
39. Bärgman, J.; Boda, C.-N.; Dozza, M. Counterfactual simulations applied to SHRP2 crashes: The effect of driver behavior models on safety benefit estimations of intelligent safety systems. *Accid. Anal. Prev.* **2017**, *102*, 165–180. [CrossRef] [PubMed]
40. Davis, G.A.; Hourdos, J.; Xiong, H.; Chatterjee, I. Outline for a causal model of traffic conflicts and crashes. *Accid. Anal. Prev.* **2011**, *43*, 1907–1919. [CrossRef]
41. Bärgman, J.; Lisovskaja, V.; Victor, T.; Flannagan, C.; Dozza, M. How does glance behavior influence crash and injury risk? A 'what-if' counterfactual simulation using crashes and near-crashes from SHRP2. *Transp. Res. Part F Traffic Psychol. Behav.* **2015**, *35*, 152–169. [CrossRef]
42. Wille, J.; Zatloukal, M. rateEFFECT—Effectiveness evaluation of active safety systems. In Proceedings of the 5th International Conference on ESAR "Expert Symposium on Accident Research", Hanover, Germany, 7–8 September 2012; pp. 1–41.
43. Eichberger, A.; Rohm, R.; Hirschberg, W.; Tomasch, E.; Steffan, H. RCS-TUG Study: Benefit Potential Investigation of Traffic Safety Systems with Respect to Different Vehicle Categories. In Proceedings of the 22th International Conference on the Enhanced Safety of Vehicles (ESV), Washington, DC, USA, 13–16 June 2011; pp. 1–13.
44. *ISO/TS 21934-2:2024*; Road Vehicles—Prospective Safety Performance Assessment of Pre-Crash Technology by Virtual Simulation: Part 2: Guidelines for Application. International Organization for Standardization: Geneva, Switzerland, 2024.
45. Zauner, C.; Tomasch, E.; Sinz, W.; Ellersdorfer, C.; Steffan, H. Assessment of the effectiveness of Intersection Assistance Systems at urban and rural accident sites. In Proceedings of the 6th International Conference on ESAR "Expert Symposium on Accident Research", Hanover, Germany, 20–21 June 2014.
46. Erbsmehl, C.T. Simulation of Real Crashes as a Method for Estimating the Potential Benefits of Advanced Safety Technologies. In Proceedings of the The 21st ESV Conference Proceedings, NHTSA, Stuttgart, Germany, 15–18 June 2009.

47. Johannsen, H. *Unfallmechanik und Unfallrekonstruktion: Grundlagen der Unfallaufklärung*, 3rd ed.; Springer Vieweg: Wiesbaden, Germany, 2013.
48. Wagner, H.-J. *Verkehrsmedizin*; Springer: Berlin/Heidelberg, Germany, 1984.
49. Burg, H.; Moser, A. *Handbuch Verkehrsunfallrekonstruktion: Unfallaufnahme, Fahrdynamik, Simulation*, 3rd ed.; Vieweg: Wiesbaden, Germany, 2017.
50. Steffan, H. PC-CRASH, A Simulation Program for Car Accidents. In Proceedings of the 26th International Symposium on Automotive Technology and Automation, Aachen, Germany, 13–17 September 1993.
51. Steffan, H.; Moser, A. The Collision and Trajectory Models of PC-CRASH. In *International Congress & Exposition*; SAE International: Warrendale, PA, USA, 1996.
52. Cliff, W.E.; Montgomery, D.T. *Validation of PC-Crash—A Momentum-Based Accident Reconstruction Program*; SAE Technical Papers; SAE International: Warrendale, PA, USA, 1996.
53. Moser, A.; Hoschopf, H.; Steffan, H.; Kasanicky, G. Validation of the PC-Crash Pedestrian Model. *SAE Trans.* **2000**, *109*, 1316–1339.
54. Hugemann, W. (Ed.) *Unfallrekonstruktion*; Verl. Autorenteam: Münster, Germany, 2007.
55. Schubert, A.; Erbsmehl, C.T.; Hannawald, L. Standardized pre-crash-scenarios in digital format on the basis of the VUFO simulation. In Proceedings of the 5th International Conference on ESAR "Expert Symposium on Accident Research", Hanover, Germany, 7–8 September 2012.
56. Gruber, M.; Kolk, H.; Klug, C.; Tomasch, E.; Feist, F.; Schneider, A.; Roth, F. The effect of P-AEB system parameters on the effectiveness for real world pedestrian accidents. In Proceedings of the 26th ESV Conference Proceedings, Eindhoven, The Netherlands, 10–13 June 2019.
57. Wagström, L.; Bohman, K.; Lindman, M.; Laudon, O.; Tomasch, E.; Klug, C.; Schachner, M.; Levallois, I.; Renaudin, R.F.; Salters, E.; et al. *Impact Scenarios and Pre-Crash Seated Positions for Automated Driving, EU Project VIRTUAL Deliverable D3.1*; Vehicle Safety Institute: Graz, Austria, 2020.
58. Burckhardt, M.; Burg, H.; Gnadler, R.; Näumann, E.; Schiemann, G. Die Brems-Reaktionsdauer von Pkw-Fahrern. *Der Verkehrsunfall* **1981**, *12*, 224–235.
59. Burckhardt, M. Zur Analyse und Synthese von Reaktionszeiten. *Verkehrsunfall* **1980**, *18*, 161–168.
60. Burckhardt, M. *Reaktionszeiten bei Notbremsvorgängen*; TÜV Rheinland: Köln, Germany, 1985.
61. Hugemann, W. Driver Reaction Times in Road Traffic. In Proceedings of the Annual EVU Meeting, Lisbon, Portugal, 25 November 2002.
62. Green, M. "How Long Does It Take to Stop?"—Methodological Analysis of Driver Perception-Brake Times. *Transp. Hum. Factors* **2000**, *2*, 195–216. [CrossRef]
63. Irzik, M.; Kranz, T.; Bühne, J.-A.; Glaeser, K.-P.; Limbeck, S.; Gail, J.; Bartolomaeus, W.; Wolf, A.; Sistenich, C.; Kaundinya, I.; et al. *Feldversuch mit Lang-Lkw (German Field Trial with Longer Trucks), Fachverlag NW in Carl Ed*; Schünemann KG: Bremen, Germany, 2018.
64. *Uniform Provisions Concerning the Approval of Motor Vehicles with Regard to the Blind Spot Information System for the Detection of Bicycles: UN Regulation No 151*; United Nations Economic Commission for Europe: Geneva, Switzerland, 2021.
65. *Uniform Provisions Concerning the Approval of Motor Vehicles with Regard to the Moving off Information System for the Detection of Pedestrians and Cyclists: Addendum 158—UN Regulation No. 159*; United Nations Economic Commission for Europe: Geneva, Switzerland, 2021.
66. Zink, F. Fünf Sensortypen für Fahrerassistenzsysteme im Überblick. *Krafthand Technikmagazin*, 15 June 2021.
67. Sensor Fusion Technology Can Bale Out the Automotive Industry from Innovation Constraints. Available online: www.ideapoke.com/growthleader/sensor-fusion-technology-can-bale-out-the-automotive-industry-from-innovation-constraints (accessed on 31 October 2024).
68. Westerhuis, F.; de Waard, D. Reading cyclist intentions: Can a lead cyclist's behaviour be predicted? *Accid. Anal. Prev.* **2017**, *105*, 146–155. [CrossRef]
69. Meijer, R.; de Hair, S.; Elfring, J.; Paardekooper, J.P. Predicting the intention of cyclists. In Proceedings of the 6th International Cycling Safety Conference (ICSC 2017), Davis, CA, USA, 20–23 September 2017.
70. Hemeren, P.E.; Johannesson, M.; Lebram, M.; Eriksson, F.; Ekman, K.; Veto, P. The use of visual cues to determine the intent of cyclists in traffic. In Proceedings of the 2014 IEEE International Inter-Disciplinary Conference on Cognitive Methods in Situation Awareness and Decision Support (CogSIMA), San Antonio, TX, USA, 3–6 March 2014; IEEE: New York, NY, USA, 2014; pp. 47–51.
71. Källhammer, J.-E.; Smith, K.; Hollnagel, E. An Empirical Method for Quantifying Drivers' Level of Acceptance of Alerts Issued by Automotive Active Safety Systems. In *Driver Acceptance of New Technology*; Regan, M.A., Horberry, T., Stevens, A., Eds.; CRC Press: Boca Raton, FL, USA, 2018; pp. 121–134.
72. Neumann, T. Analysis of Advanced Driver-Assistance Systems for Safe and Comfortable Driving of Motor Vehicles. *Sensors* **2024**, *24*, 6223. [CrossRef]
73. Parasuraman, R.; Riley, V. Humans and Automation: Use, Misuse, Disuse, Abuse. *Hum. Factors* **1997**, *39*, 230–253. [CrossRef]
74. Tang, T.; Guo, Y.; Zhang, G.; Wang, H.; Shi, Q. Understanding the Interaction between Cyclists' Traffic Violations and Enforcement Strategies: An Evolutionary Game-Theoretic Analysis. *Int. J. Environ. Res. Public Health* **2020**, *17*, 8457. [CrossRef] [PubMed]
75. Bacchieri, G.; Barros, A.J.D.; Santos, J.V.D.; Gigante, D.P. Cycling to work in Brazil: Users profile, risk behaviors, and traffic accident occurrence. *Accid. Anal. Prev.* **2010**, *42*, 1025–1030. [CrossRef]

76. Fraboni, F.; Puchades, V.M.; de Angelis, M.; Prati, G.; Pietrantoni, L. Social Influence and Different Types of Red-Light Behaviors among Cyclists. *Front. Psychol.* **2016**, *7*, 1834. [CrossRef]
77. *Bundesgesetz vom 6. Juli 1960, mit dem Vorschriften über die Straßenpolizei Erlassen Werden (Straßenverkehrsordnung 1960): StVO*; Österreichischer Nationalrat: Vienna, Austria, 2024.

Disclaimer/Publisher's Note: The statements, opinions and data contained in all publications are solely those of the individual author(s) and contributor(s) and not of MDPI and/or the editor(s). MDPI and/or the editor(s) disclaim responsibility for any injury to people or property resulting from any ideas, methods, instructions or products referred to in the content.

Article

The CornerGuard: Seeing around Corners to Prevent Broadside Collisions

Victor Xu [1] and Sheng Xu [2,*]

1 Texas Academy of Mathematics and Science (TAMS), University of North Texas, Denton, TX 76203, USA; victorxu@my.unt.edu
2 Math Department, Southern Methodist University, Dallas, TX 75275, USA
* Correspondence: sxu@smu.edu

Abstract: Nearly 3700 people are killed in broadside collisions in the U.S. every year. To reduce broadside collisions, we created and tested the CornerGuard, a prototype system that senses around a corner to alert a car driver of an impending collision with a pedestrian or automobile that is not in the line of sight (LOS). The CornerGuard leverages a microwave-transceiving radar sensor mounted on the car and a curved radio wave reflector installed at the corner to sense around the corner and detect a broadside collision threat. The car's speed is constantly read by an onboard diagnostics (OBD) system to allow the sensor to differentiate between static objects and objects approaching around the corner. Field testing demonstrated that the CornerGuard can effectively and consistently detect threats at a consistent range without blind spots under broad weather conditions. Our proof of concept study shows that the CornerGuard can be enhanced to be readily integrated into automobile construction and street infrastructure.

Keywords: Doppler effect; radar detection; broadside collision avoidance; radar wave reflector; seeing around corners

Citation: Xu, V.; Xu, S. The CornerGuard: Seeing around Corners to Prevent Broadside Collisions. *Vehicles* 2024, 6, 1468–1481. https://doi.org/10.3390/vehicles6030069

Academic Editors: Deogratias Eustace, Bhaven Naik, Heng Wei and Parth Bhavsar

Received: 13 June 2024
Revised: 16 August 2024
Accepted: 22 August 2024
Published: 27 August 2024

Copyright: © 2024 by the authors. Licensee MDPI, Basel, Switzerland. This article is an open access article distributed under the terms and conditions of the Creative Commons Attribution (CC BY) license (https://creativecommons.org/licenses/by/4.0/).

1. Introduction

In the summer of 2023, three children in North Texas were fatally hit by cars. Nearly 3700 people are killed in broadside collisions in the U.S. every year, a rate equivalent to one person every two and a half hours [1]. Across the world, traffic accidents cause serious health problems; nearly 1.35 million people are killed or disabled in traffic accidents every year [2]. The two largest factors that constitute over 90% of all broadside collisions, which occur everywhere from neighborhoods to city streets and involve pedestrians and other automobiles, are driver distraction and a limited field of view for the driver [1].

To prevent broadside collisions, we built the CornerGuard, a prototype system that can sense around a corner on the right to detect a collision threat that is not in the driver's line of sight (LOS) because of the visual obstruction by the corner. We consider only a right corner because a left corner is further away from the driver's side with a larger field of view for the driver. We combined a transceiving Doppler radar sensor mounted on a moving car and a stationary radar wave reflector installed at a street corner to detect impending pedestrian and automobile broadside collisions. The novelty of this work includes (1) the implementation of a stationary radar wave reflector and (2) a method to distinguish an approaching target around a corner from surrounding stationary objects.

Doppler radars are widely used in various civilian and military applications to detect and track moving objects in the line of sight (LOS). In [3], a model-free approach was presented for directly detecting and tracking moving objects in street scenes from point clouds obtained via a Doppler LiDAR. This approach can collect spatial information and Doppler images by using Doppler-shifted frequencies. Two types of Doppler LiDAR were used: a static terrestrial Doppler LiDAR and a mobile Doppler LiDAR. The static terrestrial Doppler LiDAR could achieve a maximum scanning range of 1 km with a scanning frequency of 5 Hz and a scanning angle of 4°. The mobile Doppler LiDAR could scan a maximum range of 400 m with the horizontal scanning angle of 40°. In [4], a microwave radar sensor mounted on a moving robot was used to provide Doppler information, which could be extracted and interpreted to obtain the velocities of both the detected objects and the robot itself. As pointed out in [4], the detection and tracking of moving objects in an outdoor environment by a mobile robot is a difficult task because of the wide variety of dynamic objects and the difficulty of separating moving objects from stationary objects.

Conventional imaging, vision, and detection systems require a direct LOS of the scene of interest. However, in many applications, obtaining a direct LOS may be unsafe, challenging, or even impossible. The concept of seeing around obstacles has been a popular topic in science fiction for years. There are two main ideas for detecting objects that are not in the LOS: through-wall techniques and reflective-surface techniques.

In [5], ultra-wideband (UWB) radar operating at the L-band (1–2 GHz) with a minimum bandwidth of 500 MHz, or a fractional bandwidth of at least 20%, was used to directly sense stationary human targets behind a wall. The detection distances range from 6.5 ft to 7.5 ft in the experiments with different wall materials and thicknesses. The detection would fail in the case of a thick concrete wall. In [6], a microwave Doppler radar sensor in the S-band (2–4 GHz) was used to detect humans behind visually opaque structures, such as building walls. The distance between the radar and the target was 3 m. We initially considered a through-wall radar (TWR) detection approach. However, through-wall detection requires a very strong radar signal and suffers from the limitations of large noises and low ranges. In contrast to through-wall techniques, the CornerGuard system proposed in this paper employs a reflective surface, which greatly improves the detection range for real traffic scenarios and simplifies the detection algorithm and implementation.

Many imaging techniques use reflective surfaces to detect objects hidden behind occluding structures [7–10]. In [10], a three-dimensional image of a scene hidden behind an occluding structure was reconstructed from an ordinary photograph of a matte LOS surface illuminated by the hidden scene. Such direct-vision-based detection methods usually require good lighting and weather conditions and sophisticated computer vision algorithms. In [11], measurements of objects moving around intersections in a realistic scene were made using a static radar operating at X-band (8–12 GHz), and radar waves returned after one or two wall reflections were processed to identify the target. In [12], a radio frequency (RF)-based method was proposed to provide accurate around-corner indoor localization through a novel encoding of how RF signals bounce off walls and occlusions. The encoding is fed to a neural network along with the radio signals to localize people around corners.

In contrast to many above-mentioned methods, the CornerGuard system proposed in this paper has the following advantages: (1) It is a very simple solution that does not require sophisticated detection algorithms. (2) It can achieve a sufficient detection range for real traffic scenarios. (3) It has low latency for punctuality. (4) It is consistent even in darkness and broad weather conditions. (5) It can be easily implemented in a current vehicle. (6) It is cost-friendly.

2. Materials and Methods

In this section, we detail the detection methodology, including its underlying theory and practical implementation.

2.1. System Overview

The CornerGuard consists of a radar subsystem, as shown in Figure 1, and a radar wave reflector subsystem, as shown in Figure 2. Note that the radar sensor is mounted on the right side of the car to detect only right corners, as shown in Figure 1b. The body of the car acts as a shield to avoid the detection of approaching cars from opposite lanes.

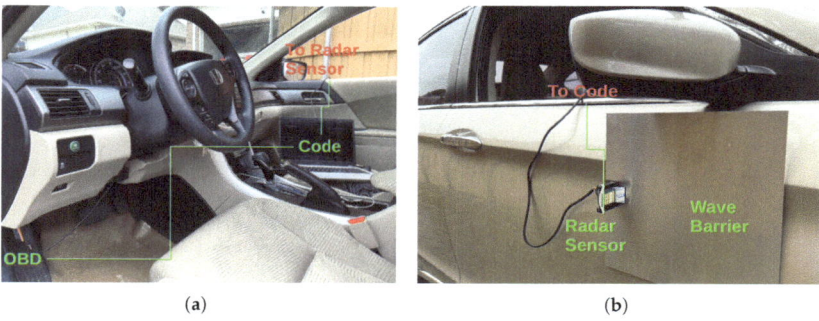

Figure 1. An overview of the CornerGuard's radar subsystem, which includes (**a**) an onboard diagnostics (OBD) device and a computer that collects and analyzes data and (**b**) a radar sensor.

Figure 2. The setup of the reflector in (**a**) blind spot tests and (**b**) field simulation tests.

The radar subsystem consists of a small, low-cost digital RFbeam 24 GHz K-LD7 microwave radar sensor, a laptop computer, and an onboard diagnostics (OBD) device, as shown in Figure 1. The car for our tests was a 2015 Honda Accord. Note that only a radar sensor is needed when a car manufacturer decides to implement the current detection approach, as a modern car already has a built-in computer system to read car sensor information, run detection algorithms, and send control signals.

2.2. Doppler Effect

The Doppler effect in radar detection is the change in the frequency of a radar wave in relation to an object that is moving relative to the source of the wave [13]. The frequency shift Δf is proportional to the relative speed Δv at which the sensor and the object approach each other. The relation is

$$\Delta f \approx 2 f_s \frac{\Delta v}{c}, \tag{1}$$

where f_s is the frequency of the radar waves emitted from the radar sensor and c is the speed of light.

The radar sensor is mounted on a moving car and approaches surrounding stationary objects. As illustrated in Figure 3a, the relative speed at which they approach each other is $v_{car} \cos \alpha$, which satisfies

$$v_{car} \cos \alpha \leq v_{car}, \tag{2}$$

where v_{car} is the ground speed of the car and α is the angle of the object relative to the moving direction of the car.

As illustrated in Figure 3b, for an object around a right corner that is not in the line of sight (LOS) of the radar sensor but seen as a virtual image by the sensor through the reflector, the relative speed at which the object and the senor approach each other is about $v_{car} + v_{obj}$, which satisfies

$$v_{car} + v_{obj} > v_{car}, \quad (3)$$

where v_{obj} is the ground speed at which the object moves toward the reflector.

Compared with the ground speed of the car, the object around the corner that moves toward the intersection can be distinguished from a surrounding stationary object, as the speed of the former relative to the car is greater than the car's speed, while the speed of the latter is slower.

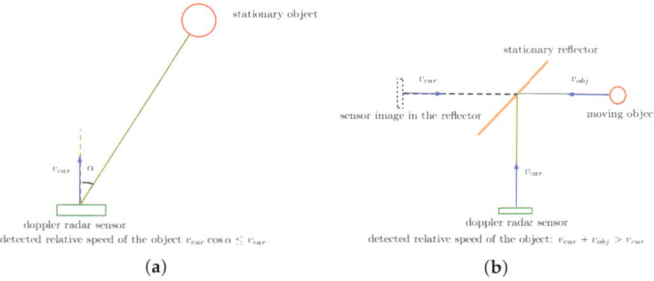

Figure 3. The relative speed at which the radar sensor and an object approach each other. (**a**) A surrounding stationary object; (**b**) an object that approaches the sensor in the reflector.

2.3. The Reflector

Metal surfaces can reflect radar waves very well. The reflector of the current Corner-Guard is made of an aluminum sheet 0.0078 inches thick, which can reflect more than 90% of a radar wave with little absorption.

The dimensions of the surface of the reflector are 48 inches by 30 inches. The center of the reflector is about the same height as the radar sensor, which is about 3 ft, the typical height of a toddler. In this way, the reflector surface does not need to be vertically curved to have no vertical blind spots for the target detection. As shown in Figure 4a, the surface of the reflector needs to be smoothly curved in the horizontal direction parallel to the ground to avoid horizontal blind spots. As shown in the inset of Figure 4b, the radar beams are directed by a 45° reflective arc to a 90° coverage if the beams hit the reflector vertically. A 45° bend, therefore, should suffice for most intersections.

The loss of radar wave energy due to reflection about the reflector may be neglected. However, the curvature of the reflector reduces the detection range of the radar sensor. Below, we estimate the range loss due to the curvature in the horizontal direction.

Without a reflector, the energy intensity received from the sensor by an object of horizontal width w at a distance $l_1 + l_2$ is proportional to w and inversely proportional to $(l_1 + l_2)^2$ as

$$I_o \approx \frac{kw}{(l_1 + l_2)^2} I_s, \quad (4)$$

where I_o is the energy intensity received by the object, I_s is the emitted energy intensity at the sensor, and k is a constant.

In the case illustrated in Figure 4b, the energy intensity received by the object is

$$I_r \approx \frac{kR}{l_1^2} I_s \cdot \frac{w}{(l_2 + R)\theta}, \quad (5)$$

where R is the horizontal dimension of the reflector and θ (in radians) is the angle defined in Figure 4b, which is about $\pi/2 \approx 3/2$.

The ratio I_r/I_o is

$$\frac{I_r}{I_o} \approx \left(\frac{l_1+l_2}{l_1}\right)^2 \cdot \frac{2R}{3(l_2+R)}, \qquad (6)$$

which is the detection range reduction factor. To have a larger detection range, we need a stronger radar sensor (with larger I_s) and a wider reflector (with larger R).

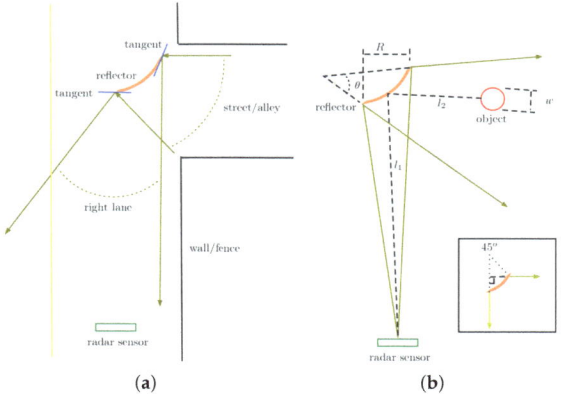

Figure 4. (**a**) The geometric design of the reflector to avoid blind spots. (**b**) The geometric analysis of radar detection range reduction due to reflector curvature.

2.4. Procedure

The following is the procedure for practical implementation:

- Programming and integrating the radar sensor. Connect a radar sensor and an OBD device to a laptop computer (Figure 1). Write a Python program to communicate with the radar sensor and OBD via USB communication ports.
- Designing and building the stationary reflector. Secure a flexible aluminum sheet to a hardboard with screws to form the reflective surface, and bend it using paracord as bowstrings to form a 45° arc surface (Figure 2). Secure the surface to a sharpened wooden plank to form a stake.
- Collecting and analyzing data. Record the speeds, distances, and angles of objects detected by the radar sensor in tests. Use an OBD package to constantly read the car's speed. Compare the reading with the speeds of the objects detected by the radar sensor in the program. Save all data to a text file. If the speed of an object detected by the radar sensor is greater than the instantaneous speed of the car read by the OBD, save the detection data to a separate text file.

The code for programming the radar sensor and OBD to collect and analyze data is given in the Appendix A.

3. Results and Discussions

The setup of the field experiment is shown in Figure 2. In addition to being tested in clear and sunny conditions, the CornerGuard was also tested in dark and misty conditions.

When an object is moving relative to the sensor, the frequency of the received radar waves reflected from the object is different from the radar waves transmitted by the sensor because of the Doppler effect. This frequency difference gives the relative speed of the object. In Figure 5, the fast Fourier transform (FFT) is used to convert raw sensor readings into a frequency difference spectrum plotted as signal amplitude versus speed (frequency difference). Additionally, the constant offset of 20 dB is plotted as a threshold line to filter out unwanted noise.

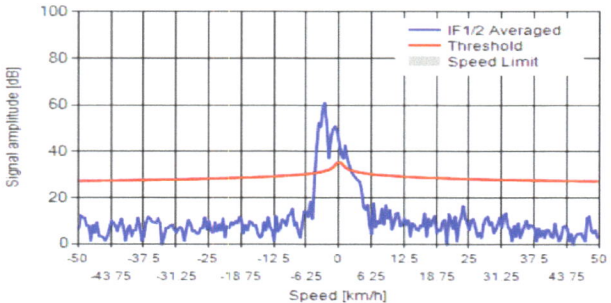

Figure 5. A spectrum obtained by the FFT (fast Fourier transform), plotted as signal magnitude vs. object speed, including a red 20 dB noise threshold line.

In the figures below, negative velocities indicate approaching targets, and positive velocities indicate receding targets. For example, in Figure 6a, from the time 0 to 10 s, the targets were receding from the sensor. Detection was then lost from time 10 to 30 s. Afterward, during the time 30 to 50 s, the targets were approaching the sensor.

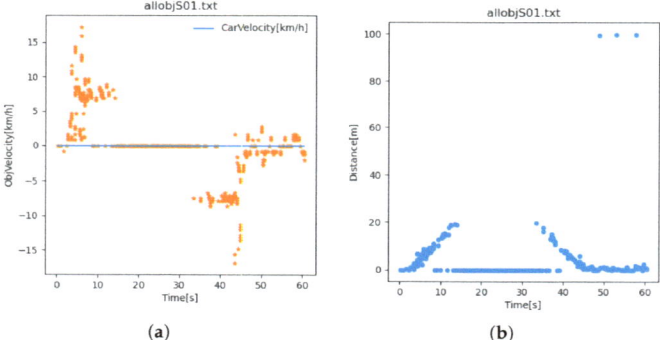

Figure 6. Range assessment through a straightaway test: (**a**) speed vs. time; (**b**) range vs. time.

3.1. Range and Blind Spot Assessment

Before starting simulation scenario tests, we ran tests to assess the range of the sensor and the coverage of the reflector.

In the range assessment test, an author walked away from the sensor in the direct LOS of the sensor until detection was lost and then walked back toward the sensor, as shown in Figure 6a. As shown in Figure 6b, the maximum detection range to detect a human being was about 20 m. We then placed the sensor perpendicular to the alleyway behind a corner fence and staked the reflector in front of the sensor to direct radar waves down the alleyway, as shown in Figure 2a and Figure 4. A similar range assessment test was run. To avoid false positives, we made sure not to come into the direct LOS of the sensor, so lines were drawn on the ground to mark the end of the sensor's direct LOS coverage, as shown in Figure 2a. The maximum detection range was reduced to about 10 m, as shown in Figure 7b.

Using the latter setup with the reflector above, we also ran a test to assess the area coverage of the reflector and identify any blind spots. We divided the alleyway into five lanes, as shown in Figure 2a, and one author walked back and forth in each lane in front of the reflector within the detection range. Each point peak in Figure 7a represents a detection in a specific lane. There were no blind spots.

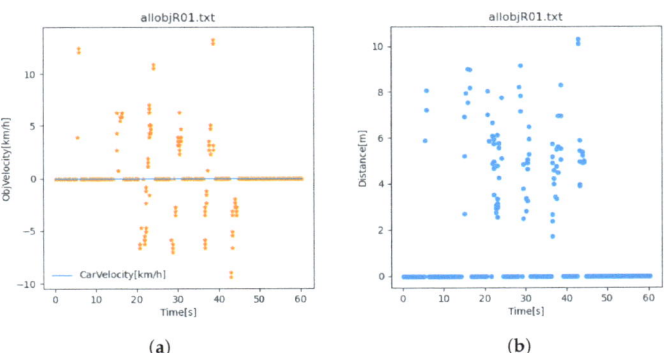

Figure 7. Blind spot assessment through a reflector test: (**a**) speed vs. time; (**b**) range vs. time.

3.2. Simulation Assessment

We first drove the car on a road to test the detection of stationary objects on the right side of the road. Shown in Figure 8a are the instantaneous velocity of the car read by the OBD and the velocities of multiple surrounding stationary objects detected by the radar sensor. As indicated in Figure 3a, the detected speeds of surrounding stationary objects should be less than or equal to the car's speed in theory. However, this may not be the case in practice because of errors in the readings of the OBD and radar sensor, as shown in Figure 8b. As indicated by Figure 8b, the speed of a surrounding stationary object detected by the radar could be up to 2 km/h larger than the car's speed read by the OBD, so to account for the errors, a collision threat as illustrated in Figure 3b was detected only when its speed was at least 3 km/h larger than the car's speed.

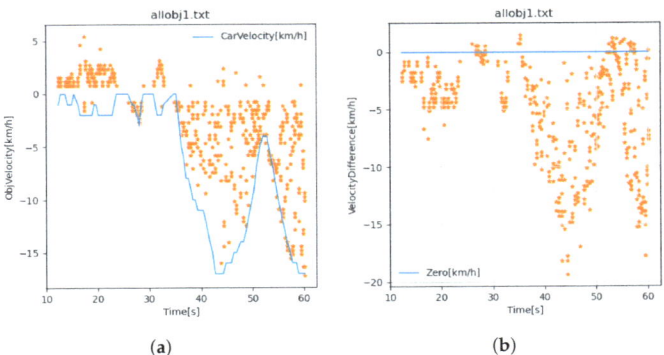

Figure 8. Error assessment of the radar sensor and OBD mounted on a car driving in stationary surroundings: (**a**) velocity vs. time; (**b**) velocity difference vs. time, where the velocity difference is the car's velocity minus the velocity of a detected object.

With all preliminary tests complete, we began running field simulation tests. The field setup is shown in Figure 2b, which is a scaled-down simulation. The car approached the reflector from the alleyway, and one author walked back and forth in front of the reflector from the house's back driveway, orthogonal to the alleyway.

The blue line in Figure 9a is the instantaneous velocity of the car as read by the OBD. The orange points represent all detected objects. Figure 9b shows the corresponding detection ranges. Figure 10a shows two detected collision threats, as the author walked back and forth twice in front of the reflector when the car approached the reflector. Figure 10b shows that the detection distance decreased from the first to the second threat.

Because radar detection does not rely on lighting conditions, the CornerGuard worked equally well when tested in dark and misty weather conditions.

The latency of the system is the time taken for data to go from the radar source to the detection algorithm output. It depends on the radar sensor, the baud rate, and the data processing algorithm. It was difficult to measure the latency. In broad terms, one might expect 40 ms latency at the analog acquisition and buffering stage, followed by a few milliseconds of processing latency, some non-deterministic network latency of maybe 5 ms [14], and finally a few milliseconds latency at the algorithm end. These numbers combine to give a total of around 50 ms. In most situations, it was reasonable to assume a latency of around 50 ms as a good working value. If the car speed is 50 mph, the distance traveled by the car in a latency of 50 ms is about 1 m.

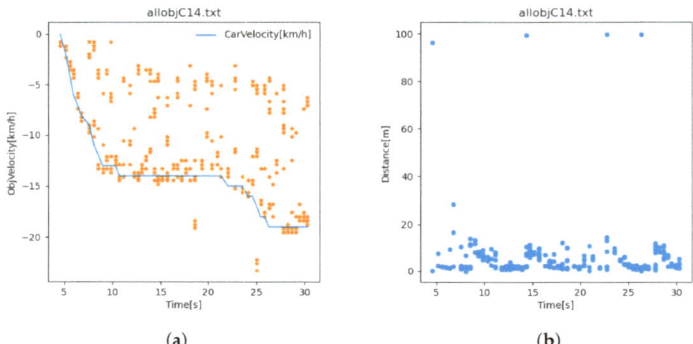

Figure 9. Detection of all objects: (**a**) velocity vs. time; (**b**) range vs. time.

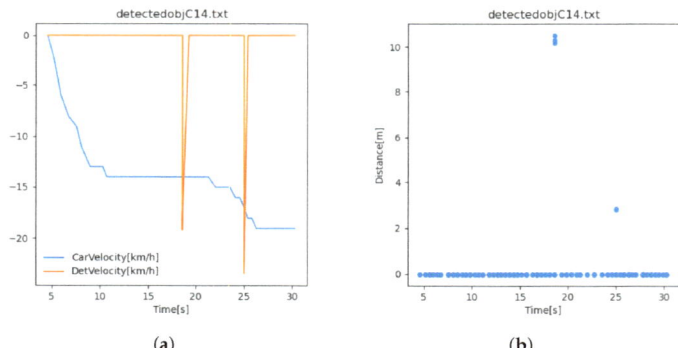

Figure 10. Detection of an approaching object: (**a**) velocity vs. time; (**b**) range vs. time.

4. Conclusions

The objective of the proof-of-concept work presented in this paper was to create a device to alert a driver of an impending collision with a non-LOS pedestrian or automobile around a corner at right. The devised solution, the CornerGuard, involved using a stationary radar wave reflector to reflect emitted radar waves to detect impending broadside collision threats.

Field trials and simulations demonstrated that the prototype CornerGuard can operate effectively and consistently in a range with no blind spots. It works well in darkness and broad weather conditions. It is also cost-friendly.

In future work, an even stronger radar sensor with a larger range and more sensitivity can be implemented, and the radar wave reflector's geometry can be further refined.

The CornerGuard can readily be implemented. By integrating the radar sensor and detection logic with car construction and strategically installing reflectors in collision-prone intersections, the rate of broadside collisions due to distraction and limited visibility can be reduced to save lives.

Author Contributions: Conceptualization, V.X.; methodology, V.X.; software, V.X. and S.X.; investigation, V.X. and S.X.; formal analysis, V.X. and S.X.; writing—original draft preparation, V.X.; writing—review and editing, S.X.; visualization, V.X. and S.X.; supervision, S.X.; project administration, S.X. All authors have read and agreed to the published version of the manuscript.

Funding: This research recieved no external funding.

Data Availability Statement: The raw data supporting the conclusions of this article will be made available by the authors on request.

Conflicts of Interest: The authors declare no conflicts of interest.

Appendix A. Code

Below is the code to communicate with the radar sensor, analyze the data, and implement the detection algorithm.

```python
# Script to read out raw target data from RFbeam K-LD7 and speed of car
$ from OBD2, detect approaching object around corner, and save data
# to files
#
# Author:   Victor Xu, Sheng Xu
# Date:     Nov and Dec 2023
# Python Version: 3
#
# Notes:    Use correct COM Port (specifed by port properties
#           in device manager (in Windows) for each serial device
#           and make sure all modules are installed before executing

import time
import serial
import matplotlib.pyplot as plt
import numpy as np
import math
import obd

import re
import winsound

print('Start tracking!')
# specify runtime, data file names
runtime = 0.5 # minutes
speederror = -3 # km/h
dur = 1000 # milisecond, duration of alerting sound
allobjfile = 'allobj282.txt' # all objects
detectedobjfile = 'detectedobj282.txt' # target object

# specify correct COM  USB ports for serial devices
COM_Port = 'COM8' # port for radar sensor
OBD_Port = 'COM9' # port for OBD2
```

```python
# create serial object with corresponding COM Port and open it
com_obj=serial.Serial(COM_Port)
print(com_obj)
com_obj.baudrate=115200
com_obj.parity=serial.PARITY_EVEN
com_obj.stopbits=serial.STOPBITS_ONE
com_obj.bytesize=serial.EIGHTBITS

# connect to sensor and set baudrate
payloadlength = (4).to_bytes(4, byteorder='little')
value = (0).to_bytes(4, byteorder='little')
header = bytes(''INIT'', 'utf-8')
cmd_init = header+payloadlength+value
com_obj.write(cmd_init)

# get response
response_init = com_obj.read(9)
if response_init[8] != 0:
    print('Error during initialisation for K-LD7')
else:
    print('K-LD7 successfully initialized!')

# delay 75ms
time.sleep(0.075)

# change to higher baudrate
com_obj.baudrate = 115200

# change max speed to 50km/h
value = (2).to_bytes(4, byteorder='little')
header = bytes(''RSPI'', 'utf-8')
cmd_frame = header+payloadlength+value
com_obj.write(cmd_frame)

# get response
response_init = com_obj.read(9)
if response_init[8] != 0:
    print('Error: Command not acknowledged')
else:
    print('Max speed successfully set!')

# change max range to 100m
value = (3).to_bytes(4, byteorder='little')
header = bytes(''RRAI'', 'utf-8')
cmd_frame = header+payloadlength+value
com_obj.write(cmd_frame)

# get response
response_init = com_obj.read(9)
if response_init[8] != 0:
    print('Error: Command not acknowledged')
else:
    print('Max distance successfully set!')
```

```python
# create figure for real-time plotting
fig = plt.figure(figsize=(10,5))
plt.ion()
plt.show()

starttime=time.time()
connection = obd.OBD(OBD_Port) # create connection with USB 0
print('OBD2 successfully connected!')
detobj = open(detectedobjfile,'w')
allobj = open(allobjfile,'w')

# readout and plot PDAT data continuously
# for ctr in range(100):
while 1:

    # request next frame data
    PDAT = (4).to_bytes(4, byteorder='little')
    header = bytes(''GNFD'', 'utf-8')
    cmd_frame = header+payloadlength+PDAT
    com_obj.write(cmd_frame)

    # get acknowledge
    resp_frame = com_obj.read(9)
    if resp_frame[8] != 0:
        print('Error: Command not acknowledged')

    # get header
    resp_frame = com_obj.read(4)

    # get payload len
    resp_len = com_obj.read(4)

    # initialize arrays
    distances_x = np.zeros(100)
    distances_y = np.zeros(100)
    speeds = np.zeros(100)
    distances = np.zeros(100)
    angles = np.zeros(100)
    i = 0

    length = resp_len[0]

    # get data, until payloadlen is zero
    while length > 0:
        PDAT_Distance = np.frombuffer(com_obj.read(2), dtype=np.uint16)
        PDAT_Speed = np.frombuffer(com_obj.read(2), dtype=np.int16)/100
        PDAT_Angle = math.radians(np.frombuffer(com_obj.read(2),\
                            dtype=np.int16)/100)
        PDAT_Magnitude = np.frombuffer(com_obj.read(2), dtype=np.uint16)

        distances_x[i] = -(PDAT_Distance * math.sin(PDAT_Angle))/100
        distances_y[i] = PDAT_Distance * math.cos(PDAT_Angle)/100
        distances[i] = PDAT_Distance/100
        speeds[i] = PDAT_Speed
```

```
            angles[i] = math.degrees(PDAT_Angle)

            i = i + 1

            # subtract stored datalen from payloadlen
            length = length - 8

    # current time
    lapsedtime = time.time()-starttime

    # read car speed from obd and convert it to float
    cmd = obd.commands.SPEED # select an OBD command (sensor)
    response = connection.query(cmd) # send the command,
                                      # and parse the response
    speedstring = str(response.value)
    print(speedstring) # in km/h
    speedstr=re.findall(r"[-+]?\d*\.?\d+|[-+]?\d+",speedstring)[0]
    carspeed=-float(speedstr)
    print(carspeed)

    # clear figure
    plt.clf()

    # plot speed/distance
    if np.count_nonzero(distances)==0:
        print(lapsedtime,carspeed,0.0,0.0,0.0,sep=',',end='\n',file=allobj)
        print(lapsedtime,carspeed,0.0,0.0,0.0,sep=',',end='\n',file=detobj)

    sub1 = plt.subplot(121)
    for j in range(np.count_nonzero(distances)):
        print(lapsedtime,carspeed,speeds[j],distances[j],angles[j],\
            sep=',',end='\n',file=allobj)
        if speeds[j]<carspeed+speederror:
            print(''Approaching object around corner detected!'')
            freq = 1000
            if distances[j]<5:
                freq = 3000 # Hz, sound frequency
            winsound.Beep(freq,dur)
            print(lapsedtime,carspeed,speeds[j],distances[j],angles[j],\
                sep=',',end='\n',file=detobj)
        else:
            print(lapsedtime,carspeed,0.0,0.0,0.0,sep=',',end='\n',\
                file=detobj)
        point_Sub1, = sub1.plot(speeds[j],distances[j],\
                marker='o',markersize=15, markerfacecolor='b',\
                markeredgecolor='k')

    plt.grid(True)
    plt.axis([-75, 75, 0, 100])
    plt.title('Distance / Speed')
    plt.xlabel('Speed [km/h]')
    plt.ylabel('Distance [m]')

    # plot distance/distance
```

```
            sub2 = plt.subplot(122)
            for y in range(np.count_nonzero(distances_x)):
                if speeds[y] > 0 :
                    point_Sub2, = sub2.plot(distances_x[y], distances_y[y],\
                    marker='o', markersize=15,markerfacecolor='g',\
                    markeredgecolor='k')
                else:
                    point_Sub2, = sub2.plot(distances_x[y], distances_y[y],\
                    marker='o',markersize=15,markerfacecolor='r',\
                    markeredgecolor='k')

            plt.grid(True)
            plt.axis([-10, 10, 0, 100])
            plt.title('Distance / Distance \n (Green: Receding, Red: Approaching)')
            plt.xlabel('Distance [m]')
            plt.ylabel('Distance [m]')

            # draw no. of targets
            plt.text(0.8, 0.95,'No. of targets: ' +
                \str(np.count_nonzero(distances)), horizontalalignment='center',\
                verticalalignment='center', transform = sub2.transAxes)

            # draw figure
            fig.canvas.draw()
            fig.canvas.flush_events()

            # reset arrays
            distances_x = np.zeros(100)
            distances_y = np.zeros(100)
            speeds = np.zeros(100)
            distances = np.zeros(100)
            i = 1

            # exit when time is up
            if lapsedtime>runtime*60: #lapsed time>runtime in minutes
                print('Trial ends!')
                break

# close files
detobj.close()
allobj.close()

# disconnect from sensor
payloadlength = (0).to_bytes(4, byteorder='little')
header = bytes(''GBYE'', 'utf-8')
cmd_frame = header+payloadlength
com_obj.write(cmd_frame)

# get response
response_gbye = com_obj.read(9)
if response_gbye[8] != 0:
    print(''Error during disconnecting with K-LD7'')
```

```
# close connection to COM port
com_obj.close()
```

References

1. Kilduff, E. Where Do Broadside Collisions Most Commonly Occur? 2021. Available online: https://emrochandkilduff.com/where-broadside-collisions-occur/ (accessed on 12 June 2024).
2. Ahmed, S.K.; Mohammed, M.G.; Abdulqadir, S.O.; El-Kader, R.G.A.; El-Shall, N.A.; Chandran, D.; Rehman, M.E.U.; Dhama, K. Road traffic accidental injuries and deaths: A neglected global health issue. *Health Sci. Rep.* **2023**, *6*, e1240. [CrossRef]
3. Ma, Y.; Anderson, J.; Crouch, S.; Shan, J. Moving Object Detection and Tracking with Doppler LiDar. *Remote Sens.* **2019**, *11*, 1154. [CrossRef]
4. Vivet, D.; Checchin, P.; Chapuis, R.; Faure, P.; Rouveure, R.; Monod, M.O. A mobile ground-based radar sensor for detection and tracking of moving objects. *EURASIP J. Adv. Signal Process.* **2012**, *2012*, 45. [CrossRef]
5. Singh, S.; Liang, Q.; Chen, D.; Sheng, L. Sense through wall human detection using UWB radar. *EURASIP J. Wirel. Commun. Netw.* **2011**, *2011*, 20. [CrossRef]
6. Gennarelli, G.L.; Soldovieri, F. Real-Time Through-Wall Situation Awareness Using a Microwave Doppler Radar Sensor. *Remote Sens.* **2016**, *8*, 621. [CrossRef]
7. Gariepy, G.; Tonolini, F.; Henderson, R.; Leach, J.; Faccio, D. Detection and tracking of moving objects hidden from view. *Nat. Photonics* **2016**, *10*, 23–26. [CrossRef]
8. Chan, S.; Warburton, R.E.; Gariepy, G.; Leach, J.; Faccio, D. Non-line-of-sight tracking of people at long range. *Opt. Express* **2017**, *25*, 10109–10117. [CrossRef] [PubMed]
9. Faccio, D.; Velten, A.; Wetzstein, G. Non-line-of-sight imaging. *Nat. Rev. Phys.* **2020**, *2*, 318–327. [CrossRef]
10. Czajkowski, R.; Murray-Bruce, J. Two-edge-resolved three-dimensional non-line-of-sight imaging with an ordinary camera. *Nat. Commun.* **2024**, *15*, 1162. [CrossRef] [PubMed]
11. Johansson, T.; Örbom, A.; Sume, A.; Rahm, J.; Nilsson, S.; Herberthson, M.; Gustafsson, M.; Andersson, Å. Radar measurements of moving objects around corners in a realistic scene. *Proc. SPIE* **2014**, *9077*, 90771Q. [CrossRef]
12. Yue, S.; He, H.; Cao, P.; Zha, K.; Koizumi, M.; Katabi, D. CornerRadar: RF-Based Indoor Localization Around Corners. *Proc. ACM Interact. Mob. Wearable Ubiquitous Technol.* **2022**, *6*, 1–24. [CrossRef]
13. Wolf, G.; Gasper, E.; Stoke, J.; Kretchman, J.; Anderson, D.; Czuba, N.; Oberoi, S.; Pujji, L.; Lyublinskaya, I.; Ingram, D.; et al. Section 17.4: The Doppler Effect. In *College Physics 2e. for AP Courses*; OpenStax, Rice University: Houston, TX, USA, 2022; ISBN-13: 978-1-951693-61-9. Available online: https://openstax.org/details/books/college-physics-ap-courses-2e (accessed on 12 June 2024).
14. Helliar, R. Measuring Latency from Radar Interfacing to Display. 2024. Available online: https://www.unmannedsystemstechnology.com/feature/measuring-latency-from-radar-interfacing-to-display/ (accessed on 12 June 2024).

Disclaimer/Publisher's Note: The statements, opinions and data contained in all publications are solely those of the individual author(s) and contributor(s) and not of MDPI and/or the editor(s). MDPI and/or the editor(s) disclaim responsibility for any injury to people or property resulting from any ideas, methods, instructions or products referred to in the content.

Article

Impacts of a Toll Information Sign and Toll Lane Configuration on Queue Length and Collision Risk at a Toll Plaza with a High Percentage of Heavy Vehicles

Farnaz Zahedieh and Chris Lee *

Department of Civil and Environmental Engineering, University of Windsor, Windsor, ON N9B 3P4, Canada; farnaz.zh75@gmail.com
* Correspondence: cclee@uwindsor.ca

Abstract: This study assessed the impacts of a toll information sign with different toll lane configurations on queue length and collision risk at a toll plaza with an estimated high percentage of heavy vehicles (HVs). The toll information sign displays information about different toll payment methods for cars and HVs upstream of the toll booth. The impacts were assessed for the toll plaza of the Gordie Howe International Bridge under construction at the Windsor–Detroit international border crossing using a traffic simulation model. Results show that the toll information sign upstream of the toll plaza and converting the toll lanes with multiple toll payment methods to electronic toll collection (ETC)-only lanes reduced queue length and collision risk. However, increasing the number of HV-only lanes for a higher percentage of HVs increased lane-change collision risk. Thus, it is recommended that toll lane configurations be changed based on the percentage of HVs to reduce collision risk at a toll plaza.

Keywords: toll information; toll lane configuration; road sign; surrogate safety measure; heavy vehicle; lane change; traffic simulation

1. Introduction

Toll plazas are a critical component of a roadway system for capital financing and ongoing infrastructure maintenance revenue. Although toll plazas have been designed and constructed for a long time, there are no widely accepted design standards for toll plaza uniformity or safety. Due to a lack of standards, there is a growing concern about safety at toll plazas. For instance, some crashes have occurred on toll roads in Canada. They were mostly high-speed-related crashes and lane-change-related crashes at toll plazas which caused death and injury.

According to the U.S. National Traffic Safety Board, toll plazas are the most dangerous locations on highways. In 2006, 49% of crashes on expressways in Illinois occurred at toll plazas and the fatality of these crashes was three times higher than the fatality of crashes on the rest of the expressways [1]. In addition, 30% of crashes on the Pennsylvania Turnpike and 38% of crashes on New Jersey toll highways occurred at toll plazas [1]. A noticeable increase in the number of crashes at toll plazas, particularly upstream of toll plazas, has generated the need to study drivers' behavior as drivers approach toll plazas [2].

To reduce the delay at toll plazas, new tolling technologies such as electronic toll collection (ETC) have been in operation at toll plazas. ETC is an automated system that allows drivers to pay tolls without stopping. ETC consists of a transponder placed inside the vehicle and is activated when the vehicle passes a roadside sensor at the toll booth [3]. ETC has numerous benefits such as lower transaction time, improved throughput, and reduced air pollution and fuel consumption [4]. However, some drivers still manually pay tolls using cash or credit cards. These drivers may be distracted when they search for cash or cards and take time to change to toll lanes which accept manual payment. These behaviors affect drivers' perception and reaction time, and consequently road safety [5]. Moreover, when there are both ETC and manual toll collection lanes in the toll plaza, drivers are more likely to abruptly change lanes to select the toll lane of their preference.

In this regard, a toll information sign that displays toll lane configurations (e.g., the method of toll payment and the type of vehicle) ahead of the toll plaza can help drivers prepare to move to the correct lane or path to the open toll booths with their preferred payment method. Thus, it is important to examine whether the toll information signs can help drivers choose the toll lane of their preference in advance and avoid abrupt speed reductions and lane changes near the toll booth, which disrupt traffic flow and increase collision risk [6].

In particular, the toll information signs and toll lane configuration might have differential effects on traffic performance and safety when the percentage of heavy vehicles (HVs) is high. Since HVs are longer in length and have lower speed and acceleration than cars, car drivers are more likely to change lanes to avoid following HVs while trying to choose toll lanes with shorter queue lengths (e.g., toll lanes with a lower number of HVs). However, there is a lack of studies on the impacts of toll information signs and toll lane configurations on traffic performance and the conflicts between cars and HVs at toll plazas with a high percentage of HVs.

To fill this research gap, this study analyzes the movements of cars and HVs at a toll plaza with toll information signs at different locations and different toll lane configurations. This study also predicts lane-change collision risk by the type of lane-changing vehicle and trailing vehicle in the target lane (car and HV) to investigate the impact on the severity of conflicts. This study could contribute to the development of operational strategies of a toll plaza to improve efficiency and safety based on the varying traffic demands of cars and HVs. Thus, the objective of this study is to assess the impacts of toll information signs and toll lane configuration on the queue length and collision risk at a toll plaza with various toll payment methods.

2. Literature Review

This section reviews the past studies on the impacts of toll information systems and toll lane configuration on lane changes, collisions, the risk of collision, and traffic performance, and describes their research gaps.

2.1. Impacts on Lane Changes

Past studies assessed the impact of static and variable toll information signs on lane changes at a toll plaza. For instance, Valdés et al. [5] found that showing the manual toll collection (MTC) and ETC lanes in overhead static signs at a toll plaza in Puerto Rico allowed smoother lane changes and reduced the number of lane changes at lower speeds. Saad et al. [7] assessed how real-time information for ramp traffic provided via a portable variable message sign (VMS) affected driver behavior at a toll plaza using a driving simulator. They found that VMS effectively kept the vehicles from the on-ramp to the toll plaza in the rightmost lane and reduced lane changing before the toll plaza.

2.2. Impacts on Collisions

Several studies assessed the safety of toll plazas using historical crash data. For instance, Abdelwahab and Abdel-Aty [2] studied traffic safety at toll plazas using the 1999 and 2000 traffic crash reports of the Central Florida expressway system. The results showed that vehicles equipped with ETC devices, especially trucks, were more likely to be involved in crashes at the toll plaza than vehicles without ETC devices. This is potentially because ETC users cannot avoid crashes when ETC lanes are blocked by non-ETC users, while they do not anticipate that they should reduce speed or stop at the toll booth. The study also found that ETC users are more likely to be severely injured than non-ETC users. Abuzwidah and Abdel-Aty [8] found that the risk of crashes was 19% higher for ETC lanes in the mainline and separate manual toll collection lanes to the side than manual toll collection on the mainline and separate ETC lanes to the side. Chakraborty et al. [9] reported that converting Hybrid Toll Plazas to an All-Electronic Toll Collection system considerably reduced the number of crashes.

2.3. Impacts on Risk of Collision

Some researchers analyzed the safety of toll plazas based on the risk of collision predicted using vehicle trajectories and surrogate safety measures. For instance, Xing et al. [10] investigated traffic conflicts in the upstream diverging area of a toll plaza using trajectory data extracted from unmanned aerial vehicle (UAV) videos and extended time-to-collision (TTC). They found that a mix of vehicles with different toll payment methods (MTC and ETC) increased collision risk upstream of the toll plaza. Jehad et al. [11] also assessed the safety impacts of different toll lane configurations on collision risk using trajectory data from a VISSIM simulation and a Surrogate Safety Assessment Model (SSAM). They found that all ETC lanes were safer than toll lane configurations with different toll payment methods because they showed the lowest number of crossing and lane-changing conflicts.

2.4. Impacts on Traffic Performance

Past studies analyzed the impacts of toll plaza configurations on traffic performance. McKinnon [12] found from VISSIM simulation results that toll lanes with multiple forms of toll payment helped disperse traffic demand during peak hours. However, accepting both manual and electronic payment degraded the level of service and increased delays for all drivers. Moreover, drivers were sensitive to slower-moving vehicles and tried to avoid queued HVs in both cash and ETC lanes. Bains et al. [13] also found from VISSIM simulation results that separate lanes for cars and HVs decreased throughput volume and increased queue length at toll plazas. Although the separation of HVs from cars is generally expected to reduce conflicts between different vehicle types and improve traffic performance, this benefit was not observed due to the high volume of cars in the studied toll plaza. Moreover, it was found that traffic volumes and types of toll service affected traffic operations at the toll plaza [14]. Mittal and Sharma [15] found from VISSIM simulation results that queue length more significantly increased as the traffic volume increased for MTC lanes than ETC lanes. However, the difference in queue length between MTC and ETC lanes was relatively smaller for HVs. Bari et al. [16] also found from VISSIM simulations that an ETC system reduced the queue delay by up to 95% compared to an MTC system as the percentage of HVs increased from 25% to 45% at a toll plaza in India. However, this study did not consider the difference in service time between cars and HVs in the simulation. Moreover, the study only calibrated differences in car-following behavior between cars and HVs in the simulation, not lane-changing behavior.

2.5. Research Gaps

In summary, these studies did not explicitly consider differences in both car-following and lane-changing behaviors between cars and HVs for evaluating the impacts of toll information signs and lane configuration. Since the percentage of HVs is generally high at toll plazas on freeways, it is important to consider the interactions between cars and HVs while they approach toll booths and their impacts on the queue length at different toll lanes and the conflicts between cars and HVs. Moreover, although microsimulation models can be used for testing various traffic conditions at a toll plaza, these past simulation-based studies did not examine the effects of the varying demand of HV traffic on the effectiveness of toll information signs and toll lane configuration in reducing conflicts and delay. Thus, this study will use a simulation-based approach to reflect the differences in behaviors between cars and HVs, and extensively analyze the effects of HVs on performance and safety in various operational strategies regarding toll information signs and toll lane configurations.

3. Methods and Data

3.1. Description of Data

In this study, the impacts of toll information signs on traffic performance and safety were assessed for the toll plaza at the Gordie Howe International Bridge between Windsor, Ontario, Canada and Detroit, Michigan, USA. The Gordie Howe International Bridge is a cable-stayed international bridge across the Detroit River and the bridge is currently under construction. The preliminary design plans, including preliminary drawings of the design layout of the toll plaza for both Canada-bound and U.S.-bound bridges, were provided by the Windsor-Detroit Bridge Authority (WDBA) in 2021. The preliminary design plans for the Canada-bound toll plaza propose a four-lane entry road for passenger cars and a three-lane entry road for heavy vehicles upstream of the toll plaza, and these entry roads merge to the toll lane as shown in Figures 1 and 2. The distance between the merge point and the entry gate of the toll plaza is 160 m. The entry gate is located 75 m upstream of the toll booth. The entry gate will be closed only when the toll booth is closed due to low traffic volume. These details are subject to change.

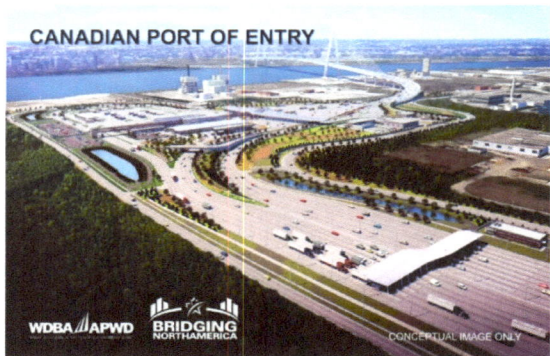

Figure 1. Conceptual design of the toll plaza at the Canada-bound Gordie Howe International Bridge. (Source: Windsor-Detroit Bridge Authority [17]).

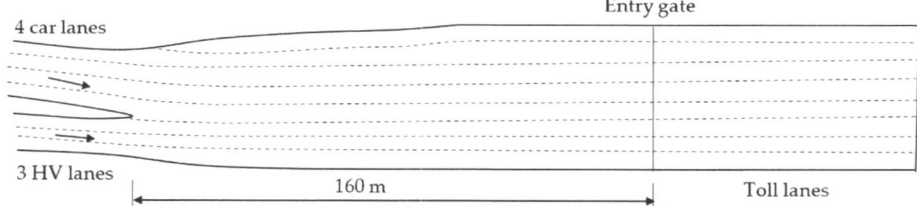

Figure 2. Schematic drawing of the Canada-bound toll plaza.

According to the WDBA, it is anticipated that the following three methods of toll payment will be accepted at the tollbooth: (1) Manual toll collection (MTC)—payment in cash, (2) Electronic toll collection (ETC)—payment with a transponder, and (3) Automatic toll collection (ATC)—non-cash payment without a transponder (e.g., credit card). In the case of ETC, as the toll payment is processed via wireless communication between a transponder and the toll booth, vehicles can pass through the toll booth without reducing their speed or stopping. In each toll lane, only specific toll payment methods (e.g., ETC and ATC only) or all toll payment methods will be accepted.

A proposed lane assignment scenario (subject to change) including eight toll lanes with the assigned payment method and type of vehicle at the Canada-bound toll plaza is shown in Figure 3. The same lane configuration scenario is also proposed for the U.S.-bound toll plaza. The lane number starts from the innermost lane. Lanes 1–3 are only open to cars whereas Lanes 7–8 are only open to HVs for all toll payment methods. Lanes 4 and 5 are only open to ETC and ATC for both cars and HVs, where the presence of toll collectors is not required. Only Lane 6 is open to all vehicles and all toll payment methods.

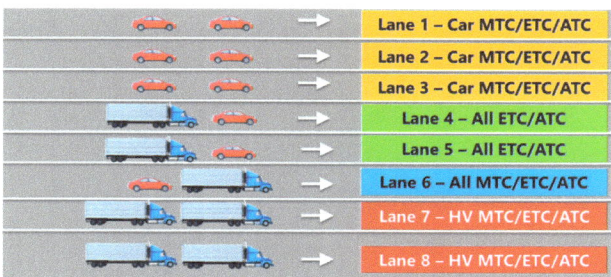

Figure 3. Current proposed configuration of toll lanes at the Canada-bound toll plaza.

3.2. VISSIM Traffic Simulation

As the Gordie Howe International Bridge is currently under construction, there is no observed traffic data for the toll plaza on the bridge. Thus, the VISSIM microscopic traffic simulation model [18] was used to replicate traffic at the bridge toll plaza with the proposed toll lane configuration shown in Figure 3. Among different microsimulation models, the VISSIM model was selected because the model is an effective tool to assess the operation at toll plazas [13] and it has been used to simulate vehicle movements at a toll plaza in previous studies [11,12,14–16]. The VISSIM model was used to predict the changes in driver behavior due to the toll information sign and assess the impacts of the sign on queue length and collision risk. The VISSIM model was developed as follows.

A VISSIM road network consists of various links, stop signs at the toll booth, and reduced speed areas. In the reduced speed areas, vehicles reduce their speed to 5 km/h after passing the entry gate and then stop at the toll booth. Peak-hour traffic demand of 300 cars and 300 HVs was used based on the assumption that 10% of total daily traffic volume, i.e., the forecasted daily traffic demand of approximately 3000 cars and 3000 HVs in 2025 according to the 2018 border crossing origin-destination surveys [19], occurs during the peak hour.

Service time is the time during which a vehicle pays a toll at the tollbooth and exits from the toll plaza, not including the waiting time in the queue. The actual service time depends on the method of toll payment. First, the service time for MTC is generally longer than ATC because of the longer transaction time for cash payments compared to non-cash payments. Bari et al. [16] observed that the median service time for MTC vehicles was about 12 s. Wang et al. [20] also found that the mean service times for cars and HVs in MTC lanes were about 11 s and 15 s, respectively, at a toll plaza in China. Similarly, Al-Deek et al. [21] found that the service time was relatively longer for HVs than cars (about 2 s) because HVs accelerate more slowly than cars after toll payment.

Second, the variability of service time is larger for MTC than ATC because the service time varies with the toll collector's experience [22]. Therefore, a higher standard deviation was assumed for the service time of manual payments. Traffic congestion also affects the service time because when toll collectors are under greater pressure from a growing queue, they tend to process transactions faster [21]. Thus, the service time for MTC will be shorter in peak hours than off-peak hours.

Based on these observed service times from past studies, the service times were determined for MTC and ATC for cars and HVs as follows: Mean service times for MTC and ATC for cars are 10 s (standard deviation (SD) = 10 s) and 5 s (SD = 5 s), respectively, and mean service time for MTC and ATC for HVs are 12 s (SD = 10 s) and 7 s (SD = 5 s), respectively. The distribution of service time was assumed to be normal distribution and the ranges of service time were generated by VISSIM.

Different proportions of toll payment methods were also assumed for cars and HVs. For HVs, a significantly higher proportion of electronic toll payments (80%) compared to manual and automatic toll payments (10% and 10%, respectively) was assumed because they are more likely to be equipped with transponders due to the law—HVs without a valid transponder are charged under the Highway Traffic Act [23]. In addition, according to the 2008 survey study for the new Windsor–Detroit border crossing, car drivers considered electronic toll payments without stopping at a toll plaza less important than HV drivers [24]. This is potentially because car drivers generally use the toll plaza less frequently than HV drivers. Due to the lower likelihood of electronic toll payments by car drivers than HV drivers and a similar likelihood of manual and automatic toll payments by car drivers, the proportions of manual, automatic, and electronic toll payments were assumed to be the same (33%).

Proportions of toll lane use for different toll payment methods were assumed for cars and HVs separately. It was assumed that cars and HVs are more likely to use the toll lanes that are exclusively open to cars and HVs, respectively—Lanes 1, 2, and 3 for cars and Lanes 7 and 8 for HVs. In the case of manual toll payments, this tendency is particularly higher for cars than HVs because car drivers are less likely to choose the toll lane with a higher proportion of HVs [16] or a higher number of large vehicles [25]. Thus, the proportions of toll lane use for cars and HVs with different toll payment methods were determined, as shown in Table 1.

Table 1. Proportions of toll lane use by vehicle type and toll payment method.

	Toll Lane							
	Car MTC/ETC/ATC			All ETC/ATC		All	HV MTC/ETC/ATC	
Vehicle Type	1	2	3	4	5	6	7	8
MTC Car		95%		N/A		5%	N/A	
ETC/ATC Car		35%		60%		5%	N/A	
MTC HV		N/A		N/A		30%	70%	
ETC/ATC HV		N/A		40%		10%	50%	

To restrict each vehicle type (i.e., car or HV with specific toll payment methods) to using only the above-designated toll lanes, the vehicle routes for each vehicle type were separately created in VISSIM. To reflect the fact that drivers are more likely to choose the toll lane with a shorter queue length to reduce waiting time, the "queue counter" was placed at each toll booth in VISSIM. This allows drivers to compare the queue length among different toll lanes and choose the lane with the shortest queue length. This reflects drivers' actual lane choice behavior at toll plazas in the real world—they are more likely to choose the toll lane with a shorter queue length, as observed in past studies [24–26].

As the real-world behaviors of drivers at the bridge could not be observed, the existing calibrated VISSIM driving behavior parameters (car-following and lane-changing) from previous studies were used. A set of 10 car-following parameters calibrated using the observed vehicle trajectories from the US-101 freeway in California [27] and a set of nine calibrated lane-changing parameters [28] were used, as shown in Table 2. The description of each parameter is also shown in Table 2. Note that the car-following parameters (CC0-CC9) are the parameters used in the Wiedemann 99 car-following model. In particular, these car-following parameters were separately set for cars and HVs to reflect the difference in their car-following behaviors. For instance, HVs maintained longer distances from the lead vehicle and applied lower acceleration and deceleration during car-following compared to cars [27].

Table 2. VISSIM calibrated parameters.

(a) Car-following parameters (Source: Durrani et al. [27])

Model Parameters	Unit	Car	Heavy Vehicle
CC0	m	4.15	4.69
CC1	m	1.5	2.7
CC2	m	11.58	14.02
CC3	s	−4	−4.55
CC4	m/s	−1.65	−2.07
CC5	m/s	1.65	2.07
CC6	m/s	11.44	11.44
CC7	m/s^2	0.09	0.1
CC8	m/s^2	0.49	0.27
CC9	m/s^2	0.45	0.25

(b) Description of car-following parameters (Source: PTV AG [18])

Parameters	Description
CC0	This is the average desired standstill distance between two vehicles and it has no variation.
CC1	Time distribution of the speed-dependent part of the desired safety distance. Shows the number and name of the time distribution. Each time distribution may be empirical or normal. Each vehicle has an individual random safety variable which is considered as CC1.
CC2	This restricts the distance difference (longitudinal oscillation) or how much further than the desired safety distance a driver allows before he intentionally moves closer to the car in front.
CC3	This controls the start of the deceleration process, i.e., the number of seconds before reaching the safety distance. At this stage, the driver recognizes a preceding slower vehicle.
CC4	This defines the negative speed difference during the following process. Low values result in a more sensitive driver reaction to the acceleration or deceleration of the preceding vehicle.
CC5	This defines the positive speed difference. Enter a positive value for CC5 which corresponds to the negative value of CC4. Low values result in a more sensitive driver reaction to the acceleration or deceleration of the preceding vehicle.
CC6	This is the influence of distance on speed oscillation. For value 0, the speed oscillation is independent of the distance and a larger value leads to greater speed oscillation with increasing distance.
CC7	This is the oscillation during acceleration.
CC8	This is the desired acceleration when starting from a standstill (limited by maximum acceleration defined within the acceleration curves).
CC9	This is the desired acceleration at 80 km/h (limited by maximum acceleration defined within the acceleration curves).

(c) Lane-changing parameters (Source: CDM Smith [28])

	Unit	Lane-change vehicle	Trailing vehicle in the target lane
Maximum deceleration	ft/s^2	−10	−8
−1 ft/s^2 per distance	ft	100	100
Accepted deceleration	ft/s^2	−3.28	−3.28
Waiting time before diffusion	s	60	
Minimum front-to-rear headway	ft	1.64	
Safety distance reduction factor	-	0.65	

Table 2. *Cont.*

(d) Description of lane-changing parameters (Source: PTV AG [18])	
Parameters	Description
Maximum Deceleration	Thi is the maximum deceleration for changing lanes based on the specified routes for own vehicle overtaking and the trailing vehicle.
Cooperative Lane changing	If vehicle A observes that a leading vehicle B on the adjacent lane wants to change to his lane A, then vehicle A will try to change lanes itself to the next lane in order to facilitate lane changing for vehicle B.
Front-to-rear headway	This is the minimum distance between two vehicles that must be available after a lane change, so that the change can take place (default value 0.5 m). A lane change during normal traffic flow might require a greater minimum distance between vehicles in order to maintain the speed-dependent safety distance.
Safety distance reduction factor	This parameter concerns the safety distance of the trailing vehicle on the new lane for determining whether a lane change will be carried out, the safety distance of the lane changer itself, and the distance to the preceding, slower lane changer. During the lane change, Vissim reduces the safety distance to the value that results from the following multiplication:Original safety distance × safety distance reduction factorThe default value of 0.6 reduces the safety distance by 40%. Once a lane change is completed, the original safety distance is taken into account again.
Waiting time before diffusion	This period of time is defined as the time a car sits waiting for a gap to change lanes in order to stay on its route before it is removed from the network.

The same range of desired speed distribution (48–58 km/h) was used for both cars and HVs as they were required to substantially reduce their speed when approaching the toll plaza. However, different default desired acceleration/deceleration functions were used for cars and HVs. Default functions were used since the functions could not be calibrated using the observed data. Similarly, default values were used for all other VISSIM parameters (e.g., vehicle models).

Considering drivers' workload for comprehending information and making decisions, it is recommended that the toll information sign be located within a half-mile (805 m) of the toll plaza [5]. Thus, to determine candidate locations for the toll information sign, it is important to ensure that drivers have enough time to decide which tollbooth or toll lane they want to use after they see messages and before they arrive at the tollbooth [7].

To evaluate the impacts of toll information signs and toll lane configurations on traffic performance and safety, three simulation experiments were conducted as shown in Figure 4. First, in Experiment 1, the impacts of the presence and location of the toll information sign were assessed for the current proposed toll lane configuration and traffic demand. With the best scenario found in Experiment 1, the impacts of alternative toll lane configurations for the current traffic demand were assessed in Experiment 2 and the impacts of alternative toll lane configuration for different percentages of HVs were assessed in Experiment 3. Different toll lane configurations were considered because of potential safety problems with the proposed toll lane configuration. For instance, as Lane 6 is open to all vehicles with all toll payment methods, it may increase conflicts between cars and HVs. In addition, Lanes 4 and 5 are open to both ETC and ATC vehicles, although ETC vehicles are not required to stop, unlike ATC vehicles. Moreover, ATC vehicles are not likely to use the innermost and outermost toll lanes (Lanes 1 and 8, respectively) because many lanes near the center of the road are open to ATC vehicles.

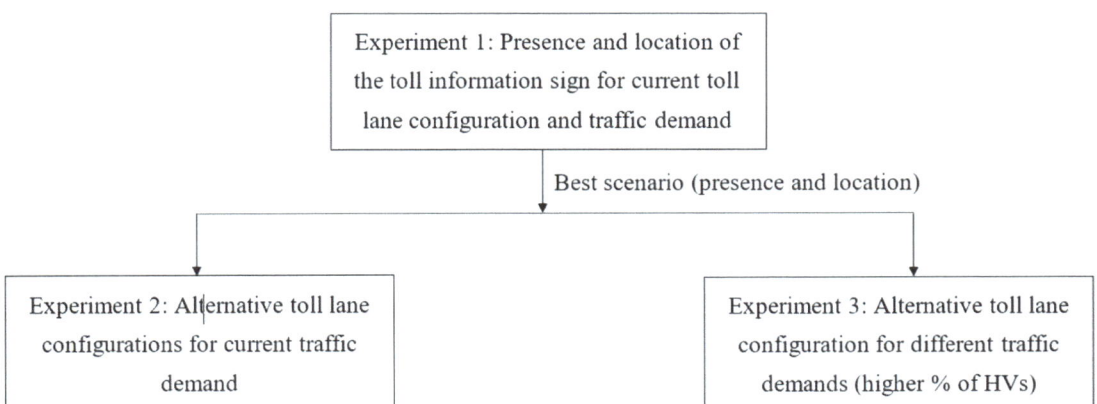

Figure 4. Study process of assessing the impacts of the toll information sign and toll lane configurations.

In Experiment 1, three scenarios (Scenario 1-1: No sign, Scenario 1-2: Sign 140 m from the entry gate, and Scenario 1-3: Separate signs for cars and HVs 75 m before the merge point) were compared, as shown in Figure 5. The purpose of this comparison is to assess whether placing the sign further upstream of the toll plaza and providing toll information in advance can reduce delays and improve the safety of the toll plaza. In the case of no sign, it was assumed that drivers could only see the toll lane sign at each toll booth and choose the toll lane 40 m before the entry gate. To reflect driver reactions to the sign in different locations, the locations of "route decision points" were changed in different scenarios. Although the drivers can see the sign before the location of the sign, in reality, it takes time to understand the messages on the sign and choose the toll lane [29]. Thus, it was assumed that drivers made route decisions at the location of the sign in this study. It was also assumed that all drivers can see the sign and choose the toll lane accordingly, although some drivers may not see the sign or may not choose the toll lane based on the sign even if they see the sign. However, due to uncertainty regarding the proportion of these drivers, this effect was not considered in the simulation.

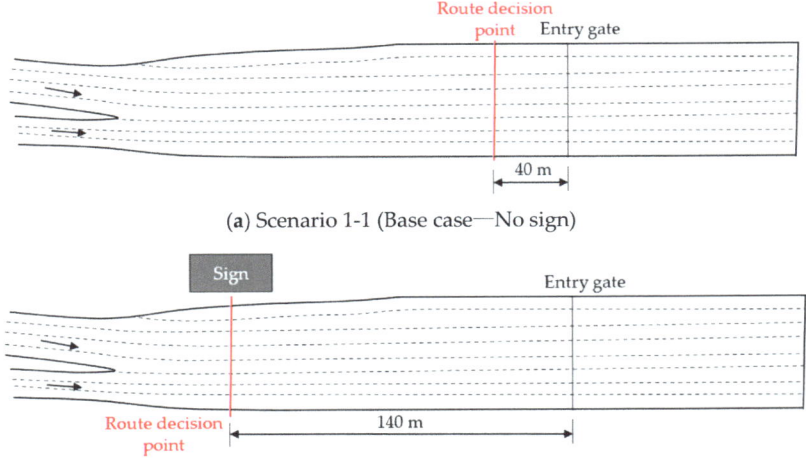

Figure 5. *Cont.*

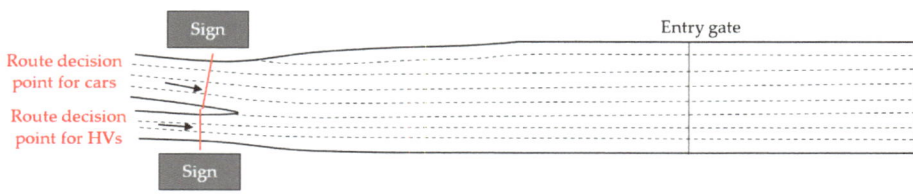

(c) Scenario 1-3 (Separate signs for cars and HVs 75 m before the merge point)

Figure 5. Scenarios on the presence and location of toll information signs in Experiment 1.

In Experiment 2, the following two alternative toll lane configurations—(1) Scenario 2-1: Convert Lanes 4 to 6 to ETC-only lanes and (2) Scenario 2-2: Convert Lanes 1 and 8 to MTC/ETC-only lanes—were compared with the best scenario in Experiment 1 (Base case), as shown in Figure 6. The toll lane configuration in Scenario 2-1 can reduce the delay more effectively as ETC vehicles do not have to stop at the toll booth and they can pass through the toll plaza without having to wait behind the lead MTC or ATC vehicles. It was assumed that 70% of ETC cars and 80% of ETC trucks will use ETC-only lanes. The toll lane configuration in Scenario 2-2 provides a higher number of lanes for MTC and ETC vehicles since the traffic demand for ATC vehicles is not much higher than ETC vehicles.

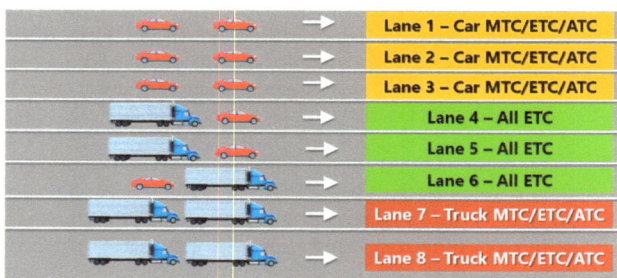

(a) Scenario 2-1 (Convert Lanes 4 to 6 to ETC-only lanes)

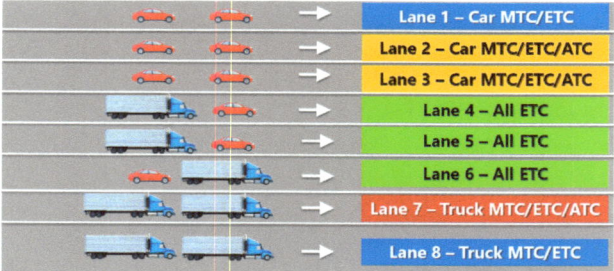

(b) Scenario 2-2 (Convert Lanes 1 and 8 to MTC/ETC-only lanes)

Figure 6. Scenarios of different lane configurations for the current traffic demand in Experiment 2.

In Experiment 3, the impact of converting Lane 6 to an HV-only lane was assessed for two different percentages of HVs (60% and 70%), which are higher than the current percentage of HVs (50%), as shown in Figure 7. The following scenarios were tested: Scenario 3-1: Current toll lane configuration for 60% HVs, Scenario 3-2: Convert Lane 6 to an HV-only lane for 60% HVs, Scenario 3-3: Current toll lane configuration for 70% HVs, and Scenario 3-4: Convert Lane 6 to an HV-only lane for 70% HVs. These scenarios were compared because the number of HVs on the bridge is expected to increase faster than the number of cars by the year 2040 according to the forecasted travel demand provided by the 2018 border crossing origin-destination surveys [19].

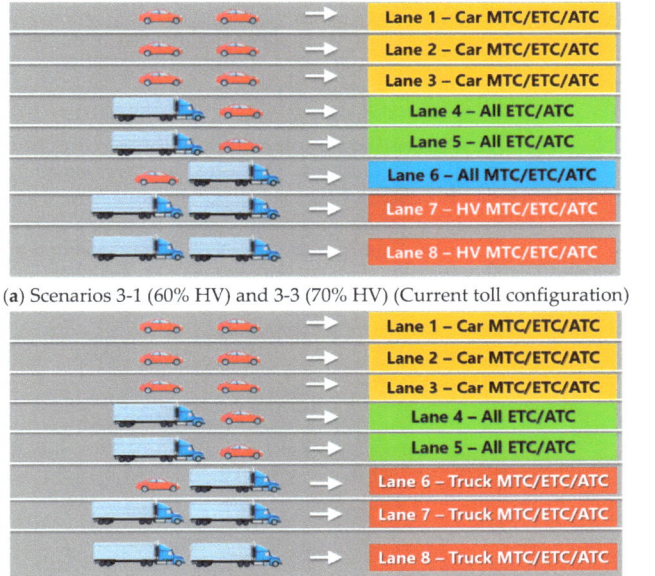

(a) Scenarios 3-1 (60% HV) and 3-3 (70% HV) (Current toll configuration)

(b) Scenarios 3-2 (60% HV) and 3-4 (70% HV) (Convert Lane 6 to an HV-only lane)

Figure 7. Scenarios of different lane configuration for 60% and 70% HVs in Experiment 3.

Each simulation scenario was run five times to consider random variations in the results. The input and output data of the VISSIM model are summarized in Table 3.

Table 3. VISSIM input and output data.

Input Data	Output Data
• Demand: 300 cars and 300 HVs for one hour • Simulation duration: 1 h • Vehicle composition: 50% cars and 50% HVs (base case) • Traffic assignment/Route choice model: The VISSIM model assumes that car and HV drivers choose the route with the shortest queue length at toll lanes with their preferred method of toll payment.	• Average and maximum queue length in each toll lane • Individual vehicle trajectories—these were used to calculate surrogate safety measures.

3.3. Estimation of Collision Risk

The results from the above scenarios were compared in terms of queue length and collision risk. Individual vehicle trajectories from VISSIM and surrogate safety measures were used to determine two types of collision risk: (1) rear-end collision risk and (2) lane-change collision risk. First, rear-end collision risk was estimated using Time-to-collision (TTC) [30] during car-following conditions. TTC is the minimum time for the following vehicle to reach the position of the lead vehicle with the initial constant speed at the time instant when the following vehicle begins braking to avoid collision with the lead vehicle. TTC is calculated using the following equation:

$$\text{TTC}(t) = \frac{S_i(t)}{V_i(t) - V_{i-1}(t)}, \quad \text{if } V_i(t) \geq V_{i-1}(t) \tag{1}$$

where $S_i(t)$ is the spacing between the rear of the lead vehicle $i - 1$ and the front of the following vehicle i at time t, and $V_i(t)$ and $V_{i-1}(t)$ = speed of the following vehicle i and the lead vehicle $i - 1$, respectively, at time t. A lower value of TTC indicates a higher rear-end collision risk.

Lane-change collision risk was estimated using surrogate safety measures developed by Wang and Stamatiadis [31] called the "Aggregate Conflict Propensity Metric (ACPM)". The ACPM assumes that lane-change conflicts may lead to a sideswipe collision or a rear-end collision. For instance, the Required Braking Rate (RBR) to avoid a sideswipe collision during lane changes ($RBR_{LC\text{-}SS}$) is calculated using the following equation:

$$RBR_{LC-SS} = \frac{\frac{2V_2 l_1}{V_1} + l_2 - l_1 \times \cos\theta - \frac{w_1}{\sin\theta} - \frac{w_2}{\tan\theta}}{\left(TTC + \left(\frac{l_1}{V_1}\right) - x\right)^2} \tag{2}$$

where TTC = time to collision, x = reaction time of driver, V_2 = speed of the trailing vehicle in the target lane, V_1 = speed of the lane-changing vehicle, l_2 = length of the trailing vehicle in the target lane, l_1 = length of the lane-changing vehicle, w_2 = width of the trailing vehicle in the target lane, w_1 = width of the lane changing-vehicle, and θ = the conflict angle, which is illustrated in Figure 8.

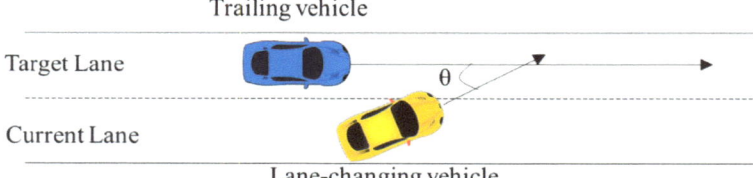

Figure 8. Conflict angle during lane changes.

The RBR to avoid a rear-end collision during lane changes ($RBR_{LC\text{-}RE}$) is calculated as follows:

$$RBR_{LC-RE} = \frac{(V_2 - V_1)^2}{2\left[V_2 \times (TTC - x) + V_1 \times x + \frac{w_1}{2\sin\theta} + \frac{w_2}{2\tan\theta} + \frac{(l_1 * \cos\theta - l_2)}{2} - l_1\right]} \tag{3}$$

The ACPM predicts that a sideswipe crash will occur if the $RBR_{LC\text{-}SS}$ is greater than the Maximum Available Braking Rate (MABR) of a given vehicle. The model also predicts that a rear-end crash will occur if the $RBR_{LC\text{-}RE}$ is greater than the MABR and $RBR_{LC\text{-}SS}$. The predicted conflicts by the ACPM were compared with annual crash frequencies by type (crossing, rear-end, and lane-change). A higher value of RBR indicates higher lane-change collision risk.

It was found that the predicted conflicts by the ACPM were strongly correlated with annual crash frequencies by type (crossing, rear-end, and lane-change) [31]. Thus, the ACPM is a reliable surrogate safety measure that can accurately predict the number of actual crashes by type. In this study, the conflict angle was assumed to be 45 degrees given that the range of conflict angles for lane-changing conflicts is 30 to 85 degrees [32]. The reaction time of the driver was assumed to be 2 s because the range of perception–reaction times for various types of highways was 1.5 to 3 s in past studies [33]. The length and width of cars were assumed to be 4 m and 2 m, respectively, and the length and width of HVs were assumed to be 10 m and 2.5 m, respectively.

4. Results and Discussion

VISSIM simulations were run for the three experiments. For each experiment, the averages of the values from five simulation runs were calculated for each scenario. The results for the three experiments are presented and discussed as follows.

4.1. Experiment 1—Impacts of the Presence and Location of the Toll Information Sign

Average queue length and maximum queue length for the entire road network were compared among the three scenarios, as shown in Table 4a. The table shows that Scenario 1-3 (separate signs for cars and HVs 75 m before the merge point) had a shorter average queue length than the other two scenarios for all toll lanes, which resulted in less delay.

Table 4. Comparison of scenarios in Experiment 1.

	Scenario 1-1 No Sign	Scenario 1-2 Sign 140 m before Gate	Scenario 1-3 Signs 75 m before Merge

(a) Queue length by toll lane

Toll lanes	Scenario 1-1		Scenario 1-2	Toll lanes	Scenario 1-3	
	Average queue length	Maximum queue length	Average queue length		Average queue length	Maximum queue length
Toll lanes	14.3	94.8	16.8	Toll lanes	12.9	81.7
Car MTC/ETC/ATC lanes (Lanes 1, 2, 3)	35.0	108.7	35.2		31.9	100.7
All ETC/ATC lanes (Lanes 4, 5)	7.0	93.5	8.2		6.2	71.8
All MTC/ETC/ATC lanes (Lane 6)	27.4	113.6	27.1		24.1	101.4

(b) Average TTC by toll lane (s)

Toll lane	Scenario 1-1	Scenario 1-2	Scenario 1-3
Lane 1	6.6	6.1	7.9
Lane 2	13.5	12.8	9.4
Lane 3	9.4	14.9	13.9
Lane 4	11.5	12.9	13.7
Lane 5	12.5	12.9	14.8
Lane 6	14.3	18.2	23.6
Lane 7	15.8	23.4	20.9
Lane 8	13.4	12.2	10.7
Average	12.4	14.8	14.6

(c) Average required braking rates during lane changes (m/s^2)

Lane-changing/ trailing vehicles	Sideswipe conflict			Rear-end conflict		
	Scenario 1-1	Scenario 1-2	Scenario 2-2	Scenario 1-1	Scenario 1-2	Scenario 2-2
Car–Car	0.24	0.21	0.26	0.23	0.14	0.27
Car–HV	0.64	0.53	0.56	0.44	0.29	0.21
HV–HV	0.47	0.26	0.40	0.26	0.28	0.24
HV–Car	0.64	0.22	0.48	1.73	0.19	0.34
Average	0.38	0.25	0.34	0.45	0.22	0.19

Table 4b shows that Scenarios 1-2 and 1-3 had longer average TTCs than Scenario 1-1 in most lanes (lanes 3, 4, 5, 6, and 7). This indicates that the toll information sign can reduce rear-end collision risk. Table 4c shows that Scenarios 1-2 and 1-3 had lower average RBRs to avoid sideswipe collisions and rear-end collisions during lane changes than Scenario 1-1.

The RBR was also compared for the following four types of lane-changing vehicles and trailing vehicles in the target lane: (1) Car–Car: lane-changing car and trailing car, (2) Car–HV: lane-changing car and trailing HV, (3) HV–Car: lane-changing HV and trailing car, and (4) HV–HV: lane-changing HV and trailing HV. The conflicts between cars and HVs (Car–HV or HV–Car) are considered the more severe conflicts because the impact of a collision between vehicles of different sizes and weights on the vehicle body is generally higher. In particular, a Car–HV collision is more severe than an HV–Car collision because cars are more likely to be severely damaged when a lead car is hit by a following HV in the rear.

It was found that Scenarios 1-2 and 1-3 generally showed lower RBRs for both sideswipe and rear-end conflicts during lane changes compared to Scenario 1-1 for all vehicle types except Car–Car. This is mainly because the total number of lane changes was lower for Scenarios 1-2 and 1-3 than for Scenario 1-1. As the number of lane changes increases, the risk of lane-change collisions also increases. In particular, Scenarios 1-2 and 1-3 showed much lower RBRs for Car–HV and HV–Car collisions. Thus, the toll information sign is effective in reducing the number of severe conflicts during lane changes.

In summary, Scenario 1-3 showed the benefits of a shorter average queue length and a lower risk of rear-end and lane-change collisions than Scenario 1-1, unlike Scenario 1-2. Thus, separate toll information signs for cars and HVs before the merge point (Scenario 1-3) was selected as the best scenario in this experiment. This indicates that providing information on toll lane configuration further upstream of the toll plaza not only helps drivers' toll lane choice but also reduces collision risk.

4.2. Experiment 2—Impacts of New Toll Lane Configurations on Estimated Traffic Demand

In this experiment, two new toll lane configuration designs were tested and compared with Scenario 1-3 (Base Case), which was the best scenario in Experiment 1. Table 5a shows that Lanes 1 to 3 and Lane 6 had longer average queue lengths in Scenario 2-1 (Convert Lanes 4 to 6 to ETC-only lanes) and Scenario 2-2 (Convert Lanes 1 and 8 to MTC/ETC-only lanes) than the Base Case. On the other hand, Lanes 4, 5, 7, and 8 had shorter average queue lengths in Scenarios 2-1 and 2-2 than the Base case. Overall, the proposed two toll lane configurations resulted in a more even distribution of queue length across different lanes than the Base case. Similar queue lengths in different toll lanes are more likely to reduce the number of lane changes. This result suggests that the current toll lane configuration needs to be changed to reduce the delay for the estimated traffic demand.

Table 5b also shows that Scenario 2-1 had a longer average TTC than Scenarios 1-3 and 2-2. This indicates that converting Lanes 4 and 6 to ETC-only lanes can more effectively reduce rear-end collision risk for all lanes except Lane 3 compared to the current toll lane configuration and converting Lanes 1 and 8 to MTC/ETC-only lanes.

Table 5c shows that both Scenarios 2-1 and 2-2 had lower RBRs for sideswipe conflict during lane changes than Scenario 1-3. In particular, the RBR for severe sideswipe conflict (i.e., Car–HV and HV–Car) was lower for Scenario 2-1 than Scenario 2-2. This indicates that converting Lanes 4 and 6 to ETC-only lanes can more effectively reduce lane-change collision risk than the current toll lane configuration and converting Lanes 1 and 8 to MTC/ETC-only lanes. However, both new toll lane configurations increased the average RBR for rear-end conflict during lane changes, although they still reduced the RBR for HV–Car conflicts compared to the current toll lane configuration.

In summary, Scenario 2-1 made the distribution of queue length across toll lanes more even and reduced both rear-end and lane-change collision risk (for sideswipe conflict only). This shows that more ETC-only lanes can enhance the safety benefit of the toll information sign by separating ETC vehicles from MTC and ATC vehicles, reducing their conflicts.

Table 5. Comparison of scenarios in Experiment 2.

	Base Case (Scenario 1-3) Current Toll Lane Configuration		Scenario 2-1 Convert Lanes 4 to 6 to ETC-Only Lanes		Scenario 2-2 Convert Lanes 1 to 8 to MTC/ETC-Only Lanes	

(a) Queue length by toll lane

	Base case (Scenario 1-3)		Scenario 2-1		Scenario 2-2	
Toll lanes	Average queue length	Maximum queue length	Average queue length	Maximum queue length	Average queue length	Maximum queue length
Lanes 1, 2, 3	12.9	81.7	16.3	76	13.1	68.8
Lanes 4, 5	31.9	100.7	24.0	92.7	24.6	90.7
Lane 6	6.2	71.8	10.4	81.8	7.5	67.5
Lanes 7, 8	24.1	101.4	15.0	87.4	20.1	83.6

(b) Average TTC by toll lane

Toll lane	Scenario 1-3	Scenario 2-1	Scenario 2-2
Lane 1	7.9	25.9	15.7
Lane 2	9.4	28.2	27.3
Lane 3	13.9	11.3	21
Lane 4	13.7	20.2	11.1
Lane 5	14.8	22	12
Lane 6	23.6	33.8	20.5
Lane 7	20.9	21.9	24.9
Lane 8	10.7	13.7	11.4
Average	14.6	21.3	14.9

(c) Average required braking rates during lane changes (m/s²)

Lane-changing/ trailing vehicles	Sideswipe conflict			Rear-end conflict		
	Scenario 1-3	Scenario 2-1	Scenario 2-2	Scenario 1-3	Scenario 2-1	Scenario 2-2
Car–Car	0.26	0.16	0.13	0.27	0.14	0.12
Car–HV	0.56	0.17	0.32	0.21	0.11	0.22
HV–HV	0.40	0.41	0.32	0.24	0.49	0.48
HV–Car	0.48	0.24	0.38	0.34	0.16	0.22
Average	0.34	0.24	0.30	0.19	0.21	0.31

4.3. Experiment 3—Impacts of Current and New Toll Lane Configurations for Different Percentages of HVs

This scenario assessed the impacts of the current and new toll lane configurations with the toll information sign for 60% and 70% HVs, which are higher than the current percentage (50%). In the new toll lane configuration, Lane 6 was converted to an HV-only lane with all toll payment methods to accommodate the higher traffic demand of HVs. Table 6a shows that Lane 6 had a longer average queue length in Scenario 3-2 than in Scenario 3-1 for 60% HVs. Similarly, Mahdi et al. [34] also found that the queue length at a toll plaza increased as the percentage of HVs increased. This is due to higher HV demand for Lane 6 in Scenario 3-2 as this lane was open to only one type of vehicle. The average queue length in the other lanes was similar in the two scenarios. A similar result was found for 70% HVs (Scenarios 3-3 and 3-4). In spite of the longer average queue length in Lane 6, the average delay per vehicle was similar between the current and new toll lane configurations for 60% and 70% HVs.

Table 6. Comparison of scenarios in Experiment 3.

	Scenario 3-1 Current Toll Lane Configuration for 60% HV		Scenario 3-2 Convert Lanes 6 to HV-only Lanes for 60% HV		Scenario 3-3 Current Toll Lane Configuration for 70% HV		Scenario 3-4 Convert Lanes 6 to HV-only Lanes for 70% HV	
(a) Queue length by toll lane								
	Scenario 3-1 Current toll lane configuration for 60% HV		Scenario 3-2 Convert Lane 6 to an HV-only lane for 60% HV		Scenario 3-3 Current toll lane configuration for 70% HV		Scenario 3-4 Convert Lane 6 to an HV-only lane for 70% HV	
Toll lanes	Ave. queue length	Max. queue length	Ave. queue length	Max. queue length	Ave. queue length	Max. queue length	Ave. queue length	Max. queue length
Lanes 1, 2, 3	10.7	81	10.6	81.2	10	83.1	10.2	83
Lanes 4, 5	32.7	103.2	31.2	107.6	34.6	108.4	31.4	109.8
Lane 6	7	75.7	10	86	8	81.4	12.4	93.2
Lanes 7, 8	29.1	105.7	30.5	114.6	32.7	111.3	33.2	118
(b) Average TTC by toll lane								
Toll lane	Scenario 3-1		Scenario 3-2		Scenario 3-3		Scenario 3-4	
Lane 1	8		8		7.8		7.7	
Lane 2	10.1		10.6		11.1		11	
Lane 3	10.2		12		8.6		8.5	
Lane 4	12		12		11.9		12.5	
Lane 5	15.6		11.6		16.8		16.8	
Lane 6	20.6		19.2		25.2		26.3	
Lane 7	26.2		30		29.4		32.5	
Lane 8	14.6		14.7		19.5		20.3	
Average	16.0		16.4		19.1		20.8	
(c) Average required braking rates during lane changes (m/s^2)								
Lane-changing/ trailing vehicles	60% HV				70% HV			
	Sideswipe conflict		Rear-end conflict		Sideswipe conflict		Rear-end conflict	
	Scenario 3-1	Scenario 3-2	Scenario 3-1	Scenario 3-2	Scenario 3-3	Scenario 3-4	Scenario 3-3	Scenario 3-4
Car–Car	0.12	0.13	0.07	0.09	0.12	0.07	0.12	0.07
Car–HV	0.62	0.89	0.24	0.83	0.83	0.45	0.72	0.3
HV–HV	0.48	0.37	0.45	0.26	0.54	0.56	0.87	1.12
HV–Car	0.40	0.41	0.32	0.42	0.29	0.34	0.22	0.77
Average	0.42	0.37	0.30	0.30	0.42	0.44	0.56	0.60

Table 6b shows that the average TTC for different lanes was similar between Scenarios 3-1 and 3-2 for 60% HVs. However, the average TTC was relatively longer for Scenario 3-4 than Scenario 3-3 for 70% HVs, particularly in HV-only lanes (Lanes 6 to 8). This indicates that opening additional HV-only lanes can reduce rear-end collision risk for a higher percentage of HVs.

Table 6c shows that the average RBR for sideswipe conflict during lane changes was lower for Scenario 3-2 than Scenario 1, particularly for HV–HV conflict, whereas the average RBR for rear-end conflicts was similar between the two scenarios for 60% HV. Thus, converting Lane 6 to an HV-only lane was effective in reducing lane-change collision risk for 60% HVs.

Table 6c also shows that Scenario 3-4 has higher average RBRs for both sideswipe and rear-end conflicts during lane changes than Scenario 3-3 for 70% HVs. In particular, RBRs for HV–Car and HV–HV conflicts were higher for Scenario 3-4 than Scenario 3-3. This indicates that converting Lane 6 to an HV-only lane increased lane-change collision risk for 70% HVs, unlike 60% HVs. This indicates that as the percentage of HVs increases, the number of lane-changing HVs and the risk of lane-change collision with the trailing car or HV in the target lane also increases. Thus, converting Lane 6 to an HV-only lane has mixed effects on severe lane-change conflicts as it reduces the collision risk for Car–HV collisions but increases the collision risk for HV–Car collisions.

In summary, converting Lane 6 to an HV-only lane increased the queue length for higher percentages of HVs than the current toll lane configuration as Lane 6 was closed for cars and more HVs used HV-only lanes. However, converting Lane 6 to an HV-only lane reduced rear-end collision risk but increased lane-change collision risk for higher percentages of HVs. This indicates that opening additional HV-only lanes to accommodate higher HV demand can have a negative effect on queue length and lane-change collision risk.

5. Conclusions and Recommendations

This study investigated the impacts of a toll information sign on queue length and collision risk using preliminary designs and lane configurations for a toll plaza on the Gordie Howe International Bridge—a new bridge under construction at the Windsor–Detroit international border crossing. This study also investigated the impacts of different toll lane configurations with the toll information sign for estimated traffic demand and different percentages of heavy vehicles (HVs). The proposed toll information sign displays the toll lane configuration with different toll payment methods and vehicle types (cars and HVs) upstream of the toll booth. There are three toll payment methods—manual toll collection (MTC), automatic toll collection (ATC), and electronic toll collection (ETC). To evaluate the impacts, the traffic flow at the toll plaza was simulated using VISSIM microscopic traffic simulation. The main findings are summarized as follows:

First, the toll information sign reduced queue length and collision risk at the toll plaza when the sign was located further upstream of the toll plaza. Separate signs for cars and HVs 75 m before the merge point at the Canada-bound bridge led to shorter average queue lengths and lower rear-end and lane-change collision risks than placing the sign after the merge point or closer to the toll booth. In particular, the sign led to a lower risk of collision between lane-changing cars and trailing HVs in the target lane. This indicates that the toll information sign helped drivers make an earlier decision to choose the toll booth with their preferred toll payment methods, avoid abrupt lane changes, and avoid severe lane-change conflicts.

Second, the effectiveness of the toll information sign in distributing queue length across toll lanes more evenly and reducing collision risk was further increased by implementing different toll lane configurations. With the toll information sign, the installation of ETC-only lanes significantly reduced rear-end and lane-change collision risk. This shows that ETC-only lanes not only allow ETC vehicles to pass the toll booth more smoothly without stopping but also decrease risky car-following and lane-change behaviors.

Third, the effectiveness of the toll information sign in reducing the queue length and collision risk varied as the percentage of HVs increased. With the toll information sign, increasing the number of HV-only lanes to accommodate the increased HV traffic demand reduced rear-end collision risk but increased queue length and lane-change collision risk at higher percentages of HVs. This shows that when the toll lane configuration is changed for varying traffic demand, it is important to consider the effects of the change on queue length and collision risk.

This study demonstrates that the toll information sign can potentially reduce the queue length and collision risk, particularly regarding more severe conflicts involving HVs during lane changes, at a toll plaza by helping drivers make earlier route decisions to choose the toll lane. In addition, the toll information sign with changeable toll lane configurations, which accommodate different traffic demands, can improve traffic performance and safety more effectively. In practice, the best toll lane configurations by time of day can be identified based on hourly car and HV traffic patterns and the toll lanes can be adjusted to match the expected demand by time of day, minimizing both rear-end and lane-change collision risk.

However, this study has some limitations. First, only a limited number of scenarios were tested in this study. Thus, more scenarios (different traffic demand for cars and HVs and different toll lane configurations) need to be tested to observe the general pattern of impacts of traffic demand and toll lanes on traffic performance and safety. Second, surrogate safety measures used for the assessment of lane-change collision risk in this study have some limitations such as not considering the trailing vehicles that have lower speed than the lane-changing vehicles. Thus, lane-change collision risk for various lane-change situations needs to be captured using a new surrogate safety measure. Third, as this study only focused on the traffic upstream of the toll booth and immediately after the toll booth, the effect of traffic conditions downstream of the toll booth on drivers' lane choice was not considered. Lastly, the severity of collisions was evaluated only based on the types of vehicles involved in conflicts (car or HV), not the speeds of vehicles at the time of collision.

In future studies, it is recommended to collect real-world driver behavior data after the bridge is open and use the validated simulation model to evaluate the impacts of the toll information sign on queue length and collision risk. It is also recommended to analyze the effect of drivers' familiarity with toll lane configuration on their toll lane choice. Since the drivers who regularly or frequently cross the bridge (e.g., commuters) will have better knowledge of the toll lane configuration from their experience, their lane choice behaviors are likely to be different from the other drivers. It is also worth investigating the impacts of car and HV drivers' different compliance rates with the toll information sign and different HV classes on traffic performance and safety. Lastly, economic analysis is recommended to examine the cost-benefit ratio of implementing the recommended toll information sign and toll lane configuration, including the cost of ETC systems and the economic impact of collisions.

Author Contributions: Conceptualization, F.Z. and C.L.; methodology, F.Z.; software, F.Z.; investigation, F.Z. and C.L.; data curation, F.Z.; formal analysis, F.Z.; writing—original draft, F.Z.; writing—review & editing, C.L.; supervision, C.L.; funding acquisition, C.L. All authors have read and agreed to the published version of the manuscript.

Funding: This research was funded by Natural Sciences and Engineering Research Council of Canada [Grant number: RGPIN-2019-04430].

Data Availability Statement: The primary research data can be accessed upon request.

Acknowledgments: The authors thank Anas Abdulghani for developing the VISSIM simulation model and Windsor-Detroit Bridge Authority (WDBA) for providing the data used in this study.

Conflicts of Interest: The authors declare no conflicts of interest. The funders had no role in the design of the study; in the collection, analyses, or interpretation of data; in the writing of the manuscript; or in the decision to publish the results.

References

1. National Transportation Safety Board. 2023. Available online: http://www.ntsb.gov (accessed on 22 June 2022).
2. Abdelwahab, H.T.; Abdel-Aty, M. Artificial neural networks and logit models for traffic safety analysis of toll plazas. *Transp. Res. Rec. J. Transp. Res. Board* **2022**, *1784*, 115–125. [CrossRef]
3. Coelho, M.C.; Farias, T.L.; Rouphail, N.M. Measuring and modeling emission effects for toll facilities. *Transp. Res. Rec. J. Transp. Res. Board* **2005**, *1941*, 136–144. [CrossRef]
4. Yang, H.; Ozbay, K.; Bartin, B.; Ozturk, O. Assessing the safety effects of removing highway mainline barrier toll plazas. *J. Transp. Eng.* **2014**, *140*, 04014038. [CrossRef]
5. Valdés, D.; Colucci, B.; Fisher, D.L.; Ruiz, J.; Colón, E.; García, R. Driving simulation of the safety and operation performance at a freeway toll plaza. *Transp. Res. Rec. J. Transp. Res. Board* **2016**, *2602*, 129–137. [CrossRef]
6. Zhang, H.; Zhang, C.; Zhang, Y.; Ma, J.; He, J.; Liu, Z.; Yan, X.; Hong, Q. Calculation model of preposition distance of electronic toll collection signs based on lane changing behavior. *MATEC Web Conf.* **2020**, *325*, 01003. [CrossRef]
7. Saad, M.; Abdel-Aty, M.; Lee, J. Analysis of driving behavior at expressway toll plazas. *Transp. Res. Part F Traffic Psychol. Behav.* **2019**, *61*, 163–177. [CrossRef]
8. Abuzwidah, M.; Abdel-Aty, M. Crash risk analysis of different designs of toll plazas. *Saf. Sci.* **2018**, *107*, 77–84. [CrossRef]
9. Chakraborty, M.; Stapleton, S.Y.; Ghamami, M.; Gates, T.J. Safety effectiveness of all-electronic toll collection systems. *Adv. Transp. Stud.* **2020**, *2*, 127–142.

10. Xing, L.; He, J.; Abdel-Aty, M.; Cai, Q.; Li, Y.; Zheng, O. Examining traffic conflicts of upstream toll plaza using vehicles' trajectory data. *Accid. Anal. Prev.* **2019**, *125*, 174–187. [CrossRef]
11. Jehad, A.E.; Ismail, A.; Borhan, M.N.; Ishak, S.Z. Modelling and optimizing of Electronic Toll Collection (ETC) at Malaysian toll plazas using microsimulation models. *Int. J. Eng. Technol.* **2018**, *7*, 2304–2308. [CrossRef]
12. McKinnon, I. Operational and Safety-Based Analyses of Varied Toll Lane Configurations. Master's Thesis, University of Massachusetts, Amherst, MA, USA, 2013.
13. Bains, M.S.; Arkatkar, S.S.; Anbumani, K.S.; Subramaniam, S. Optimizing and modeling tollway operations using microsimulation. *Transp. Res. Rec. J. Transp. Res. Board* **2017**, *2615*, 43–54. [CrossRef]
14. Hamid, A.H.A. Simulation of traffic operation and management at Malaysian toll plazas using VISSIM. *Malays. Univ. Transp. Res. Forum Conf.* **2011**, 13–21.
15. Mittal, H.; Sharma, N. Operational optimization of toll plaza queue length using microscopic simulation VISSIM model. *J. Algebr. Stat.* **2022**, *13*, 418–425.
16. Bari, C.; Gupta, U.; Chandra, S.; Antoniou, C.; Dhamaniya, A. Examining the effect of the Electronic Toll Collection (ETC) system on queue delay using a microsimulation approach at toll plaza—A case study of Ghoti toll plaza, India. In Proceedings of the 7th IEEE International Conference on Models and Technologies for Intelligent Transportation Systems (MT-ITS), Online, 16–17 June 2021; pp. 1–6.
17. Windsor-Detroit Bridge Authority. Annual Report 2019–2020. 2022. Available online: https://www.gordiehoweinternationalbridge.com/en/corporate-reports (accessed on 22 June 2022).
18. PTV AG. *PTV VISSIM 2021 User Manual*; PTV AG: Karlsruhe, Germany, 2021.
19. WSP. *Supplemental Travel Demand Modeling Technical Report Gordie Howe International Bridge*; Michigan Department of Transportation: Lansing, MI, USA, 2018.
20. Wang, H.; Zou, F.; Tian, J.; Guo, F.; Cai, Q. Analysis on lane capacity for expressway toll station using toll data. *J. Adv. Transp.* **2022**, *2022*, 9277000. [CrossRef]
21. Al-Deek, H.M.; Mohammed, A.; Radwan, A.E. Operational benefits of electronic toll collection: Case study. *J. Transp. Eng.* **1997**, *123*, 467–477. [CrossRef]
22. Woo, T.H.; Hoel, L.A. Toll plaza capacity and level of service. *Transp. Res. Rec.* **1991**, *1320*, 119–127.
23. 407 Express Toll Route. 407 ETR Heavy Multiple Unit Vehicle Rate Chart. 2022. Available online: https://www.407etr.com/en/tolls/tolls/rate-chart-multi.html (accessed on 22 June 2022).
24. IBI Group. *Traffic and Revenue Forecaster: Windsor Gateway Project Origin-Destination Travel Surveys Summary Report*; Transport Canada: Ottwa, ON, Canada, 2008.
25. Liang, H.; Yang, S.; Pan, B. Probability model of driver's choice behavior in toll plaza. *IOP Conf. Ser. Earth Environ. Sci.* **2019**, *371*, 052051. [CrossRef]
26. Parmar, H.; Chakroborty, P.; Kunduc, D. Modelling automobile drivers toll-lane choice behaviour at a toll plaza using mixed logit model. In Proceedings of the 2nd Conference of Transportation Research Group of India (2nd CTRG), Kolkata, India, 12–15 December 2013.
27. Durrani, U.; Lee, C.; Maoh, H. Calibrating the Wiedemann's vehicle-following model using mixed vehicle-pair interactions. *Transp. Res. Part C Emerg. Technol.* **2016**, *67*, 227–242. [CrossRef]
28. CDM Smith. *US 231 Scottsville Road Scoping and Traffic Operations Study. Appendix B: VISSIM Development and Calibration Report*; KYTC Item No. 3-8702.00; KKentucky Transportation Cabinet, Division of Planning: Warren County, KY, USA, 2014.
29. Chrysler, S.T.; Williams, A.A.; Fitzpatrick, K. *Driver Comprehension of Signing and Marking for Toll Facilities*; Report No. FHWA/TX-08/0-5446-2; Texas Transportation Institute, The Texas A&M University System: College Station, TX, USA, 2008.
30. Hydén, C. *The Development of a Method for Traffic Safety Evaluation: The Swedish Traffic Conflict Techniques*; Lund Institute of Technology, Department of Traffic Planning and Engineering: Lund, Sweden, 1987.
31. Wang, C.; Stamatiadis, N. Surrogate safety measure for simulation-based conflict study. *Transp. Res. Rec. J. Transp. Res. Board* **2013**, *2386*, 72–80. [CrossRef]
32. Gettman, D.; Pu, L.; Sayed, T.; Shelby, S. *Surrogate Safety Assessment Model and Validation: Final Report*; Report No. FHWA-HRT-08-051; Federal Highway Administration: McLean, VA, USA, 2008.
33. Neuman, T.R. New approach to design for stopping sight distance. *Transp. Res. Rec.* **1989**, *1208*, 14–22.
34. Mahdi, M.B.; Leong, L.V.; Sadullah, A.F.M. Use of Microscopic Traffic Simulation Software to Determine Heavy-Vehicle Influence on Queue Lengths at Toll Plazas. *Arab. J. Sci. Eng.* **2019**, *44*, 7297–7311. [CrossRef]

Disclaimer/Publisher's Note: The statements, opinions and data contained in all publications are solely those of the individual author(s) and contributor(s) and not of MDPI and/or the editor(s). MDPI and/or the editor(s) disclaim responsibility for any injury to people or property resulting from any ideas, methods, instructions or products referred to in the content.

Article

Radar-Based Pedestrian and Vehicle Detection and Identification for Driving Assistance

Fernando Viadero-Monasterio [1,*], Luciano Alonso-Rentería [2], Juan Pérez-Oria [2] and Fernando Viadero-Rueda [3]

1. Mechanical Engineering Department, Advanced Vehicle Dynamics and Mechatronic Systems (VEDYMEC), Universidad Carlos III de Madrid, Avda. de la Universidad 30, 28911 Leganés, Spain
2. Control Engineering Group, Universidad de Cantabria, 39005 Santander, Spain; luciano.alonso@unican.es (L.A.-R.); oria@teisa.unican.es (J.P.-O.)
3. Structural and Mechanical Engineering Department, Universidad de Cantabria, 39005 Santander, Spain; viaderof@unican.es
* Correspondence: fviadero@ing.uc3m.es

Abstract: The introduction of advanced driver assistance systems has significantly reduced vehicle accidents by providing crucial support for high-speed driving and alerting drivers to imminent dangers. Despite these advancements, current systems still depend on the driver's ability to respond to warnings effectively. To address this limitation, this research focused on developing a neural network model for the automatic detection and classification of objects in front of a vehicle, including pedestrians and other vehicles, using radar technology. Radar sensors were employed to detect objects by measuring the distance to the object and analyzing the power of the reflected signals to determine the type of object detected. Experimental tests were conducted to evaluate the performance of the radar-based system under various driving conditions, assessing its accuracy in detecting and classifying different objects. The proposed neural network model achieved a high accuracy rate, correctly identifying approximately 91% of objects in the test scenarios. The results demonstrate that this model can be used to inform drivers of potential hazards or to initiate autonomous braking and steering maneuvers to prevent collisions. This research contributes to the development of more effective safety features for vehicles, enhancing the overall effectiveness of driver assistance systems and paving the way for future advancements in autonomous driving technology.

Keywords: vehicle safety; road transport; vehicle safety; intelligent traffic vehicle; radar; adas; urban traffic; neural network

Citation: Viadero-Monasterio, F.; Alonso-Renteria, L.; Pérez-Oria, J.; Viadero-Rueda, F. Radar-Based Pedestrian and Vehicle Detection and Identification for Driving Assistance. *Vehicles* **2024**, *6*, 1185–1199. https://doi.org/10.3390/vehicles6030056

Academic Editors: Deogratias Eustace, Bhaven Naik, Heng Wei and Parth Bhavsar

Received: 3 June 2024
Revised: 2 July 2024
Accepted: 8 July 2024
Published: 9 July 2024

Copyright: © 2024 by the authors. Licensee MDPI, Basel, Switzerland. This article is an open access article distributed under the terms and conditions of the Creative Commons Attribution (CC BY) license (https://creativecommons.org/licenses/by/4.0/).

1. Introduction

Radar technology is increasingly used to detect both moving and stationary objects [1]. The word "radar", from "radio detection and ranging", means not only the detection of objects, but also the evaluation of certain parameters of these objects at the same time.

Radars emit a radio pulse, which is reflected by the target and typically received at the same position as the emitter. This "echo" allows for the extraction of a great deal of information [2]. The reflection of radar waves varies according to their wavelength and the shape and properties of the target. When the object is much smaller than the wavelength, it becomes invisible to the wave, that is, the wave passes through it as if it did not exist. When the sizes are similar, a part of the wave energy is reflected and another portion passes through the object, resulting in the diffraction effect [3]. Early radars used very long wavelengths, larger than the targets, which resulted in weak echo signals. Today's radars use small wavelengths (a few centimetres or less) that allow objects the size of a human arm to be detected. Short-wave signals (3 kHz–30 MHz) reflect off curves and edges, just as light flashes off a curved piece of glass. The radar cross section (RCS) of an object is a key factor that determines the degree of reflectance of radio waves [4].

Radar sensors can also be used to measure velocities thanks to the Doppler effect [5]. By taking advantage of the fact that the return signal from a moving target is frequency shifted, it is possible to measure the relative velocity of the object with respect to the radar. The velocity components perpendicular to the radar line of sight cannot be estimated by the Doppler effect alone and would require memory to calculate them by tracking the evolution of the target's position [6].

It is not uncommon to find radars integrated with other sensors in order to achieve complex applications. Some of the most noticeable uses of this technology integration include the following:

- Object tracking and classification [7–12].
- Non-contact heart and breathing rate estimation [13–16].
- Vehicle platoon control [17–20].
- Human gait recognition [21–23].

Specifically in the vehicle research field, the development of advanced driver assistance systems (ADASs) has led to a significant decrease in the number of traffic accidents [24,25]. ADASs commonly incorporate radar sensors to facilitate multiple tasks, such as cruise control, to automatically slow down or speed up the vehicle to maintain a set gap with the vehicle ahead [26]; emergency braking, where a vehicle may decelerate sharply without driver involvement in order to avoid a potential collision [27]; blind spot detection, where radar sensors are used to monitor the blind spots and alerts the driver in the event of a potential collision when changing lanes [28]; parking assistance, to precisely detect an open parking space nearby [29]; etc.

The integration of multiple sensors in ADASs will result in a significant computational burden, which may not be feasible in real-time applications with low-cost architectures [30,31]. Although it is now relatively inexpensive to include additional sensors in vehicles, we were curious as to whether a single sensor would be sufficient for some applications. Moreover, concerning radar features, most are based solely on the analysis of the distance and velocity measured from the radar, which fail to utilise the full potential of radar sensors. While it is often forgot that the RCS of an object determines how waves are reflected, numerous studies can been found on RCS reduction, useful for military and defence applications [32–34]. If the RCS is sufficiently low, the object cannot be detected. However, in detectable objects, the reflected power at a given distance will differ according to the object properties. In [35], the RCS and Doppler signature of targets are used to differentiate pedestrians and vehicles; however, targets can be stationary or there can be no observable Doppler signature, which limits the practical application. In [36], machine learning techniques are applied for target classification under static conditions. For some advanced radars that are capable of imaging, targets can be represented by radar images, as in [37], where targets are visualized as radar point clouds, discarding RCS and Doppler data. Nevertheless, the aforementioned sensors are prohibitively expensive and therefore unsuitable for inclusion in series production vehicles. The results of these works have led to the formulation of the following hypothesis: is it possible to detect and identify vehicles and pedestrians solely through the use of low-cost radar information such as RCS and distance?

The objective of this paper is to design a system for detecting and identifying objects in front of a vehicle, exclusively using two radar measurements: distance to the target and reflected power, which is correlated with the RCS of the target. A frequency-modulated continuous-wave (FCMW) radar was mounted on a vehicle during the experiments. A neural network was designed to classify each pair of measurements with the appropriate object. This information is provided to the user in order to assist them in driving. It can be used as part of ADAS systems in order to perform the appropriate response to a given stimulus. This may involve adapting the vehicle speed or performing emergency maneuvers in hazardous situations, such as when a pedestrian crosses the road unexpectedly, thereby enhancing safety. Potential applications of the proposed system include, but are not limited to, adaptive cruise control, automated emergency braking systems, and collision avoidance.

The manuscript is structured as follows: In Section 2, the fundamental principles of radar are formulated. In Section 3, a brief description of neural networks is provided. In Section 4, the experimental setup employed in this study is described. In Section 5, the experimental results are processed and a neural network is trained to identify the objects detected. In Section 6, the conclusions and future works related to this research are presented.

2. Radar Sensor

Radars emit a radio pulse, which is reflected when a target is hit. The reflected power to the radar receiving antenna is given by the following expression:

$$P_r = \frac{P_t G_t A_r \sigma F^4}{(4\pi)^2 R_t^2 R_r^2} \tag{1}$$

where P_r is the reflected power, P_t is the transmitter power, G_t is the gain of the transmitting antenna, A_r is the effective area of the receiving antenna, σ is the radar cross section of the target (typical RCS values are presented in Table 1), F is the pattern propagation factor (as reference, $F = 1$ in a vacuum), R_t is the distance from the transmitter to the target, and R_r is the distance from the receiver to the target. In most common applications, transmitting and receiving antennas are located together, then $R_t = R_r = R$, where R is the distance to the target.

Table 1. Typical RCS values [38,39].

Target	σ (m^2)
Bug	0.00001
Large bird	0.01
F-117 fighter	0.1
Human	1
Automobile	10

The use of separate transmit and receive antennas is recommended as it provides greater sensitivity and isolation. In the case of limited space and the only option being one common antenna, the receiver antenna can be removed. However, the received signal must be decoupled from the shared transmit/receive path, which results in a deterioration of data reception, with reduced sensitivity, as the received signal is fed into two ports: receive and transmit, where it is lost.

2.1. Pulse Radar

One method for measuring the distance between a radar and an object is to transmit a small electromagnetic pulse and subsequently measure the time taken for the echo to return (Figure 1) [40].

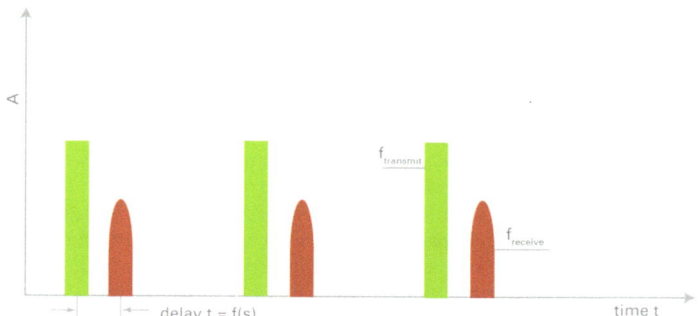

Figure 1. Time-dependent shape of transmit and receive signal of a pulse radar [41].

In order to have a good resolution, especially for objects at close range, these pulses must be very short. The distance can be calculated as half the transit time multiplied by the propagation speed of the pulse. The accurate estimation of distance necessitates the utilisation of high-performance electronic components. The majority of radars utilise the same antenna for both sending and receiving; therefore, during the transmission of the pulse, no echo can be received. This establishes the so-called "blind distance" of the radar, below which the radar is rendered ineffective [42].

2.2. FMCW Radar

Frequency-modulated continuous-wave (FMCW) radar represents a different approach to detect stationary objects [43–45]. The comparison of frequencies is a more accurate and simpler method than the comparison of times. To achieve this, a sinusoidal signal is emitted at a frequency that varies continuously over time. Consequently, when the echo arrives, its frequency will differ from that of the original signal. By comparing the two signals, it is possible to ascertain the elapsed time and therefore the distance to the target (see Figure 2). The greater the frequency offset, the greater the distance, calculated using the following formula:

$$R = \frac{c_0}{2} T \frac{f_D}{\Delta f} \qquad (2)$$

where c_0 is the speed of light, T is the sawtooth repetition time period, f_D is the differential frequency and Δf is the frequency deviation.

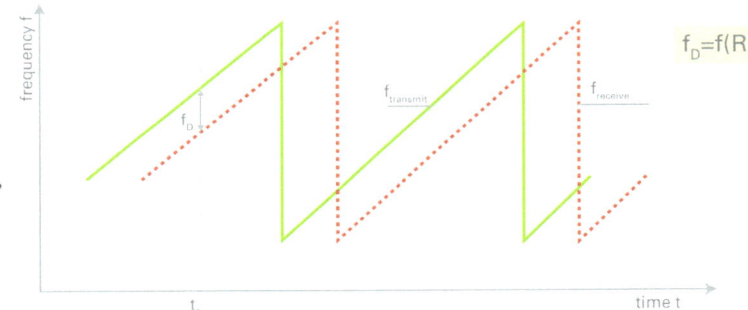

Figure 2. Time-dependent shape of transmit and receive signal of a FMCW radar with sawtooth modulation scheme [41].

The accuracy of the measurement is dependent upon the bandwidth utilised. Furthermore, it is important to note that the laws of each country define which frequencies are permitted and which are prohibited. In the event that the bandwidth is insufficient, two distinct objects may be erroneously identified as a single entity, see Figure 3. A bandwidth higher than 250 MHz is not allowed because of regulation reasons in Europe (ETSI 300-440) and US (FCC 15.245). Therefore the best achievable resolution for the commercial radar iSYS-4004 used in this work is limited to 60 cm [41].

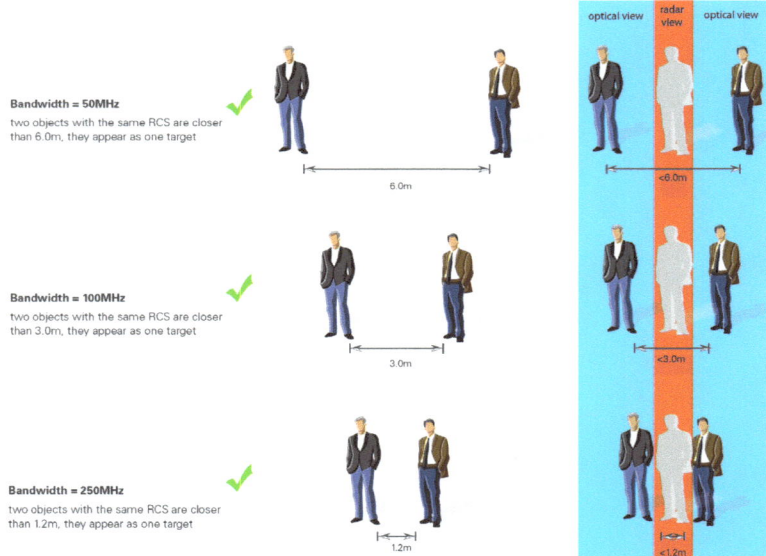

Figure 3. Example of the bandwidth effect on the commercial radar iSYS-4004 [41].

3. Neural-Network-Based Identification

Artificial neural networks are computational systems inspired by the biological neural networks that are part of animal brains. Such systems are capable of learning to perform tasks through the feeding of a large set of examples, typically without the need for any task-specific rules to be programmed into them [46]. A neural network is based on a collection of interconnected units, or nodes, which are analogous to the neurons in a biological brain. Each connection functions in a manner analogous to synapses in a biological brain, transmitting a signal from one artificial neuron to another. An artificial neuron that receives a signal can process it and then signal additional artificial neurons that are connected to it.

In typical implementations, the signal at a connection between artificial neurons is a real number, and the output of each artificial neuron is calculated by some nonlinear function of the sum of its inputs. The connections between artificial neurons are designated as "edges". The weights of the artificial neurons and edges are typically adjusted as the learning process progresses. The weight of the connection affects the strength of the signal transmitted at that point. It is possible for artificial neurons to have a threshold, whereby the signal is only transmitted if the aggregated signal crosses that threshold. Artificial neurons are typically aggregated in layers. The various layers are capable of implementing distinct types of transformations on their inputs. Signals are transmitted from the initial layer, designated the input layer, to the final layer, the output layer. This transmission can be performed in multiple stages, with signals passing through one or more intermediate layers.

An artificial neural network is composed of the following:

- **Neurons**. A neuron j (see Figure 4) that receives an input $p_j(k)$ from the predecessor neurons has the following components:
    - An activation $a_j(k)$.
    - A threshold Θ_j, which is usually fixed, unless a learning function updates it.
    - An activation function f, which evaluates the new activation in the following instant $k+1$, using $a_j(k)$, Θ_j, and the new input $p_j(k)$, leading to

$$a_j(k+1) = f(a_j(k), p_j(k), \Theta_j) \qquad (3)$$

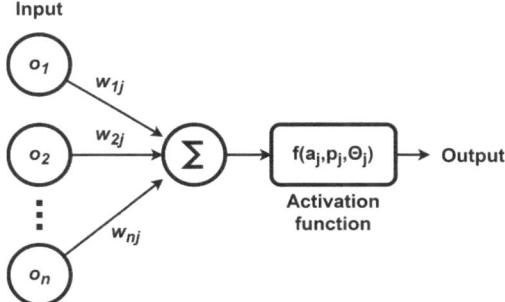

Figure 4. Artificial neuron.

The most commonly used activation functions are as follows [47]:

* Sigmoid. The sigmoid activation function converts an input from range $(-\infty, +\infty)$ to the range $[0, 1]$. The sigmoid function is usually used in the output layer for classification purposes. One of the benefits of the sigmoid function is that it has a smooth derivative. The sigmoid activation function is defined as follows:

$$\sigma(x) = \frac{1}{1+e^{-x}} \quad (4)$$

* Hyperbolic tangent. This has a similar structure to the sigmoid function; however, the output is within the range $[-1, +1]$. Compared to the sigmoid function, the hyperbolic tangent has a higher derivative. The hyperbolic tangent function is defined as follows:

$$\tanh(x) = \frac{2}{1+e^{-2x}} - 1 \quad (5)$$

* Rectified linear unit (ReLU). This is a frequently employed activation function that returns the value of the input if it is positive; otherwise, it returns zero. The ReLU function is defined as follows:

$$\text{ReLU}(x) = \max(0, x) \quad (6)$$

* Parametric leaky version of a ReLU (PReLU). In this case, instead of the function being zero for negative inputs, it returns a small negative slope α. The PReLU function is defined as follows:

$$\text{PReLU}(x) = \max(0, x) + \alpha \cdot \min(0, x) \quad (7)$$

* Exponential linear unit (ELU). This function provides some improvement to ReLU. The ELU activation function is defined as follows:

$$\text{ELU}(x) = \max(0, x) + \min(0, \alpha(e^x - 1)) \quad (8)$$

* Scaled exponential linear unit (SELU). Another variation to ReLU. The SELU activation function is defined as follows:

$$\text{SELU}(x) = \gamma \cdot \left(\max(0, x) + \min(0, \alpha(e^x - 1)) \right) \quad (9)$$

* Swish. The Swish function does not have an upper bound. The Swish function is defined as follows:

$$\text{Swish}(x) = \frac{x}{1+e^{-x}} \quad (10)$$

* Mish. A variant with a similar shape to the Swish function. The Mish function is defined as follows:

$$\text{Mish}(x) = x \cdot \tanh\left(\log(1 + e^x)\right) \tag{11}$$

- An output signal that computes the activation output

$$o_j(t) = f_{out}(a_j(k)) \tag{12}$$

In the majority of cases, the output function is the identity function.

- **Connections and weights**. The neural network is based on connections. Each connection transmits the output of the neuron i to the input of the neuron j. Each connection is assigned the weight w_{ij}.
- **Propagation functions**. These calculate the input $p_j(k)$ to the neuron j depending on the outputs $o(k)$ from the predecessor neurons. A common propagation function is

$$p_j(k) = \sum w_{ij} o_i(k) \tag{13}$$

- **Learning rules**. The learning rule is a rule or algorithm that modifies the parameters of the neural network in order to produce a desired outcome when presented with a specific input. This learning process involves modifying the weights and thresholds of variables within the network.

The network topology indicates the existence of two broad categories of artificial neural networks, which can be distinguished by the following characteristics:

- **Feedforward neural network**. This is the first and simplest type. In this network, the transfer of information occurs in a unidirectional manner, from the input layer to the output layer, without any loops. The process of learning occurs through the updating of connection weights in response to the processing of each piece of data, with the subsequent evaluation of the error between the actual and expected results.
- **Recurrent neural network**. These networks propagate data forward, but also backwards, from later processing stages to earlier stages. It is possible that recurrent neural networks may exhibit chaotic behaviour due to the backpropagation process.

The addition of further hidden layers to a neural network can enhance its performance, enabling it to learn more complex and abstract data representations, which are beneficial for tasks such as image recognition and natural language processing. However, this increases the number of parameters, computational requirements, and training time. The addition of excessive layers can result in overfitting, whereby the network performs well on training data but poorly on test data.

The capabilities of neural networks can be broadly categorised into the following areas:

- Function approximation or regression analysis, including time series prediction, fitness approximation and modelling.
- Classification, pattern and sequence recognition.
- Data processing, filtering and clustering.
- Robotics and control.

In order to detect and differentiate objects in the course of traffic, a neural network is to be designed which, by means of two input parameters (distance to the target and intensity of the signal returned by the object), is capable of determining whether the object detected is a person, a car, or nothing relevant.

The network is then constructed with two inputs: distance to target in metres and signal intensity in decibels; and three logical outputs: one for people, another for vehicles, and a third for irrelevant objects. The neural network will have two layers: one in which the neurons interpret the input values, and another, the output layer, which will provide the logic outputs based on the output values of the previous layer. In order to obtain a binary value for the object identification, a sigmoid activation function is to be used in the output layer. The network will be configured in a feedforward topology at each layer in order to reduce its overall complexity and the time required for learning. The learning process is supervised, which requires the training of the network with a substantial number of previously acquired and personally processed datasets derived from experimental tests. The greater the quantity of data utilised in the design of the network, the more reliable the resulting model will be. Two potential outcomes may be observed:

- In the event that the network functions as intended, with a low error rate, it can be applied to develop an autonomous driving system.
- In the event that the network exhibits a high number of errors, it is not reliable. In such instances, it is necessary to attempt to modify the number of neurons in the network, the data set with which it is trained, or the additional input parameters.

4. Experimental Setup

Figure 5 presents the architecture mounted on the vehicle. The experiments are recorded using a Logitech C270 camera. A commercial radar model, the iSYS-4004 from InnoSent, is employed for detecting objects. The technical specifications of the radar iSYS-4004 are presented in Table 2. Both the camera and the radar are connected to a laptop, on which the algorithms for object detection and identification are executed.

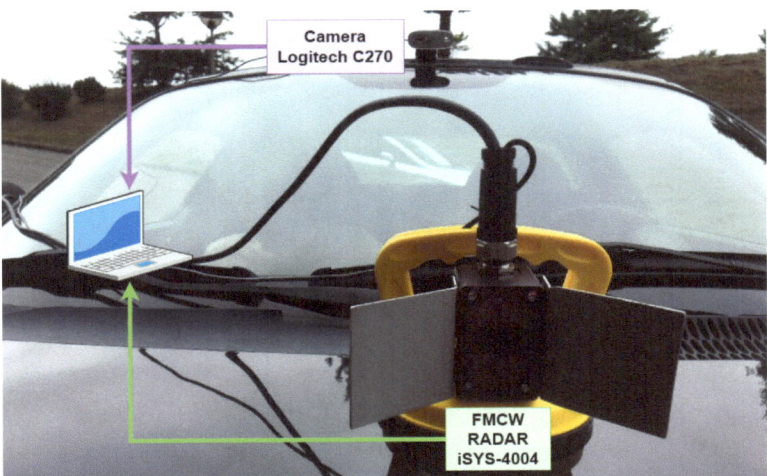

Figure 5. Mounting of the radar on the vehicle.

The configuration presented here is designed to detect and identify vehicles and pedestrians in front of the vehicle. It should be noted that the radar is mounted on the vehicle bonnet and not on the front bump, in order to ensure that the minimum detection distance of 1.1 m is always met. The radar system is only capable of detecting the first object in front of the vehicle, and thus, the density of vehicles and pedestrians during driving does not affect the radar measurements.

For each object detected by the radar, two measurements must be analysed: the distance to the target and the reflected power, which is related to the object RCS. From that, a neural network will be designed so that it is capable of detecting and identifying pedestrians and vehicles. Any other object must be ignored. The procedure for processing the radar-measured information is illustrated Figure 6.

Table 2. Technical specifications of the radar iSYS-4004 [41].

Parameter	Conditions	Min	Max	Units
Radar				
Transmit frequency		24.000	24.250	GHz
Occupied bandwidth	EU-Version		250	MHz
	US/UK/France-Version		100	MHz
Output power (EIRP)	25 °C		20	dBm
Sensor				
Detection distance	EU-Version	1.1	35	m
	US/UK/F-Version	2.7	35	m
Accuracy	250 MHz bandwidth (EU)	−3	3	cm
	100 MHz bandwidth (US)	−7.5	7.5	cm
Resolution	250 MHz bandwidth (EU)		60	mm
	100 MHz bandwidth (US)		150	mm
Operating temperature		−25	60	°C

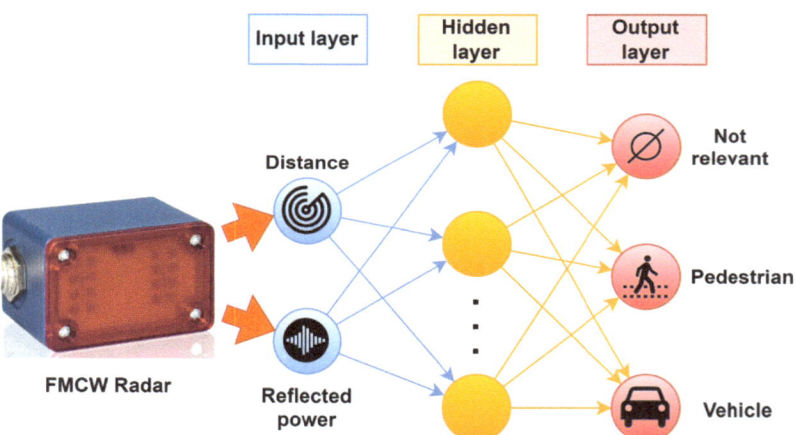

Figure 6. Processing of radar information by a neural network for object detection and identification.

A series of experiments were conducted at different times of the day (morning, afternoon, evening, night) in the city of Santander, Spain. The experiment route is presented in Figure 7. From the initial processing of radar data, it has been demonstrated that the intensity of the signal returned by a vehicle is significantly greater than that returned by a person, and that these signals are distinct from those returned by other types of objects found on the road, such as traffic signs and rubbish bins. This is why the use of a neural network to differentiate each detection and classify it according to the type of object that produces it is an appropriate solution for the aim of the work. Further details will be provided in the subsequent section.

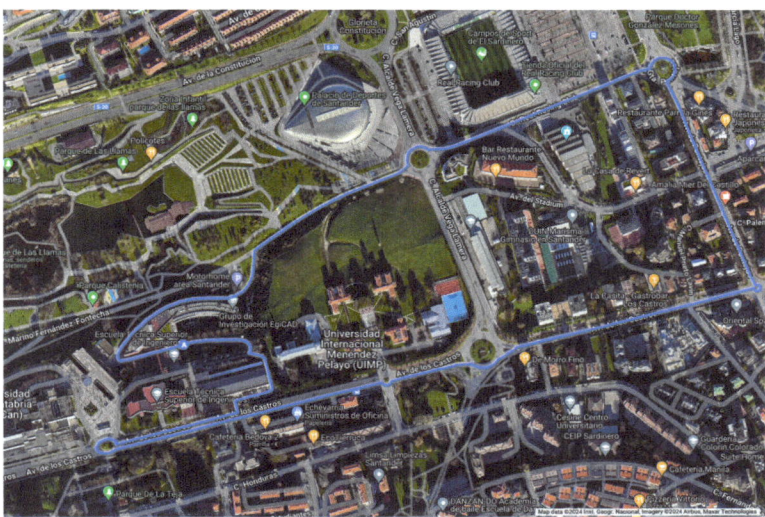

Figure 7. Route followed during the experiments.

5. Experimental Results

A set of 2131 different objects were detected during driving using the experimental setup presented in Section 4. While the radar measured the distance to the target and reflected power, the user must first identify the objects in order to train the neural network for object detection and identification. Table 3 presents a sample of the data collected, which will constitute the input–output data employed in the training of the network. The first two columns are the input vectors, with the distance to the object and the intensity of the signal reflected. The final three columns represent the desired output vectors, which comprise three binary values indicating the type of object detected (pedestrian, vehicle or none of the above).

Table 3. Sample input–output data from the neural network neural network.

Distance (m)	Power Reflected	Irrelevant Object	Pedestrian	Vehicle
8.65	80.58	1	0	0
8.73	65.23	1	0	0
9.75	65.46	1	0	0
2.69	82.52	0	1	0
3.09	82.17	0	1	0
5.07	79.04	0	1	0
5.10	77.52	0	1	0
5.75	91.48	0	0	1
5.92	88.20	0	0	1
6.09	89.50	0	0	1
6.2	89.09	0	0	1
6.82	73.81	0	0	1

The collected data are then interpreted using the Deep Learning Toolbox from Matlab version 2024a, resulting in the generation of a neural network that contains 30 neurons in a hidden layer and 3 neurons in the output layer. A training set with 70% of the data is used to train the network, a validation set with 20% of the data is used to validate the generated network, and a test set with the remaining 10% of the data is used to test the performance of the network. During the training process, the data division is random, the training method chosen is the scaled conjugate gradient, and cross-entropy is selected as a performance indicator. Classes 1, 2 and 3 denote irrelevant objects, pedestrians, and vehicles, respectively. The network constructed from the dataset returned the confusion matrices presented in Figure 8, which indicates an overall 91.1% identification performance. Subsequent trials employed a residual neural network, resulting in an accuracy of 81%. Notably, this value is considerably inferior to the proposed solution, and thus, the use of a residual neural network was discarded.

Figure 8. Training, validation, test, and global confusion matrices.

The true positive rate (TPR) against the false positive rate (FPR) of the proposed network is presented using the receiver operating characteristic curve (ROC), shown in Figure 9. It is important to note that pedestrians and vehicles are never misidentified as irrelevant objects. Furthermore, there are no irrelevant objects misidentified as pedestrians or vehicles. Although pedestrians can be misidentified as vehicles, this is not a severe issue: in the event that a pedestrian is erroneously identified as a vehicle, a possible safety protocol would adopt a cautious approach, such as slowing down or stopping, to avoid collisions. This conservative approach ensures that safety is maintained despite occasional classification errors.

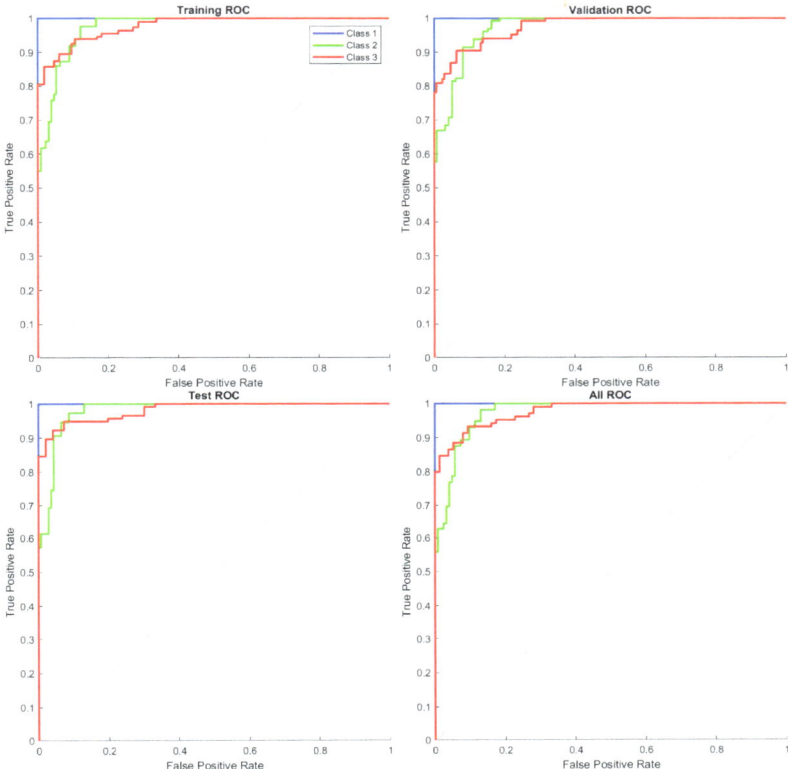

Figure 9. ROC curve of the neural network.

After the network had been designed, it was implemented on the vehicle architecture in order to detect objects in real time and to assist the driver, as seen in Figure 10. The aforementioned results have demonstrated the veracity of the hypothesis that it is possible to detect and identify vehicles and pedestrians solely through the use of low-cost radar information such as RCS and distance, which validates our work.

Figure 10. On-screen display of radar data. Distance and type of objects.

6. Discussion and Conclusions

A pedestrian and vehicle detection and identification system has been developed based on a FCMW radar and a neural network. The radar provides the distance and intensity of the signal reflected by the nearest object, which constitute the input vector to the network. The network outputs a vector of three binary components, each corresponding to one of the possible classes (pedestrian, vehicle or irrelevant objects).

A high success rate in identification has been achieved (91.1% overall); additionally, the low false positive rate observed in the experiments reflects the robustness of the radar-based object detection system in avoiding incorrect hazard alerts. However, there is still room for improvement, and it would be beneficial to conduct further research before the system can be implemented commercially. One possible modification to the network structure would be to increase the number of layers and/or neurons per layer. Furthermore, deep learning techniques could even be used to process the raw signal provided by the radar (with the consequent computational cost), which would significantly improve the identification capability.

The successful implementation of the neural network model to process radar data signifies a step forward in developing autonomous driving systems that do not solely depend on driver intervention. This advancement paves the way for more sophisticated safety features, such as autonomous braking and steering maneuvers, which could significantly reduce the risk of collisions. Furthermore, this research highlights the potential for further enhancements in radar-based detection systems through the refinement of neural network algorithms and the expansion of the range of detectable objects and scenarios.

As a part of future work, the acquired data will be employed by ADAS to facilitate the generation of an appropriate response, such as deceleration, cessation of motion, or the execution of an evasive manoeuvre in the event of potential collisions according to the object type. Furthermore, event-triggering and fault detection mechanisms should be designed to filter the data and detect potential errors [48–50]. In addition, to increase the versatility of the proposed method, the neural network can be improved by including weather data such as temperature as an input.

Author Contributions: Conceptualization, F.V.-M., L.A.-R. and J.P.-O.; methodology, F.V.-M., L.A.-R., J.P.-O. and F.V.-R.; software, F.V.-M.; validation, F.V.-M., L.A.-R., J.P.-O. and F.V.-R.; formal analysis, F.V.-M., L.A.-R., J.P.-O. and F.V.-R.; investigation, F.V.-M., L.A.-R., J.P.-O. and F.V.-R.; resources, F.V.-M., L.A.-R. and J.P.-O.; data curation, F.V.-M.; writing, F.V.-M., L.A.-R., J.P.-O. and F.V.-R.; visualization, F.V.-M.; supervision, L.A.-R., J.P.-O. and F.V.-R.; project administration, L.A.-R. and J.P.-O. All authors have read and agreed to the published version of the manuscript.

Funding: This work has been supported by the Madrid Government (Comunidad de Madrid-Spain) under the Multiannual Agreement with UC3M in the line of Excellence of University Professors (EPUC3M20), and in the context of the V PRICIT (Regional Programme of Research and Technological Innovation). The APC was funded by MDPI.

Data Availability Statement: The original contributions presented in the study are included in the article, further inquiries can be directed to the corresponding author.

Conflicts of Interest: The authors declare no conflicts of interest.

Abbreviations

The following abbreviations are used in this manuscript:

ADAS	Advanced Driver Assistance Systems
FMCW	Frequency-Modulated Continuous-Wave
FPR	False Positive Rate
PReLU	Parametric Rectified Linear Unit
Radar	Radio Detection And Ranging
RCS	Radar Cross Section
ReLU	Rectified Linear Unit

	ROC	Receiver Operating Characteristic
	SELU	Scaled Exponential Linear Unit
	TPR	True Positive Rate

References

1. Song, H.; Shin, H.C. Classification and Spectral Mapping of Stationary and Moving Objects in Road Environments Using FMCW Radar. *IEEE Access* **2020**, *8*, 22955–22963. [CrossRef]
2. Feghhi, R.; Oloumi, D.; Rambabu, K. Tunable Subnanosecond Gaussian Pulse Radar Transmitter: Theory and Analysis. *IEEE Trans. Microw. Theory Tech.* **2020**, *68*, 3823–3833. [CrossRef]
3. Bauer, S.; Tielens, E.K.; Haest, B. Monitoring aerial insect biodiversity: A radar perspective. *Philos. Trans. R. Soc. B* **2024**, *379*, 20230113. [CrossRef] [PubMed]
4. Ramachandran, T.; Faruque, M.R.I.; Singh, M.S.J.; Khandaker, M.U.; Salman, M.; Youssef, A.A. Reduction of Radar Cross Section by Adopting Symmetrical Coding Metamaterial Design for Terahertz Frequency Applications. *Materials* **2023**, *16*, 1030. [CrossRef] [PubMed]
5. Gilson, L.; Imad, A.; Rabet, L.; Van Roey, J.; Gallant, J. Real-time measurement of projectile velocity in a ballistic fabric with a high-frequency Doppler radar. *Exp. Mech.* **2021**, *61*, 533–547. [CrossRef]
6. Liu, H.; Wang, P.; Lin, J.; Ding, H.; Chen, H.; Xu, F. Real-Time Longitudinal and Lateral State Estimation of Preceding Vehicle Based on Moving Horizon Estimation. *IEEE Trans. Veh. Technol.* **2021**, *70*, 8755–8768. [CrossRef]
7. Sengupta, A.; Cheng, L.; Cao, S. Robust Multiobject Tracking Using Mmwave Radar-Camera Sensor Fusion. *IEEE Sens. Lett.* **2022**, *6*, 5501304. [CrossRef]
8. Bai, J.; Li, S.; Huang, L.; Chen, H. Robust Detection and Tracking Method for Moving Object Based on Radar and Camera Data Fusion. *IEEE Sens. J.* **2021**, *21*, 10761–10774. [CrossRef]
9. Huang, X.; Tsoi, J.K.; Patel, N. mmWave radar sensors fusion for indoor object detection and tracking. *Electronics* **2022**, *11*, 2209. [CrossRef]
10. Ravindran, R.; Santora, M.J.; Jamali, M.M. Camera, LiDAR, and Radar Sensor Fusion Based on Bayesian Neural Network (CLR-BNN). *IEEE Sens. J.* **2022**, *22*, 6964–6974. [CrossRef]
11. Liu, P.; Yu, G.; Wang, Z.; Zhou, B.; Chen, P. Object Classification Based on Enhanced Evidence Theory: Radar–Vision Fusion Approach for Roadside Application. *IEEE Trans. Instrum. Meas.* **2022**, *71*, 5006412. [CrossRef]
12. Song, Y.; Xie, Z.; Wang, X.; Zou, Y. MS-YOLO: Object Detection Based on YOLOv5 Optimized Fusion Millimeter-Wave Radar and Machine Vision. *IEEE Sens. J.* **2022**, *22*, 15435–15447. [CrossRef]
13. Rong, Y.; Dutta, A.; Chiriyath, A.; Bliss, D.W. Motion-tolerant non-contact heart-rate measurements from radar sensor fusion. *Sensors* **2021**, *21*, 1774. [CrossRef]
14. Xia, W.; Li, Y.; Dong, S. Radar-Based High-Accuracy Cardiac Activity Sensing. *IEEE Trans. Instrum. Meas.* **2021**, *70*, 4003213. [CrossRef]
15. Yuan, S.; Fan, S.; Deng, Z.; Pan, P. Heart Rate Variability Monitoring Based on Doppler Radar Using Deep Learning. *Sensors* **2024**, *24*, 2026. [CrossRef] [PubMed]
16. Gharamohammadi, A.; Pirani, M.; Khajepour, A.; Shaker, G. Multibin Breathing Pattern Estimation by Radar Fusion for Enhanced Driver Monitoring. *IEEE Trans. Instrum. Meas.* **2024**, *73*, 8001212. [CrossRef]
17. Log, M.M.; Thoresen, T.; Eitrheim, M.H.; Levin, T.; Tørset, T. Using Low-Cost Radar Sensors and Action Cameras to Measure Inter-Vehicle Distances in Real-World Truck Platooning. *Appl. Syst. Innov.* **2023**, *6*, 55. [CrossRef]
18. Lazar, R.G.; Pauca, O.; Maxim, A.; Caruntu, C.F. Control Architecture for Connected Vehicle Platoons: From Sensor Data to Controller Design Using Vehicle-to-Everything Communication. *Sensors* **2023**, *23*, 7576. [CrossRef] [PubMed]
19. Pipicelli, M.; Gimelli, A.; Sessa, B.; De Nola, F.; Toscano, G.; Di Blasio, G. Architecture and potential of connected and autonomous vehicles. *Vehicles* **2024**, *6*, 275–304. [CrossRef]
20. Hussain, S.A.; Shahian Jahromi, B.; Cetin, S. Cooperative highway lane merge of connected vehicles using nonlinear model predictive optimal controller. *Vehicles* **2020**, *2*, 249–266. [CrossRef]
21. Shi, Y.; Du, L.; Chen, X.; Liao, X.; Yu, Z.; Li, Z.; Wang, C.; Xue, S. Robust Gait Recognition Based on Deep CNNs With Camera and Radar Sensor Fusion. *IEEE Internet Things J.* **2023**, *10*, 10817–10832. [CrossRef]
22. Li, J.; Li, B.; Wang, L.; Liu, W. Passive Multiuser Gait Identification Through Micro-Doppler Calibration Using mmWave Radar. *IEEE Internet Things J.* **2024**, *11*, 6868–6877. [CrossRef]
23. He, X.; Zhang, Y.; Dong, X. Extraction of Human Limbs Based on Micro-Doppler-Range Trajectories Using Wideband Interferometric Radar. *Sensors* **2023**, *23*, 7544. [CrossRef] [PubMed]
24. Viadero-Monasterio, F.; Nguyen, A.T.; Lauber, J.; Boada, M.J.L.; Boada, B.L. Event-Triggered Robust Path Tracking Control Considering Roll Stability Under Network-Induced Delays for Autonomous Vehicles. *IEEE Trans. Intell. Transp. Syst.* **2023**, *24*, 14743–14756. [CrossRef]
25. Meléndez-Useros, M.; Jiménez-Salas, M.; Viadero-Monasterio, F.; Boada, B.L. Tire slip H∞ control for optimal braking depending on road condition. *Sensors* **2023**, *23*, 1417. [CrossRef] [PubMed]
26. Eichelberger, A.H.; McCartt, A.T. Toyota drivers' experiences with dynamic radar cruise control, pre-collision system, and lane-keeping assist. *J. Saf. Res.* **2016**, *56*, 67–73. [CrossRef]

27. Sidorenko, G.; Thunberg, J.; Sjöberg, K.; Fedorov, A.; Vinel, A. Safety of Automatic Emergency Braking in Platooning. *IEEE Trans. Veh. Technol.* **2022**, *71*, 2319–2332. [CrossRef]
28. Kim, W.; Yang, H.; Kim, J. Blind Spot Detection Radar System Design for Safe Driving of Smart Vehicles. *Appl. Sci.* **2023**, *13*, 6147. [CrossRef]
29. Kumar, K.; Singh, V.; Raja, L.; Bhagirath, S.N. A Review of Parking Slot Types and their Detection Techniques for Smart Cities. *Smart Cities* **2023**, *6*, 2639–2660. [CrossRef]
30. Viadero-Monasterio, F.; Boada, B.L.; Zhang, H.; Boada, M.J.L. Integral-Based Event Triggering Actuator Fault-Tolerant Control for an Active Suspension System Under a Networked Communication Scheme. *IEEE Trans. Veh. Technol.* **2023**, *72*, 13848–13860. [CrossRef]
31. Viadero-Monasterio, F.; García, J.; Meléndez-Useros, M.; Jiménez-Salas, M.; Boada, B.L.; López Boada, M.J. Simultaneous Estimation of Vehicle Sideslip and Roll Angles Using an Event-Triggered-Based IoT Architecture. *Machines* **2024**, *12*, 53. [CrossRef]
32. Zaker, R.; Sadeghzadeh, A. Passive techniques for target radar cross section reduction: A comprehensive review. *Int. J. RF Microw. Comput.-Aided Eng.* **2020**, *30*, e22411. [CrossRef]
33. Andrade, L.A.D.; Santos, L.S.C.D.; Gama, A.M. Analysis of radar cross section reduction of fighter aircraft by means of computer simulation. *J. Aerosp. Technol. Manag.* **2014**, *6*, 177–182.
34. Ramachandran, T.; Faruque, M.R.I.; Islam, M.T.; Khandaker, M.U.; Tamam, N.; Sulieman, A. Design and analysis of multi-layer and cuboid coding metamaterials for radar cross-section reduction. *Materials* **2022**, *15*, 4282. [CrossRef] [PubMed]
35. Lee, S.; Yoon, Y.J.; Lee, J.E.; Kim, S.C. Human–vehicle classification using feature-based SVM in 77-GHz automotive FMCW radar. *IET Radar Sonar Navig.* **2017**, *11*, 1589–1596. [CrossRef]
36. Cai, X.; Giallorenzo, M.; Sarabandi, K. Machine Learning-Based Target Classification for MMW Radar in Autonomous Driving. *IEEE Trans. Intell. Veh.* **2021**, *6*, 678–689. [CrossRef]
37. Wang, H.; Liu, Y.; Ni, L.; Luo, Y. Micro-Doppler effect removal in inverse synthetic aperture radar imaging based on UNet. *Electron. Lett.* **2023**, *59*, e12814. [CrossRef]
38. Skolnik, M.I. *Introduction to Radar Systems*; McGraw-Hill: New York, NY, USA, 1980; Volume 3.
39. Rezende, M.C.; Martin, I.M.; Faez, R.; Miacci, M.A.S.; Nohara, E.L. Radar cross section measurements (8–12 GHz) of magnetic and dielectric microwave absorbing thin sheets. *Rev. Fis. Apl. Instrum.* **2002**, *15*, 24–29.
40. Marcum, J. A statistical theory of target detection by pulsed radar. *IRE Trans. Inf. Theory* **1960**, *6*, 59–267. [CrossRef]
41. InnoSent. iSYS-4004—Radarsystem. 2024. Available online: https://www.innosent.de/en/radar-systems/isys-4004-radarsystem (accessed on 20 May 2024).
42. Wang, R.; Hu, C.; Fu, X.; Long, T.; Zeng, T. Micro-Doppler measurement of insect wing-beat frequencies with W-band coherent radar. *Sci. Rep.* **2017**, *7*, 1396. [CrossRef] [PubMed]
43. Kim, B.S.; Jin, Y.; Lee, J.; Kim, S. FMCW radar estimation algorithm with high resolution and low complexity based on reduced search area. *Sensors* **2022**, *22*, 1202. [CrossRef] [PubMed]
44. Muslam, M.M.A. Enhancing Security in Vehicle-to-Vehicle Communication: A Comprehensive Review of Protocols and Techniques. *Vehicles* **2024**, *6*, 450–467. [CrossRef]
45. Stove, A.G. Linear FMCW radar techniques. In *IEE Proceedings F (Radar and Signal Processing)*; IET: Birmingham, UK, 1992.
46. Jwo, D.J.; Biswal, A.; Mir, I.A. Artificial neural networks for navigation systems: A review of recent research. *Appl. Sci.* **2023**, *13*, 4475. [CrossRef]
47. Rasamoelina, A.D.; Adjailia, F.; Sinčák, P. A Review of Activation Function for Artificial Neural Network. In Proceedings of the 2020 IEEE 18th World Symposium on Applied Machine Intelligence and Informatics (SAMI), Herl'any, Slovakia, 23–25 January 2020; pp. 281–286. [CrossRef]
48. Viadero-Monasterio, F.; Boada, B.; Boada, M.; Díaz, V. H∞ dynamic output feedback control for a networked control active suspension system under actuator faults. *Mech. Syst. Signal Process.* **2022**, *162*, 108050. [CrossRef]
49. Boada, B.L.; Viadero-Monasterio, F.; Zhang, H.; Boada, M.J.L. Simultaneous Estimation of Vehicle Sideslip and Roll Angles Using an Integral-Based Event-Triggered H_∞ Observer Considering Intravehicle Communications. *IEEE Trans. Veh. Technol.* **2023**, *72*, 4411–4425. [CrossRef]
50. Meléndez-Useros, M.; Jiménez-Salas, M.; Viadero-Monasterio, F.; López-Boada, M.J. Novel Methodology for Integrated Actuator and Sensors Fault Detection and Estimation in an Active Suspension System. *IEEE Trans. Reliab.* **2024**, 1–14. [CrossRef]

Disclaimer/Publisher's Note: The statements, opinions and data contained in all publications are solely those of the individual author(s) and contributor(s) and not of MDPI and/or the editor(s). MDPI and/or the editor(s) disclaim responsibility for any injury to people or property resulting from any ideas, methods, instructions or products referred to in the content.

Article

Meta-Feature-Based Traffic Accident Risk Prediction: A Novel Approach to Forecasting Severity and Incidence

Wei Sun [1], Lili Nurliynana Abdullah [2,*], Puteri Suhaiza Sulaiman [2] and Fatimah Khalid [2]

1. Computer Vision, Faculty of Computer Science and Information, Universiti Putra Malaysia, Serdang 43400, Malaysia; gs66334@student.upm.edu.my
2. Department of Multimedia, Faculty of Computer Science and Information, Universiti Putra Malaysia, Serdang 43400, Malaysia
* Correspondence: liyana@upm.edu.my

Abstract: This study aims to improve the accuracy of predicting the severity of traffic accidents by developing an innovative traffic accident risk prediction model—StackTrafficRiskPrediction. The model combines multidimensional data analysis including environmental factors, human factors, roadway characteristics, and accident-related meta-features. In the model comparison, the StackTrafficRiskPrediction model achieves an accuracy of 0.9613, 0.9069, and 0.7508 in predicting fatal, serious, and minor accidents, respectively, which significantly outperforms the traditional logistic regression model. In the experimental part, we analyzed the severity of traffic accidents under different age groups of drivers, driving experience, road conditions, light and weather conditions. The results showed that drivers between 31 and 50 years of age with 2 to 5 years of driving experience were more likely to be involved in serious crashes. In addition, it was found that drivers tend to adopt a more cautious driving style in poor road and weather conditions, which increases the margin of safety. In terms of model evaluation, the StackTrafficRiskPrediction model performs best in terms of accuracy, recall, and ROC–AUC values, but performs poorly in predicting small-sample categories. Our study also revealed limitations of the current methodology, such as the sample imbalance problem and the limitations of environmental and human factors in the study. Future research can overcome these limitations by collecting more diverse data, exploring a wider range of influencing factors, and applying more advanced data analysis techniques.

Keywords: traffic accident risk prediction; meta-features; machine learning; environmental factors; human factors; traffic safety management

1. Introduction

Traffic accidents have escalated into a significant global public health issue, resulting in a considerable number of fatalities and injuries annually. According to the 2018 Global Status Report on Road Safety by the World Health Organization (WHO), approximately 1.35 million individuals experience road accidents worldwide annually, with traffic-related injuries being the leading cause of death among individuals aged 5 to 29 years [1]. Consequently, the prevention and reduction in traffic accidents on an international scale are an imperative necessity. During our investigation into the effects of urbanization on traffic accidents, it was discerned that human factors are crucial in influencing traffic accident occurrences in numerous countries and regions. Data collected from the World Health Organization (WHO) indicate that approximately 10% of road traffic deaths are related to drink driving; this corresponds to self-reported rates of 16–21% of people admitting to drink driving in a survey conducted by the European Survey Research Association (ESRA). The same self-reports reveal that nearly 50% of drivers across 48 countries report exceeding the speed limit outside built-up areas [2]. Speeding, drink-driving, driver fatigue, distracted driving, and non-use of safety belts, child restraints and helmets are among the key behaviours contributing to road injury and death [3]. Vulnerable road users such as pedestrians, cyclists, moped riders, and motorcyclists are particularly at high risk of severe or fatal injury when motor vehicles collide with them because of their lack of protection [4].

In our investigation of the effects of urbanization on traffic accidents, we determined that human factors play a pivotal role in influencing traffic accident occurrences across various countries and regions. Particularly in Morocco, human factors have been identified as one of the primary reasons behind the nation's roads being ranked among the most perilous globally. A survey conducted in Sudan revealed that individual factors were responsible for 60.6% of traffic accidents, with suboptimal road conditions (45.5%), animal-related factors (5.6%), and vehicle scarcity (1.4%) also contributing significantly [5]. The Czech In-depth Accident Study (CzIDAS) indicates that distractions account for 40% of the analyzed accidents, highlighting the significance of this factor. Distractions may stem from a variety of causes, including attention overload (35%), distracted driving (19%), and monotonous driving (13%) [6]. Furthermore, the likelihood of road traffic accidents is directly correlated with environmental factors such as rainfall, extreme low temperatures, fog, and hot weather conditions. The incident rates of accidents are 34%, 25%, 21%, and 20%, respectively, attributable to fog, rain, temperature variances, and additional weather-related factors [7]. From a geographical standpoint, the proportion of fatal traffic collisions is notably higher in rural regions (66%) compared to urban areas (34%). Accidents predominantly occur on straight roads, succeeded by curved roads, intersections, and Y/T intersections, which witness the highest rates of traffic fatalities [8]. This paragraph accentuates the impact of human factors, environmental conditions, and geographical location on the rates of traffic accidents, factors that are especially critical in the context of urbanization. Urbanization directly influences road-use patterns and traffic flow, thereby significantly impacting traffic safety.

However, challenges remain in the realm of traffic safety research. The issue of data imbalance in traffic accident studies is a persistent concern [9,10], as is the need for greater interpretability and transparency in traffic safety risk analysis [11–13]. Additionally, while much research has focused on local attributes of traffic accidents, there is growing recognition of the importance of incorporating contextual information from the entire scene for a more explicit and classification [14,15].

In light of these findings, there is a growing need for advanced methods to analyze and predict traffic crash risk. Traditional models, while valuable, have limitations in terms of predictive accuracy and the ability to handle complex, multifaceted data. This gap highlights the need for new methods that combine the strengths of various approaches to provide more accurate analysis. This study introduces StackTrafficRiskPrediction, a predictive model of traffic risk hazard, which is a pioneering attempt in the field of traffic safety analysis. In this study, a series of classification models are first utilized to generate meta-features, which are subsequently applied to train a regression model, i.e., a meta-model. In this way, we are able to not only capture the underlying patterns of the data using classification models, but also provide greater flexibility and accuracy in predicting continuous outputs through regression models. Our results not only provide an effective framework for predicting injury severity in traffic accidents, but also offer new perspectives on the application of machine learning in the field of traffic safety.

2. Literature Review

Within the scholarly discourse on traffic accident severity classification, accidents are typically categorized into the following three distinct types: "fatal", "serious", and "minor". Fatal crashes, defined as accidents resulting in the death of one or more individuals, have a profound global impact. Research underscores this, noting that on average, 1.35 million people perish annually in traffic accidents [16,17]. Serious accidents refer to incidents that culminate in substantial injuries, albeit non-fatal in nature. The severity of these accidents is typically assessed based on the quantity of individuals injured and the extent of direct property damage incurred [18]. Minor accidents are characterized by less severe injuries, and while the direct discourse on such incidents is limited, ancillary research implicitly addresses these minor injuries through the analysis of various accident types and their influence on overall accident severity [19]. These classifications offer a foundational framework for comprehending the diverse severities of injuries sustained in traffic accidents and are pivotal in the development of tailored prevention strategies and interventions.

An exhaustive review of the literature pertaining to factors influencing traffic accidents reveals that meteorological conditions, roadway conditions, and individual factors are integral in determining the frequency and severity of traffic accidents. Meteorological conditions exert a substantial impact on traffic accidents, with varying weather conditions influencing different types of accidents in distinct manners, for instance, snowy conditions predominantly affect cycling accidents, whereas daylight glare significantly elevates the risk of multi-vehicle collisions on highways [20–24]. Roadway conditions, encompassing aspects such as traffic congestion and the state of the pavement, play a pivotal role in the incidence of accidents. Research has elucidated an inverse correlation between traffic congestion and the frequency of accidents, while the condition of the road surface has also been found to significantly influence the occurrence of accidents [25,26]. Individual factors, particularly those encompassing driver error and fatigue, exert a profound impact on the incidence of road accidents. While existing research has delved into the relationship between personal factors and traffic accidents, a notable research gap remains regarding the precise assessment of the impact of personal factors, particularly in relation to drivers' psychological and physiological states on accidents [27,28]. These studies illuminate the myriad factors influencing road accidents and underscore areas necessitating further exploration in future research endeavors to enhance overall road safety.

Conventional traffic accident data analysis methodologies have been employed to meticulously examine traffic safety issues, utilizing a spectrum of data analysis techniques including plain Bayesian classifiers, logistic regression, linear regression, K-nearest neighbours (K-NN) algorithms, K-mean clustering algorithms, auto-encoders, transfer learning, and transformer techniques. These methods are extensively utilized in road safety research, encompassing a broad spectrum of aspects ranging from road condition analysis to driving behaviour assessment and the development of collision warning systems. Plain Bayesian classifiers have gained particular prominence in applications such as pavement detection and the safety assessment of driving behaviour [29–31]. Logistic regression has been used to analyze accident severity and driving behaviour [32–34], whereas linear regression has played an important role in studies on the relationship between economic dynamics, road design improvements and traffic safety [35–37]. K-NN algorithms have shown their clustering and classification capabilities in accident prediction and case retrieval [38,39]. K-mean clustering and auto-coders have been used to extract hidden information from traffic accident data and to performing accident hotspot identification [40–42]. Transfer learning and transformer techniques have shown potential in traffic accident risk prediction and detection [43–46]. These research methodologies not only demonstrate the diversity and intricacy of data analytics within the realm of traffic safety, but also highlight potential limitations and chart out future research trajectories for the application of these techniques in real-world traffic scenarios.

Research in applied traffic accident analysis has focused on the following three areas: traffic accident prediction, real-time traffic behaviour analysis, and driver fatigue and distraction detection. Research in traffic accident prediction focuses on understanding the factors that lead to accidents and applying various machine learning models to make predictions, especially on motorways and high-class roads [47,48]. Real-time traffic behaviour analysis uses advanced techniques such as linking vehicle data for real-time assessment of traffic safety and analyzing the driving behaviour of urban bus drivers [49]. The field of driver fatigue and distraction detection, on the other hand, focuses on the development of effective detection methods and systems, including identification using machine learning techniques [50–52]. These studies elucidate the multifaceted nature and intricacy of road safety research, simultaneously identifying the limitations of current studies and outlining prospective avenues for future research. This includes refining the applicability of predictive models, converting research findings into actionable road safety measures, and augmenting the thoroughness and scalability of real-time assessment frameworks.

Research in contextual information analysis of traffic accidents focuses on understanding personality and behavioural traits in traffic accidents, utilizing nationwide traffic accident datasets, and applying advanced technologies such as the Internet of Vehicles (IoV) and artificial intelligence (AI) for accident prediction and prevention. Research has shown that driver personality and behavioural patterns have a significant impact on traffic safety [53–56]. In addition, the use of metadata and meta-features is becoming increasingly important in crash analysis, as these techniques can improve the accuracy and efficiency of crash detection, understand the relationship between driving behaviour and crash risk, and perform long-term trend analysis [57–59]. Collectively, these studies underscore the significance of comprehending contextual factors in traffic accidents and exemplify the implementation of sophisticated techniques such as artificial intelligence, machine learning, and context-aware systems in exhaustive traffic accident analysis. These studies furnish the field with novel insights, methodologies, and data resources, bearing significant practical implications for the enhancement of traffic safety and the prevention of accidents.

The application and analysis of metadata are becoming important research directions in the field of traffic accident analysis. The utilization of metadata not only improves the accuracy and efficiency of traffic accident detection, but also provides insights for understanding the context and causes of accidents. For example, a traffic accident detection model developed using a metadata registry demonstrates how the accuracy of accident detection can be improved [60]. Through meta-analysis of the relationship between traffic violations and accidents, researchers have been able to reveal biases between self-reported and archived data as well as provide insights into the link between personality traits and traffic accidents [57]. On a technical level, the development of multidimensional design methods for spatial data warehouses and geo-decision tools demonstrates the important application of metadata in spatial analysis and road accident analysis [59]. Long-term trend analyses using metadata, such as the analysis of road accidents in the Ugandan region, have revealed patterns and trends in accident occurrence [61]. These studies show that metadata play a key role in improving traffic safety and preventing accidents.

Overall, these studies not only provide insights into the meta-characterization of traffic accidents, but also provide valuable references for future traffic safety management and accident prevention strategies. By integrating multiple data and models, the application of meta-characterization shows great potential in improving traffic safety.

3. Research Methodology

Based on the detailed background provided in the previous two chapters, the experimental design in Chapter 3 focuses on developing and validating the StackTrafficRiskPrediction model as shown in Figure 1. The study began with data collection, followed by data cleaning to deal with incomplete and erroneous data. This was immediately followed by feature extraction, focusing on traffic risk features. After defining and selecting the meta-features, the meta-feature generation process was performed. Then, the meta-model was designed, and regression techniques were selected to integrate it into a complete model. In the comparison phase, the new model was compared with existing models. Finally, a training and evaluation phase was performed, which included a training process and evaluation metrics to assess model performance. The entire process emphasizes a step-by-step approach from data preprocessing to model comparison and evaluation to ensure model accuracy and validity.

3.1. Model Structural Design

3.1.1. Objective

The main goal of the StackTrafficRiskPrediction model is to improve the accuracy of traffic accident risk prediction by utilizing stacked integrated learning methods. This model aims to improve the prediction of traffic accident severity by creating meta-features through a classification-based base model. It integrates multiple factors, including environmental conditions, road characteristics, and human factors, to comprehensively analyze the complexity of traffic accidents and enhance prediction.

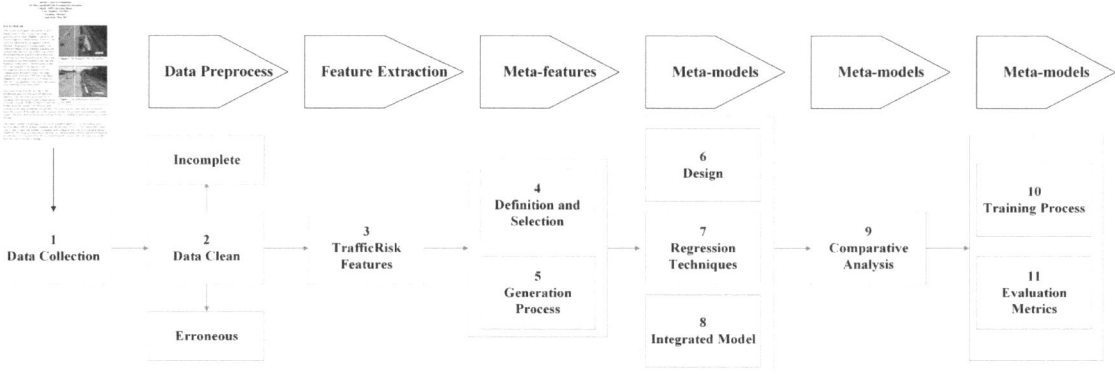

Figure 1. Flow chart of StackTrafficRiskPrediction.

3.1.2. Meta Model Structure

The StackTrafficRiskPrediction model is a sophisticated ensemble learning framework that combines multiple machine learning techniques to improve the prediction of traffic crash risks. The architecture of this model is built upon two primary layers, the base layer and the meta-model layer as shown in Figure 2.

(1) Base Layer (Classification Models):
- Composition: This layer comprises a series of different classification models. Each model is designed to capture specific aspects of traffic accident data, such as accident severity, type of accident, and contributing factors.
- Function: These models analyze various features of the data, like weather conditions, road types, and driver behaviors, to classify different aspects of traffic accidents.
- Output: The primary output of this layer is a set of meta-features. These are derived from the predictions of each classification model and represent a higher-level abstraction of the data.

(2) Meta-Model Layer (Regression Model):
- Integration: The meta-model is a regression model that takes the meta-features generated by the base layer as its input. This layer effectively synthesizes the insights gained from the base classification models into a cohesive prediction.
- Algorithm selection: Logistics regression was chosen for the regression algorithm in the meta-model.
- Objective: The purpose of the meta-model is to predict the continuous risk score of traffic accidents, providing a nuanced understanding of the likelihood and severity of accidents under various conditions.

(3) Stacking Mechanism:
- Principle: The model employs a stacking approach where the predictions of several base classifiers serve as input features for the meta-model. This approach harnesses the strengths of different models, mitigating their individual weaknesses.
- Advantage: By combining multiple models, the StackTrafficRiskPrediction model aims to capture a broader spectrum of patterns and relationships within the data, which might be missed by a single model.

(4) Integration with Classification Models:
- The output of the classification model is first converted into meta-features. These meta-features are normalized to ensure consistency in their scales and distributions, making the meta-features suitable as inputs to the meta-model. In the process of weighting and combining meta-features, different weights are assigned to each meta-feature based on their predictive power and relevance. In addition, the study employs feature selection and dimensionality reduction techniques to refine the meta-feature set. Then, in the model training and tuning phase, the meta-model is trained on the basis of these meta-features with the goal of minimizing the prediction error and optimizing the performance metrics.

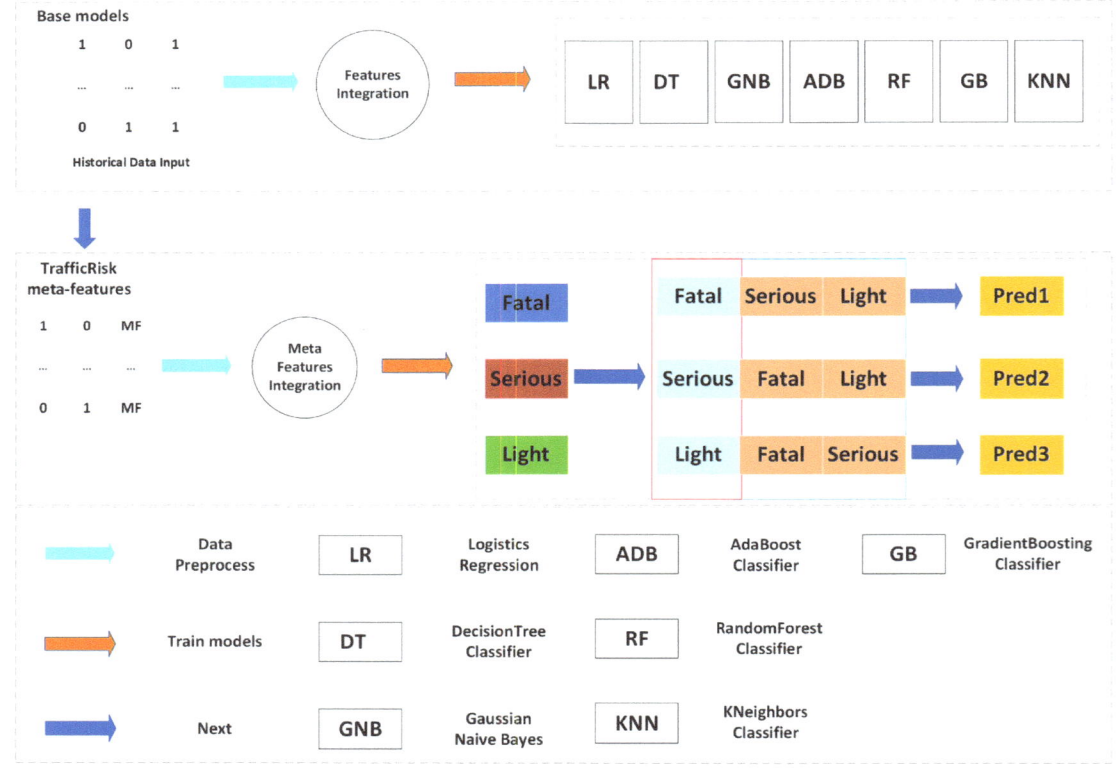

Figure 2. An overview of the StackTrafficRiskPrediction.

3.2. Data Collection and Preprocessing

3.2.1. Data Collection

The StackTrafficRiskPrediction model utilizes data from multiple sources for the analysis of factors influencing traffic accidents. These sources include data from police and transportation department reports, providing detailed information on each accident, including time, location, type of vehicles involved, nature of the accident, weather data, road condition, and casualties, as shown in Table 1. In this study, 4000 traffic accidents were selected as data sets from February 2016 to December 2020.

Table 1. Original data items.

Items	Explanation	Types
Time	Specific moment of the accident occurrence, usually indicated by hours and minutes.	Randomness
Day_of_week	The specific day of the week on which the accident occurred.	'Monday', 'Sunday', 'Friday', 'Wednesday', 'Saturday', 'Thursday', 'Tuesday'
Age_band_of_driver	A categorized range indicating the age group of the driver involved.	'18–30', '31–50', 'Under 18', 'Over 51'
Sex_of_driver	The gender of the driver involved in the accident.	'Male', 'Female'
Educational_level	The highest level of formal education attained by the driver.	'Above high school', 'Junior high school', 'Elementary school', 'High school'
Driving_experience	The total duration or years of experience the driver has in driving.	'1–2 yr', 'Above 10 yr', '5–10 yr', '2–5 yr'
Area_accident_occured	The specific location or type of area where the accident took place.	'Residential areas', 'Office areas', 'Recreational areas', 'Industrial areas', 'Church areas', 'Market areas', 'Rural village areas', 'Outside rural areas', 'Hospital areas', 'School areas', 'Rural village areas Office areas', 'Recreational areas'
Road_surface_conditions	The condition of the road at the accident spot.	'Dry', 'Wet or damp', 'Snow', 'Flood over 3 cm. deep'
Light_conditions	The level of natural or artificial lighting at the time of the accident.	'Daylight', 'Darkness-lights lit', 'Darkness-no lighting', 'Darkness-lights unlit'
Weather_conditions	The environmental weather conditions during the accident.	'Normal', 'Raining', 'Raining and Windy', 'Cloudy', 'Windy', 'Snow', 'Fog or mist'
Individual	This term could refer to any single person involved in the accident, often focusing on their specific characteristics or role.	'Drinking', 'Normal', 'Operating', 'Talking', 'Texting'
Accident_severity	The classification of the accident based on its seriousness or consequences.	'Light', 'Serious', 'Fatal'

3.2.2. Data Cleaning

Data collected from these various sources contain inconsistencies, missing values, and outliers. The cleaning process includes the following steps:

- Dealing with missing values: depending on the nature and extent of the missing data, missing values are identified, and records of missing values are removed.
- Consistency checking: this is carried out to ensure that data from different sources are consistent in terms of units, scale, and format.

3.3. Definition and Selection of Meta-Features

3.3.1. Definition

In machine learning and statistical modeling, meta-features usually refer to features derived from the original data set to enhance the predictiveness and interpretability of the model. Traditionally, these features might include statistical descriptors, model-based predictions, or be the product of feature engineering [62]. In this study, the traditional meta-feature definition is extended and applied to the context of traffic accident risk prediction. The meta-features studied are not only derived based on the raw data, but also include higher-order features derived from the predictions and internal states of the underlying classification model. These higher-order features can capture subtle patterns and relationships that cannot be observed or quantified through the raw data alone [63].

3.3.2. Selection

In terms of meta-feature selection, this study selected multiple types of meta-features to improve the accuracy and explanatory power of traffic accident risk prediction. Specifically, they include traditional statistical descriptor meta-features, meta-features based on traffic accident prediction results, and high-order meta-features derived from predictions and

internal states of classification models. These meta-features not only reflect the fundamental
properties of the original data, but also enhance the model's predictive power by capturing
deeper patterns and relationships.

3.3.3. Generation Process of Meta-Features

Traditional interactive variables or derived features, such as polynomial combination
and categorical feature intersection, belong to traditional feature engineering methods.
These methods mainly combine two or more original features through mathematical or
logical operations to create new features to reveal possible interactive effects between
these features.

- Polynomial combination: By combining features through mathematical operations
  (such as multiplication) as shown in Table 2, new features are generated, such as
  multiplying "Age" and "Driving Experience" to obtain "Age_Experience", which is
  used to reveal how these two variables jointly affect the risk of accidents.
- Categorical feature crossover: This includes combining classification features into a
  new classification feature as shown in Table 3, such as combining "road conditions"
  with "weather conditions" to generate a new feature "Road_Weather". These features
  capture direct relationships between variables by explicitly combining them in the
  original data.

Table 2. Polynomial combination of meta-features.

Items	Explanation	Types of Examples
Age_Experience	The effect of the interaction of age and experience on accident risk is revealed.	'(18–30) × (1–2 yr)', '(31–50) × (1–2 yr)', '(Under 18) × (1–2 yr)', '(Over 51) × (1–2 yr)', etc.

Table 3. Categorical feature crossover of meta-features.

Items	Explanation	Types of Examples
Road_Weather	Indicates a combination of different pavement conditions in each weather.	'Dry-Normal', 'Wet or damp-Normal', 'Snow-Normal', 'Flood over 3 cm. deep-Normal', etc.
Individual_Road	Indicates a combination of different personal factors in each roadway.	'Drinking-Dry', 'Normal-Dry', 'Operating-Dry', 'Talking-Dry', 'Texting-Dry', etc.
Individual_Weather	Indicates combinations of different personal factors in each weather.	'Drinking-Raining', 'Normal-Raining', 'Operating-Raining', 'Talking-Raining', 'Texting-Raining', etc.

Each base classification model in the StackTrafficRiskPrediction framework focuses
on predicting the severity of traffic accidents, using the classification base model output
probabilities as new features as shown in Table 4. At the same time, the input factors are
shown in Table 1. This includes extracting features from the intermediate layers of the
deep learning models, and capturing complex patterns learned by the models. Finally, it is
ensured that these meta-features are properly normalized and transformed for input into
the meta-model.

Table 4. Value of meta-features.

Items	Explanation
LogisticRegression	The output from a logistic regression model can be used as a meta-feature, representing the probability of accident_severity occurring.
DecisionTreeClassifier	The decision paths taken in a decision tree, which lead to a certain prediction, can be encoded as meta-features.
KNeighborsClassifier	For each prediction, the count or proportion of neighbors voting for each class can be used as a meta-feature.

Table 4. *Cont.*

Items	Explanation
Gaussian Naive Bayes	The posterior probabilities generated by GNB, based on the assumption of normally distributed features, can be used.
RandomForestClassifier	Random forests provide insights into feature importance, which can be used as meta-features.
AdaBoostClassifier	AdaBoost focuses on instances that are harder to classify, adjusting weights accordingly.
GradientBoostingClassifier	The outputs from gradient boosting, which builds trees in a sequential correction manner, can be used.

In contrast to the above feature engineering, the research uses transformers to obtain internal high-order features as Table 5. These features are extracted from the internal structure of the model and can reflect deeper data patterns and relationships. These outputs can reflect the contextual and deep semantic information in the text data. In the creation of internal high-level characteristics, 128 dimensions were studied to extract 128 characteristics of each traffic accident case.

Table 5. Meta-features of internal high-level characteristics.

Feature1	Feature2	...	Feature128
2.8537998	2.8497229	...	2.8519573
2.9121785	2.90959	...	2.9110537
...	...	...	...
2.781281	2.7881584	...	2.7881358

Combining the above-mentioned ways of combining the features, the meta-features of this study were obtained.

3.4. Model Training and Evaluation

3.4.1. Training Process

- Model Training: The training process involves feeding the training dataset into the model and iteratively adjusting the model parameters to minimize the loss function.
- Complexity Management: To handle the complexity of the model, especially if using a deep learning approach, techniques like dropout and early stopping are employed to prevent overfitting.
- Hyperparameter Tuning: Techniques like grid search can be used to find the optimal set of hyperparameters for the model.

3.4.2. Evaluation Metrics

For Classification Components:

- Accuracy: Measures the proportion of correctly predicted instances.
- Recall: Measures the proportion of actual positives that were correctly identified.
- F1: The F1 score is the reconciled mean of precision and recall, and is a composite of precision and recall, particularly applicable to those cases where the categories are unbalanced.

Validation Techniques

- Cross-Validation: K-fold cross-validation is used, especially for smaller datasets, to ensure that the model's performance is consistent across different subsets of the data. This technique involves dividing the data into k subsets and training the model k times, each time using a different subset as the test set and the remaining as the training set.
- Performance Benchmarking: The model's performance is compared with established benchmarks or similar models in the field to assess its relative effectiveness.

In summary, the training and evaluation of the StackTrafficRiskPrediction model require careful consideration of data handling, model complexity, and appropriate evaluation metrics. The combination of different metrics for classification and regression components will provide a greater understanding of the model's performance.

4. Results and Discussion

After experiments, the performance of the severity prediction model of traffic accidents based on the meta-based model was obtained as follows Table 6. This meta-model performs best in categorizing minor accidents with very high accuracy. It also showed some reliability in predicting serious and fatal accidents. And when comparing the model without meta characteristics, the accuracy rate is higher than other models.

Table 6. Performance of meta-model testing.

Model Type	Fatal	Serious	Light
Meta-model	0.9613	0.9069	0.7508
LogisticRegression	0.7182	0.8669	0.6289

The results of the five-fold cross-validation are shown in Table 7, which shows the performance of the meta-model on different accident severity levels (fatal, serious, and light). For fatal accidents, the accuracy of the model averages 0.8248 and reaches a maximum of 0.9396, which indicates that the model has high accuracy and stability for predicting fatal accidents. However, it performs relatively poorly in the prediction of serious accidents, with an average accuracy of 0.7336, with the lowest accuracy dropping to 0.6094, which may point out that the model has some limitations or needs further optimization in dealing with such accidents.

Table 7. Five-fold validation of meta-model testing.

Type	Fatal	Serious	Light
Accuracy	0.8283	0.7553	0.8283
	0.7381	0.6094	0.6180
	0.7339	0.7682	0.7639
	0.8841	0.7982	0.7725
	0.9396	0.7370	0.7715
Average	0.8248	0.7336	0.7503

For light accidents, the model performed similarly to fatal accidents, with an average accuracy of 0.7503, which shows that the model is relatively balanced but slightly less accurate in predicting light accidents than fatal accidents. In addition, there is a small difference in the minimum accuracy between the predictions of minor and fatal accidents, which suggests that there is some consistency in the model's performance in predicting accidents of different severities. Overall, the meta-model showed some volatility in the prediction of traffic accidents at various severity levels, especially the fluctuation of accuracy on the prediction of severe accidents, which requires targeted improvement or adjustment of the model parameters to improve the accuracy and stability of the prediction in subsequent studies.

After analyzing the data from the study, as shown in Figure 3, it was found that people between 31 and 50 years old are prone to major traffic accidents. Also, when analyzing the data on driving experience and severity of traffic risk, it was found that drivers with 2–5 years of experience were more likely to be involved in traffic accidents. Among the factors about road surface, light and weather, the study found that when drivers encounter bad road surface and weather, they instead drive more carefully and have a higher safety margin than a normal driving environment.

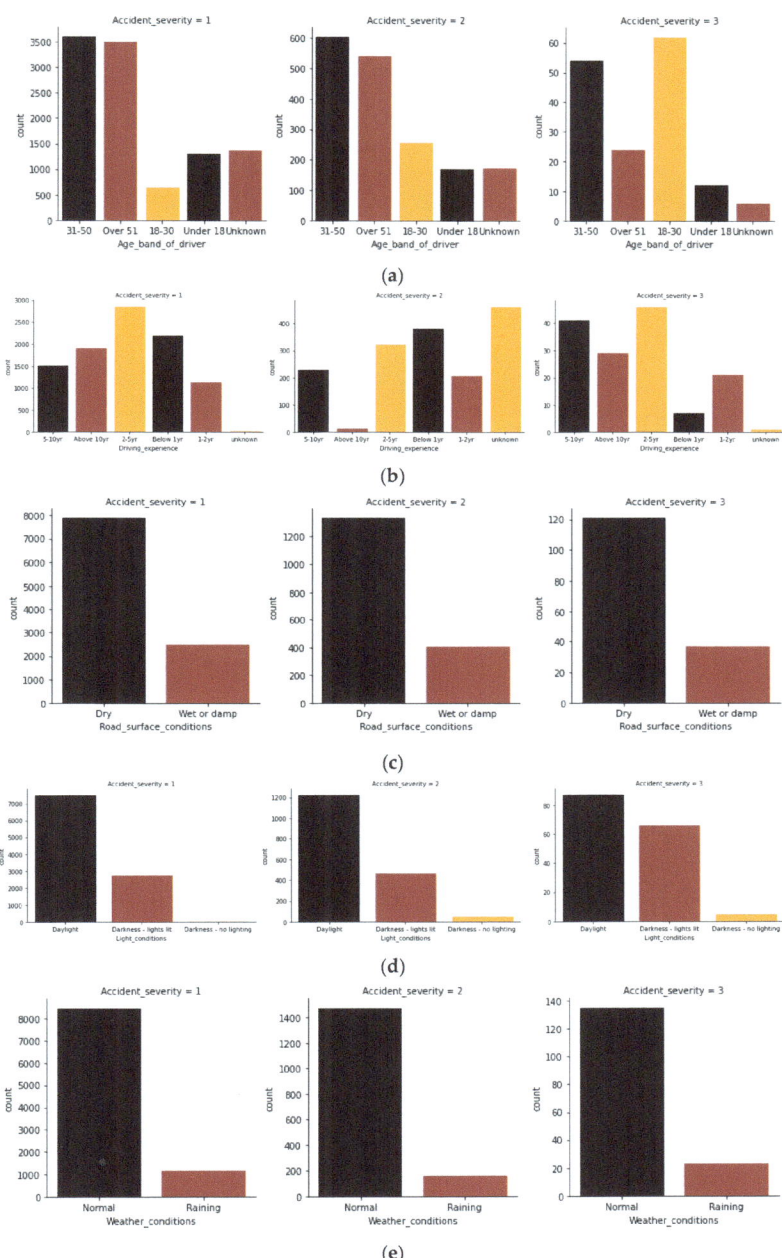

Figure 3. Incidents of severity of traffic accidents due to different factors. In summary, 1 means light, 2 means serious, and 3 means fatal. (**a**) Age_band_of_driver: Incidents of traffic accident severity due to driver age. (**b**) Driving_experience: Incidents of traffic accident severity due to driver experience. (**c**) Road_surface_conditions: Incidents of traffic accident severity due to road. (**d**) Light_conditions: Incidents of traffic accident severity due to light. (**e**) Weather_conditions: Incidents of traffic accident severity due to weather.

As shown in Figure 4, without the addition of meta-features, the study found a correlation between "Accident_severity" and several factors. In particular, "Number_of_casualties" has a significant positive correlation with accident severity, meaning that as the number of casualties in an accident increase, the severity of the accident tends to increase. In addition, 'Light_conditions' also showed some degree of correlation with accident severity, suggesting that the severity of accidents varies under different light conditions. However, factors such as 'Weather_conditions', 'Road_surface_conditions' and 'Type_of_collision' were associated with the 'Type_of_collision'. Factors such as "Accident_severity" correlate strongly with "Road_surface_conditions" and "Type_of_collision", suggesting that they are major factors in accident severity. Therefore, the meta-feature selection in the study was performed by combining these features to form a new dataset based on the base model of the study.

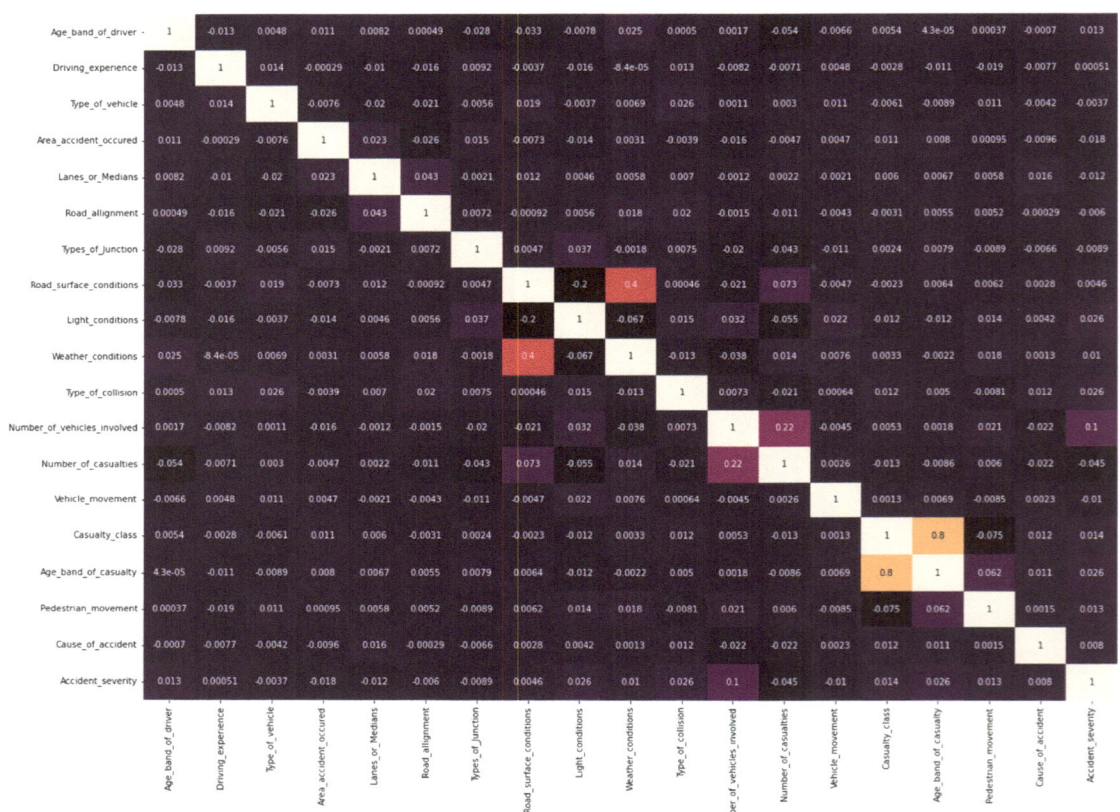

Figure 4. Heatmap of accident severity without meta-features.

The prediction results of each base model were derived after the training and evaluation of the model, as shown in Table 8. In the performance evaluation of the different base models of the StackTrafficRiskPrediction model, we find that the GradientBoostingClassifier performs the best on all the metrics, with the highest accuracy, recall, and F1 scores, and shows optimal performance on the ROC–AUC values. RandomForestClassifier and LogisticRegression follow closely, and these two models have better F1 scores and ROC–AUC values while maintaining high accuracy and recall, showing a more balanced performance. AdaBoostClassifier (AdaBoostClassifier) also shows good performance similar to logistic regression. In contrast, Gaussian Naive Bayes and KNeighborsClassifier, while performing moderately well in terms of accuracy and recall, were slightly lacking in terms of F1 scores and ROC–AUC values. The DecisionTreeClassifier performed the worst on this dataset, especially on the ROC–AUC values, possibly due to overfitting or failing to effectively capture the complexity of the data.

Table 8. Performance of basic models without meta-features.

Items	Accuracy	Recall	F1 Score	ROC–AUC
LogisticRegression	0.84375	0.84375	0.77224	0.61956
DecisionTreeClassifier	0.74472	0.74472	0.75342	0.55011
KNeighborsClassifier	0.82629	0.82629	0.76982	0.54124
Gaussian Naive Bayes	0.81452	0.81452	0.76221	0.61222
RandomForestClassifier	0.84618	0.84618	0.78371	0.68336
AdaBoostClassifier	0.84253	0.84253	0.77181	0.62343
GradientBoostingClassifier	0.84862	0.84862	0.78441	0.70539

In evaluating the predictive performance of the GradientBoostingClassifier model as shown in Table 9, it can be analyzed in terms of its precision, recall, F1 score, and overall accuracy on different categories. The model performs well in terms of overall accuracy, reaching 0.77, while its weighted avg (weighted avg) precision, recall, and F1 score are 0.75, 0.77, and 0.76, respectively, which shows a high prediction efficiency taking into account the difference in the number of samples in the categories. In particular, on category 2, the model exhibits high precision (0.85), recall (0.89), and F1 score (0.87), indicating a significant advantage in prediction in this category. However, in terms of macro avg precision, recall and F1 score, the average performance of the model on different categories is only around 0.37, reflecting a more insufficient performance on small-sample categories (especially categories 0 and 1), which may be related to the insufficient number of samples and the imbalance of categories. In summary, the GradientBoostingClassifier performs well in dealing with major categories, but still needs to be improved in terms of prediction accuracy on small-sample categories to achieve a more balanced and prediction effect.

Table 9. Prediction performance of GradientBoostingClassifier.

	Precision	Recall	F1 Score	Support
0	0.08	0.04	0.05	52
1	0.23	0.18	0.20	552
2	0.85	0.89	0.87	3091
Accuracy			0.77	3695
Macro avg	0.38	0.37	0.37	3695
Weighted avg	0.75	0.77	0.76	3695

This heat map shows the correlation between various factors and accident severity in Figure 5. The depth of the color indicates the strength of the correlation, where red represents a positive correlation and blue represents a negative correlation. Analyzing the chart reveals that no factors show a very strong positive correlation with accident severity. However, Light_conditions and Age_of_driver showed strong negative correlations with accident severity, suggesting that better lighting conditions or certain age groups of drivers may lead to lower accident severity. Weather_conditions also showed a negative correlation, but the correlation was not particularly strong.

Comparative analysis of the performance of the meta-model with several other models (including logistic regression, decision tree classifier, K nearest neighbor classifier, Gaussian Naive Bayes, random forest classifier, AdaBoost classifier and gradient boosting classifier) was carried out. Finally, we discovered some salient features of the meta-model and its advantages and disadvantages, as shown in Table 10.

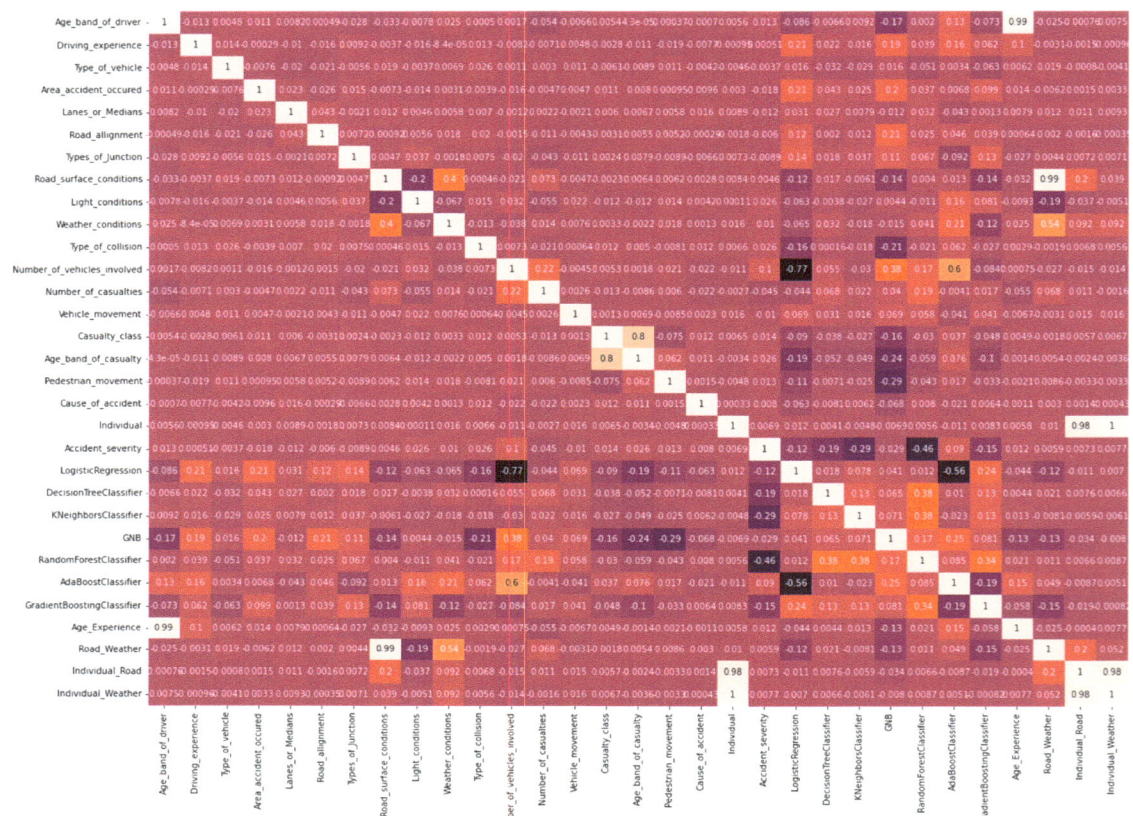

Figure 5. Heatmap of accident severity with meta-features.

Table 10. Performance of meta-model with other models.

Items	Type	Accuracy	Precision	Recall	F1 Score
LogisticRegression	Fatal	0.7165	0.6829	0.1555	0.2533
	Serious	0.8969	0.0625	0.0021	0.0041
	Light	0.6116	0.6312	0.8157	0.7117
DecisionTreeClassifier	Fatal	0.7365	0.5680	0.5787	0.5733
	Serious	0.8213	0.2087	0.2697	0.2353
	Light	0.6655	0.7277	0.6954	0.7112
KNeighborsClassifier	Fatal	0.7302	0.5852	0.4389	0.5016
	Serious	0.8918	0.3333	0.0500	0.0870
	Light	0.6478	0.6807	0.7544	0.7157
Gaussian Naive Bayes	Fatal	0.6985	0.5422	0.2466	0.3390
	Serious	0.8960	0.4286	0.0250	0.0472
	Light	0.6271	0.6313	0.8675	0.7308
RandomForestClassifier	Fatal	0.7955	0.7200	0.5838	0.6448
	Serious	0.8703	0.5455	0.2000	0.2927
	Light	0.7509	0.7623	0.8279	0.7937
AdaBoostClassifier	Fatal	0.7268	0.6354	0.3297	0.4342
	Serious	0.8669	0.2083	0.2397	0.2153
	Light	0.6976	0.6868	0.8783	0.7708
GradientBoostingClassifier	Fatal	0.7864	0.6786	0.3081	0.4238

Table 10. *Cont.*

Items	Type	Accuracy	Precision	Recall	F1 Score
	Serious	0.8535	0.5000	0.0167	0.0323
	Light	0.7208	0.6810	0.8487	0.7556
Meta-model	Fatal	0.9613	0.0344	0.1111	0.0526
	Serious	0.9069	0.0525	0.0121	0.0241
	Light	0.7508	0.8837	0.7524	0.8128

First of all, the meta-model performs outstandingly in processing "Fatal"-type events with an accuracy of 0.9613, which is much higher than the other models, showing its potential in identifying serious events. However, the performance of the meta-model in terms of precision and recall is unsatisfactory. Its precision rate is only 0.0344, the recall rate is 0.1111, and the F1 score is extremely low, only 0.0526. This shows that although the model can identify "Fatal" events well, it still needs to be greatly improved in terms of certainty and coverage.

For "Serious" and "Light"-type events, the meta-model's performance also shows certain advantages. In the "Serious"-type event, its accuracy reached 0.9069, but it also faced problems of low precision and low recall, and its corresponding F1 score was only 0.0241. In the "Light"-type event, the meta-model showed high accuracy (0.7508), precision (0.8837) and recall (0.7524), and the F1 score reached 0.8128, showing good overall performance.

Overall, the performance of the meta-model in processing different types of events varies. Its main advantage lies in its high accuracy for "Fatal"-type events, indicating that it can effectively distinguish serious events in some cases. However, this model is generally low in precision and recall, especially when dealing with "Fatal" and "Serious"-type events, which may lead to a large number of misjudgments and missed misjudgments, thus affecting the actual application effect of the model. Therefore, future work should focus on improving the precision and recall of the meta-model to achieve more balanced and reliable performance.

In the StackTrafficRiskPrediction framework, the meta-model is an advanced regression model designed to capture the complex relationships between traffic risk factors and predict the severity of traffic accidents by integrating multiple meta-features derived from different basic classification models. This model structure includes an input layer, multiple processing layers and an output layer, which is designed to process and output the level of traffic accident risk through a deep neural network. Meta-features include combined features and base model predicted probabilities, and the choice of regression technique—first classifying severity using a random forest classifier and subsequently modeling using linear regression—is based on the properties of the meta-feature and the size and complexity of the data.

However, although the meta-model shows high accuracy in the prediction of "Fatal"-type events, it performs poorly in terms of precision and recall overall, especially when dealing with "Fatal" and "Serious"-type events. This performance may be due to problems in several aspects, i.e., the integration of meta-features may not be sufficient, the model may be too simplified and fail to simulate the complex relationships between data in detail, or the model may be overfitted on specific data, resulting in insufficient generalization ability.

In response to the above problems, there are still some methods to improve the performance of the meta-model. First, we can strengthen feature engineering, which can further analyze and integrate more diverse features, such as introducing time series analysis or data features of specific locations to enhance the model's ability to handle complex predictive capabilities of traffic scenarios. Secondly, to optimize the model structure, we can consider adjusting the existing neural network architecture and explore the application of new deep learning technologies, such as convolutional neural networks (CNNs) or long short-term memory networks (LSTMs). These technologies can better handle time and spatially dependent data. Finally, the model training method can be strengthened, and more advanced cross-validation and regularization strategies can be adopted to avoid overfitting and ensure that the model has good prediction accuracy and adaptability to unseen data. By implementing these improvements, the meta-model will be able to more effectively assess and predict traffic accident risks and provide more accurate and reliable decision support for traffic safety management.

5. Conclusions

In this study, the research introduces the StackTrafficRiskPrediction model, which is a method for predicting the severity of traffic crashes by utilizing meta-features derived from environmental, human factors and traffic characteristics. The results show that the model is effective in identifying key factors affecting the risk of traffic crashes, such as driver age, driving experience, road surface conditions, lighting conditions, and weather conditions.

The innovative aspect of our work is the meta-modeling approach, in which we employ a stacked integrated learning strategy. This strategy utilizes the outputs of various underlying classification models as meta-features, which are subsequently used to train regression models aimed at predicting the severity of traffic accidents. A comparative performance analysis shows that the meta-model has an accuracy of 0.9613, 0.9069, and 0.7508 in predicting fatal, serious, and minor accidents, respectively, demonstrating high predictive effectiveness, and excels especially when dealing with fatal and serious accident prediction. This approach allows for a more detailed picture of complex patterns in the data, thus improving the overall predictive accuracy of the model. In contrast, traditional logistic regression models perform poorly in these areas, with accuracies of only 0.7182, 0.8669, and 0.6289 in predicting fatal, serious, and minor accidents. This further highlights the superiority of the StackTrafficRiskPrediction model.

Despite these advantages, we also observed that although the model performs well in predicting major categories such as accident severity, its accuracy is limited when dealing with categories with smaller sample sizes. In addition, our study highlights some limitations that need to be addressed. The problem of sample imbalance, especially in small categories, suggests the need for further data collection and integration to enhance the generalization ability of the model. In addition, although this study focused on specific environmental and human factors, it did not cover all potential factors that may affect the risk of traffic accidents. Future research could gain a more comprehensive understanding of crash risk by exploring other influencing factors such as vehicle technology and roadway design.

In conclusion, the StackTrafficRiskPrediction model demonstrates great potential in advancing the field of traffic accident risk prediction. By continually refining and extending the model, we aim to develop more robust tools and strategies for traffic safety management and accident prevention that can significantly reduce the incidence and severity of traffic accidents.

Author Contributions: Conceptualization, W.S. and L.N.A.; methodology, W.S.; software, W.S.; validation, W.S., L.N.A., P.S.S. and F.K.; formal analysis, W.S.; investigation, W.S.; resources, W.S.; data curation, W.S.; writing—original draft preparation, W.S.; writing—review and editing, L.N.A., P.S.S. and F.K.; visualization, W.S.; supervision, W.S.; project administration, L.N.A. All authors have read and agreed to the published version of the manuscript.

Funding: This research received no external funding.

Data Availability Statement: The data presented in this study are available on request from the corresponding author.

Conflicts of Interest: The authors declare no conflict of interest.

References

1. World Health Organization. *Global Status Report on Road Safety 2018*; World Health Organization: Geneva, Switzerland, 2018; Available online: https://www.who.int/publications/i/item/9789241565684 (accessed on 6 April 2024).
2. World Health Organization. *Global Status Report on Road Safety 2023*; World Health Organization: Geneva, Switzerland, 2023; Available online: https://www.who.int/publications/i/item/9789240086517 (accessed on 6 April 2024).

3. World Health Organization. *Global Plan for the Decade of Action for Road Safety 2021–2030*; World Health Organization: Geneva, Switzerland, 2021; Available online: https://www.who.int/teams/social-determinants-of-health/safety-and-mobility/decade-of-action-for-road-safety-2021-2030 (accessed on 6 April 2024).
4. UN ESCAP. Road Safety: Saving Lives beyond 2020 in the Asia-Pacific Region. 2020. Available online: https://hdl.handle.net/20.500.12870/2881 (accessed on 6 April 2024).
5. Deme, D. Review on factors causes road traffic accident in Africa. *J. Civ. Eng. Res. Technol.* **2019**, *1*, 1–8. [CrossRef]
6. Bucsuházy, K.; Matuchová, E.; Zůvala, R.; Moravcová, P.; Kostíková, M.; Mikulec, R. Human factors contributing to the road traffic accident occurrence. *Transp. Res. Procedia* **2020**, *45*, 555–561. [CrossRef]
7. Hammad, H.M.; Ashraf, M.; Abbas, F.; Bakhat, H.F.; Qaisrani, S.A.; Mubeen, M.; Fahad, S.; Awais, M. Environmental factors affecting the frequency of road traffic accidents: A case study of sub-urban area of Pakistan. *Environ. Sci. Pollut. Res.* **2019**, *26*, 11674–11685. [CrossRef] [PubMed]
8. Darma, Y.; Karim, M.R.; Abdullah, S. An analysis of Malaysia road traffic death distribution by road environment. *Sādhanā* **2017**, *42*, 1605–1615. [CrossRef]
9. Parsa, A.B.; Taghipour, H.; Derrible, S.; Mohammadian, A. Real-time accident detection: Coping with imbalanced data. *Accid. Anal. Prev.* **2019**, *129*, 202–210. [CrossRef]
10. Zhang, Z.; Niu, Z.; Li, Y.; Ma, X.; Sun, S. Research on the influence factors of accident severity of new energy vehicles based on ensemble learning. *Front. Energy Res.* **2023**, *11*, 1329688. [CrossRef]
11. Adadi, A.; Berrada, M. Peeking Inside the Black-Box: A Survey on Explainable Artificial Intelligence (XAI). *IEEE Access* **2018**, *6*, 52138–52160. [CrossRef]
12. Coeckelbergh, M. Artificial Intelligence, Responsibility Attribution, and a Relational Justification of Explainability. *Sci. Eng. Ethics* **2020**, *26*, 2051–2068. [CrossRef]
13. Gilpin, L.H.; Bau, D.; Yuan, B.Z.; Bajwa, A.; Specter, M.; Kagal, L. Explaining Explanations: An Overview of Interpretability of Machine Learning. *arXiv* **2019**, arXiv:1806.00069.
14. Kumar, S.; Mahima Srivastava, D.K.; Kharya, P.; Sachan, N.; Kiran, K. Analysis of risk factors contributing to road traffic accidents in a tertiary care hospital. A hospital based cross-sectional study. *Chin. J. Traumatol.* **2020**, *23*, 159–162. [CrossRef]
15. Panda, C.; Dash, A.K.; Dash, D.P. Assessment of Risk Factors of Road Traffic Accidents: A Panel Model Analysis of Several States in India. *Vis. J. Bus. Perspect.* **2020**, *23*, 097226292211132. [CrossRef]
16. Ahmed, S.K.; Mohammed, M.G.; Abdulqadir, S.O.; El-Kader RG, A.; El-Shall, N.A.; Chandran, D.; Rehman ME, U.; Dhama, K. Road traffic accidental injuries and deaths: A neglected global health issue. *Health Sci. Rep.* **2023**, *6*, e1240. [CrossRef] [PubMed]
17. Chand, A.; Jayesh, S.; Bhasi, A.B. Road traffic accidents: An overview of data sources, analysis techniques and contributing factors. *Mater. Today Proc.* **2021**, *47*, 5135–5141. [CrossRef]
18. Jianfeng, X.; Hongyu, G.; Jian, T.; Liu, L.; Haizhu, L. A classification and recognition model for the severity of road traffic accident. *Adv. Mech. Eng.* **2019**, *11*, 168781401985189. [CrossRef]
19. Yang, Z.; Zhang, W.; Feng, J. Predicting multiple types of traffic accident severity with explanations: A multi-task deep learning framework. *Saf. Sci.* **2022**, *146*, 105522. [CrossRef]
20. Becker, N.; Rust, H.W.; Ulbrich, U. Weather impacts on various types of road crashes: A quantitative analysis using generalized additive models. *Eur. Transp. Res. Rev.* **2022**, *14*, 37. [CrossRef]
21. Drosu, A.; Cofaru, C.; Popescu, M.V. Relationships between Accident Severity and Weather and Roadway Adherence Factors in Crashes Occurred in Different Type of Collisions. In *The 30th SIAR International Congress of Automotive and Transport Engineering: Science and Management of Automotive and Transportation Engineering*; Springer International Publishing: Cham, Switzerland, 2020; pp. 251–264.
22. Edwards, J.B. The Relationship Between Road Accident Severity and Recorded Weather. *J. Saf. Res.* **1998**, *29*, 249–262. [CrossRef]
23. Lio, C.-F.; Cheong, H.-H.; Un, C.-H.; Lo, I.-L.; Tsai, S.-Y. The association between meteorological variables and road traffic injuries: A study from Macao. *PeerJ* **2019**, *7*, e6438. [CrossRef] [PubMed]
24. Xing, F.; Huang, H.; Zhan, Z.; Zhai, X.; Ou, C.; Sze, N.N.; Hon, K.K. Hourly associations between weather factors and traffic crashes: Non-linear and lag effects. *Anal. Methods Accid. Res.* **2019**, *24*, 100109. [CrossRef]
25. Mkwata, R.; Chong, E.E.M. Effect of pavement surface conditions on road traffic accident—A Review. *E3S Web Conf.* **2022**, *347*, 01017. [CrossRef]
26. Retallack, A.E.; Ostendorf, B. Current Understanding of the Effects of Congestion on Traffic Accidents. *Int. J. Environ. Res. Public Health* **2019**, *16*, 3400. [CrossRef] [PubMed]
27. Gopalakrishnan, S. A Public Health Perspective of Road Traffic Accidents. *J. Fam. Med. Prim. Care* **2012**, *1*, 144. [CrossRef] [PubMed]
28. Paramasivan, K.; Subburaj, R.; Sharma, V.M.; Sudarsanam, N. Relationship between mobility and road traffic injuries during COVID-19 pandemic—The role of attendant factors. *PLoS ONE* **2022**, *17*, e0268190. [CrossRef] [PubMed]
29. Tijani, A.; Molyet, R.; Alam, M. Collision Warning System Using Naïve Bayes Classifier. *Tech. Rom. J. Appl. Sci. Technol.* **2022**, *4*, 39–56. [CrossRef]
30. Yang, F.J. An implementation of naive bayes classifier. In Proceedings of the 2018 International Conference on Computational Science and Computational Intelligence (CSCI), Las Vegas, NV, USA, 12–14 December 2018; IEEE: New York, NY, USA, 2018; pp. 301–306.

31. Yang, L.; Aghaabbasi, M.; Ali, M.; Jan, A.; Bouallegue, B.; Javed, M.F.; Salem, N.M. Comparative Analysis of the Optimized KNN, SVM, and Ensemble DT Models Using Bayesian Optimization for Predicting Pedestrian Fatalities: An Advance towards Realizing the Sustainable Safety of Pedestrians. *Sustainability* **2022**, *14*, 10467. [CrossRef]
32. Ashqar, H.I.; Shaheen, Q.H.; Ashur, S.A.; Rakha, H.A. Impact of risk factors on work zone crashes using logistic models and Random Forest. In Proceedings of the 2021 IEEE International Intelligent Transportation Systems Conference (ITSC), Indianapolis, IN, USA, 19–22 September 2021; IEEE: New York, NY, USA, 2021; pp. 1815–1820.
33. Eboli, L.; Forciniti, C.; Mazzulla, G. Factors influencing accident severity: An analysis by road accident type. *Transp. Res. Procedia* **2020**, *47*, 449–456. [CrossRef]
34. Otte, D.; Facius, T.; Brand, S. Serious injuries in the traffic accident situation: Definition, importance and orientation for countermeasures based on a representative sample of in-depth-accident-cases in Germany. *Int. J. Crashworth.* **2018**, *23*, 18–31. [CrossRef]
35. Aldala'in, S.A.; Sukor, N.S.A.; Obaidat, M.T. The Impact of Road Alignment toward Road Safety: A review from statistical perspective. In *Proceedings of AICCE'19: Transforming the Nation for a Sustainable Tomorrow*; Springer: Berlin/Heidelberg, Germany, 2020; Volume 4, pp. 729–735.
36. Hauer, E. *The Art of Regression Modeling in Road Safety*; Springer: New York, NY, USA, 2015; Volume 38.
37. Ranadive, M.S.; Das, B.B.; Mehta, Y.A.; Gupta, R. (Eds.) *Recent Trends in Construction Technology and Management: Select Proceedings of ACTM 2021*; Springer Nature: Singapore, 2023; Volume 260. [CrossRef]
38. Dong, X.; Lu, M. Optimal Road accident case retrieval algorithm based on k-nearest neighbor. *Adv. Mech. Eng.* **2019**, *11*, 168781401882452. [CrossRef]
39. Hatti, M. (Ed.) *Artificial Intelligence and Heuristics for Smart Energy Efficiency in Smart Cities: Case Study: Tipasa, Algeria*; Springer Nature: Singapore, 2021; Volume 361.
40. Anderson, T.K. Kernel density estimation and K-means clustering to profile road accident hotspots. *Accid. Anal. Prev.* **2009**, *41*, 359–364. [CrossRef]
41. Priyanka, G.; Jayakarthik, D.R. Road Safety Analysis by Using K-Means Algorithm. *Int. J. Pure Appl. Math.* **2020**, *119*, 253–257.
42. Puspitasari, D.; Wahyudi, M.; Rizaldi, M.; Nurhadi, A.; Ramanda, K.; Sumanto. K-Means Algorithm for Clustering the Location of Accident-Prone on the Highway. *J. Phys. Conf. Ser.* **2020**, *1641*, 012086. [CrossRef]
43. Kang, M.; Lee, W.; Hwang, K.; Yoon, Y. Vision Transformer for Detecting Critical Situations and Extracting Functional Scenario for Automated Vehicle Safety Assessment. *Sustainability* **2022**, *14*, 9680. [CrossRef]
44. Liu, X.; Lu, J.; Chen, X.; Fong YH, C.; Ma, X.; Zhang, F. Attention based spatio-temporal graph convolutional network with focal loss for crash risk evaluation on urban road traffic network based on multi-source risks. *Accid. Anal. Prev.* **2023**, *192*, 107262. [CrossRef]
45. Sohail, A.; Cheema, M.A.; Ali, M.E.; Toosi, A.N.; Rakha, H.A. Data-driven approaches for road safety: A comprehensive systematic literature review. *Saf. Sci.* **2023**, *158*, 105949. [CrossRef]
46. Tamagusko, T.; Correia, M.G.; Huynh, M.A.; Ferreira, A. Deep Learning applied to Road Accident Detection with Transfer Learning and Synthetic Images. *Transp. Res. Procedia* **2022**, *64*, 90–97. [CrossRef]
47. Silva, P.B.; Andrade, M.; Ferreira, S. Machine learning applied to road safety modeling: A systematic literature review. *J. Traffic Transp. Eng.* **2020**, *7*, 775–790. [CrossRef]
48. Pourroostaei Ardakani, S.; Liang, X.; Mengistu, K.T.; So, R.S.; Wei, X.; He, B.; Cheshmehzangi, A. Road car accident prediction using a machine-learning-enabled data analysis. *Sustainability* **2023**, *15*, 5939. [CrossRef]
49. Mussah, A.R.; Adu-Gyamfi, Y. Machine Learning Framework for Real-Time Assessment of Traffic Safety Utilizing Connected Vehicle Data. *Sustainability* **2022**, *14*, 15348. [CrossRef]
50. Dong, B.-T.; Lin, H.-Y.; Chang, C.-C. Driver Fatigue and Distracted Driving Detection Using Random Forest and Convolutional Neural Network. *Appl. Sci.* **2022**, *12*, 8674. [CrossRef]
51. Kashevnik, A.; Shchedrin, R.; Kaiser, C.; Stocker, A. Driver Distraction Detection Methods: A Literature Review and Framework. *IEEE Access* **2021**, *9*, 60063–60076. [CrossRef]
52. Koay, H.V.; Chuah, J.H.; Chow, C.-O.; Chang, Y.-L. Detecting and recognizing driver distraction through various data modality using machine learning: A review, recent advances, simplified framework and open challenges (2014–2021). *Eng. Appl. Artif. Intell.* **2022**, *115*, 105309. [CrossRef]
53. Aswad, M.; Al-Sultan, S.; Zedan, H. Context aware accidents prediction and prevention system for VANET. In Proceedings of the 3rd International Conference on Context-Aware Systems and Applications, Dubai, United Arab Emirates, 7–9 October 2014; pp. 162–168.
54. Legree, P.J.; Heffner, T.S.; Psotka, J.; Martin, D.E.; Medsker, G.J. Traffic crash involvement: Experiential driving knowledge and stressful contextual antecedents. *J. Appl. Psychol.* **2003**, *88*, 15–26. [CrossRef] [PubMed]
55. Sümer, N. Personality and behavioral predictors of traffic accidents: Testing a contextual mediated model. *Accid. Anal. Prev.* **2003**, *35*, 949–964. [CrossRef] [PubMed]
56. Zhu, X.; Yuan, Y.; Hu, X.; Chiu, Y.-C.; Ma, Y.-L. A Bayesian Network model for contextual versus non-contextual driving behavior assessment. *Transp. Res. Part C Emerg. Technol.* **2017**, *81*, 172–187. [CrossRef]
57. Af Wåhlberg, A.; Barraclough, P.; Freeman, J. Personality versus traffic accidents; meta-analysis of real and method effects. *Transp. Res. Part F Traffic Psychol. Behav.* **2017**, *44*, 90–104. [CrossRef]

58. Barraclough, P.; Af Wåhlberg, A.; Freeman, J.; Watson, B.; Watson, A. Predicting Crashes Using Traffic Offences. A Meta-Analysis that Examines Potential Bias between Self-Report and Archival Data. *PLoS ONE* **2016**, *11*, e0153390. [CrossRef] [PubMed]
59. Selmoune, N.; Derbal, K.; Alimazighi, Z. Spatial Data Warehouse Multidimensional Design Approach and Geo-Decisional Tool for Road Accidents Analysis. In Proceedings of the 2019 International Conference on Information and Communication Technologies for Disaster Management (ICT-DM), Paris, France, 18–20 December 2019; IEEE: New York, NY, USA, 2019; pp. 1–8.
60. Ki, Y.K.; Kim, J.W.; Baik, D.K. A traffic accident detection model using metadata registry. In Proceedings of the Fourth International Conference on Software Engineering Research, Management and Applications (SERA'06), Seattle, WA, USA, 9–11 August 2006; IEEE: New York, NY, USA, 2006; pp. 255–259.
61. Balikuddembe, J.K.; Ardalan, A.; Khorasani-Zavareh, D.; Nejati, A.; Munanura, K.S. Road traffic incidents in Uganda: A systematic review study of a five-year trend. *J. Inj. Violence Res.* **2017**, *9*, 17–25. [CrossRef]
62. Jomaa, H.S.; Schmidt-Thieme, L.; Grabocka, J. Dataset2vec: Learning dataset meta-features. *Data Min. Knowl. Discov.* **2021**, *35*, 964–985. [CrossRef]
63. Zhou, H.; Xiao, S.; Zhang, S.; Peng, J.; Zhang, S.; Li, J. Jump Self-attention: Capturing High-order Statistics in Transformers. *Adv. Neural Inf. Process. Syst.* **2022**, *35*, 17899–17910.

Disclaimer/Publisher's Note: The statements, opinions and data contained in all publications are solely those of the individual author(s) and contributor(s) and not of MDPI and/or the editor(s). MDPI and/or the editor(s) disclaim responsibility for any injury to people or property resulting from any ideas, methods, instructions or products referred to in the content.

Article

Connected Automated and Human-Driven Vehicle Mixed Traffic in Urban Freeway Interchanges: Safety Analysis and Design Assumptions

Anna Granà [1,*], Salvatore Curto [2], Andrea Petralia [2] and Tullio Giuffrè [2,*]

1 Department of Engineering, University of Palermo, Viale delle Scienze ed 8, 90128 Palermo, Italy
2 Department of Engineering and Architecture, University of Enna Kore, Viale delle Olimpiadi, 94100 Enna, Italy; salvatore.curto@unikorestudent.it (S.C.); andrea.petralia@unikorestudent.it (A.P.)
* Correspondence: anna.grana@unipa.it (A.G.); tullio.giuffre@unikore.it (T.G.)

Citation: Granà, A.; Curto, S.; Petralia, A.; Giuffrè, T. Connected Automated and Human-Driven Vehicle Mixed Traffic in Urban Freeway Interchanges: Safety Analysis and Design Assumptions. *Vehicles* 2024, *6*, 693–710. https://doi.org/10.3390/vehicles6020032

Academic Editors: Deogratias Eustace, Bhaven Naik, Heng Wei and Parth Bhavsar

Received: 1 March 2024
Revised: 8 April 2024
Accepted: 9 April 2024
Published: 11 April 2024

Copyright: © 2024 by the authors. Licensee MDPI, Basel, Switzerland. This article is an open access article distributed under the terms and conditions of the Creative Commons Attribution (CC BY) license (https:// creativecommons.org/licenses/by/ 4.0/).

Abstract: The introduction of connected automated vehicles (CAVs) on freeways raises significant challenges, particularly in interactions with human-driven vehicles, impacting traffic flow and safety. This study employs traffic microsimulation and surrogate safety assessment measures software to delve into CAV–human driver interactions, estimating potential conflicts. While previous research acknowledges that human drivers adjust their behavior when sharing the road with CAVs, the underlying reasons and the extent of associated risks are not fully understood yet. The study focuses on how CAV presence can diminish conflicts, employing surrogate safety measures and real-world mixed traffic data, and assesses the safety and performance of freeway interchange configurations in Italy and the US across diverse urban contexts. This research proposes tools for optimizing urban layouts to minimize conflicts in mixed traffic environments. Results reveal that adding auxiliary lanes enhances safety, particularly for CAVs and rear-end collisions. Along interchange ramps, an exclusive CAV stream performs similarly to human-driven ones in terms of longitudinal conflicts, but mixed traffic flows, consisting of both CAVs and human-driven vehicles, may result in more conflicts. Notably, when CAVs follow human-driven vehicles in near-identical conditions, more conflicts arise, emphasizing the complexity of CAV integration and the need for careful safety measures and roadway design considerations.

Keywords: urban freeways interchanges; surrogate safety measures; connected automated vehicles; VISSIM; road design

1. Introduction

Over the past few decades, escalating traffic volumes have had significant effects on road safety, traffic congestion, and fuel consumption [1]. To enhance drivers' performance and mitigate human errors, tools like adaptive cruise control (ACC) and cooperative adaptive cruise control (CACC) have been implemented [2]. However, the transformative innovation expected to profoundly impact these aspects in the coming decades is the advent of connected autonomous vehicles (CAVs). Connected autonomous vehicles integrate digital technology with automated systems to assist or replace human drivers [3]. These vehicles operate autonomously, utilizing sensors and cameras to continuously analyze their surroundings. Additionally, they establish ongoing GPS-based location tracking systems and telecommunications networks [4,5]. This autonomy enables them to execute precise maneuvers, thereby positively impacting both traffic network performance and safety [6].

Despite these advancements, the integration of CAVs in traffic introduces driving challenges, particularly at critical points such as interchanges and intersections, where they still interact with non-autonomous or manually operated vehicles. Microsimulation is a valuable tool for forecasting the potential impacts of CAV circulation [7,8]. It enables the analysis of collaborative behaviors, such as the formation of vehicle platoons, and provides opportunities for refinement based on implementation experiences in various what-if scenarios [9]. Studies have shown that human drivers in CAV platoons adjust their maneuvers, frequently decreasing their time headways [10,11]. Nevertheless, scenarios characterized by diminished reaction times of human drivers may lead to collisions [12].

The effectiveness of precision maneuvers executed by CAVs is intricately tied to the capabilities of their detection systems and the quality of surrounding infrastructure. Hence, the operational efficiency of CAVs experiences a notable improvement when the roadway infrastructures adhere to high-quality standards [13]. This becomes especially critical at interchanges, where vehicles interact at elevated speeds. Consequently, there is a pressing need to examine potential adaptations and impacts stemming from the collaboration between CAVs and human-driven vehicles.

Country-specific design standards have led to a wide range of geometric layouts for interchanges worldwide [14–16]. Varied design standards may result in differences in safety features and impact factors such as capacity, throughput, and congestion, influencing overall traffic flow dynamics, safety performance, and the efficiency of interchanges.

Building upon this, the study aims to address the question: What is the impact of the coexistence of CAVs and human-driven vehicles on operational and safety performance at freeway interchanges? The research involved collecting and analyzing various geometric interchange configurations in Italy and the US to explore both advantages and disadvantages in the transition to complete CAV driving.

To achieve this goal, the evaluation utilized VISSIM (Version 10) [17] in modeling urban freeway interchanges; the microscopic traffic simulation tool has been coupled with the surrogate safety assessment model (SSAM) [18] to identify potential traffic conflicts.

The study transitions from an interest in understanding the potential evolution of road traffic parameters along the ring roads under mixed traffic conditions involving CAVs and human-driven vehicles. Specifically, the focus is on the geometry of urban freeway interchanges, which exhibit significant variations influenced by both traffic parameters and landscape features. The study delves into the complexity of these interchanges by examining a sample of existing road infrastructures characterized by high heterogeneity. Despite some limitations in the approach used, the research findings underscore the need to improve road policies to optimize and adapt road interchanges for accommodating the next generation of traffic vehicles.

2. Related Research Studies

The efficiency of CAV driving at freeway interchanges may be affected or compromised by human-driven vehicles in traffic [19]. In this context, a potential solution could involve adjusting the access configurations of dedicated lanes, especially at the "turning points" where traffic flow shifts or turns. These areas often play a crucial role in influencing the overall performance and efficiency of the interchange. Updating intersection management and enhancing access configurations at these turning points could potentially serve as a solution to address challenges related to mixed traffic [20,21]. Considering the potential to reduce human errors through CAVs, there is an outlook of decreasing crashes [22]. For high-occupancy vehicles alone, the removal of human mistakes is estimated to eliminate 93–97% of crashes [23].

In this context, to facilitate the transition to CAV driving, the implementation of autonomous vehicle/toll (AVT) lanes could be realized, where CAVs have free access, while human-driven vehicles must pay a toll [24]. To understand the impact of driving next to CAV platoons on the behavior of both human-driven vehicle drivers and CAVs, it is crucial to examine the behavioral adaptation. This assessment should be based on the interaction between CAVs and human-driven vehicles (HdVs), providing insights into the implications of segregating CAVs and human-driven vehicles through dedicated lanes. The examination of behavioral adaptation and its effects on traffic efficiency and safety performances, stemming from the experience of driving with CAVs, drivers' inclination to transition between automated and manual modes and vice versa, the consequences of such transitions (i.e., the transition of control), and drivers' choices in car following and lane changing while CAV driving has been defined as "any change of driver, traveller, and travel behaviours that occurs following user interaction with a change to the road traffic system, in addition to those behaviours specifically and immediately targeted by the initiators of the change" by Kulmala and Rämä [25]. Thus, behavioral adaptation encompasses behaviors not only specifically and immediately targeted by the initiators of the change but also those resulting from the interaction among vehicles during driving. Additionally, there is a notable lack of understanding concerning the effects on performance and safety resulting from various design setups of road segments featuring dedicated lanes during maneuvers (such as merging, splitting, transitioning between manual and automated control, and entering or exiting dedicated lanes). The influence of diverse lane utilization policies on driver behavior and, consequently, on traffic performance and safety remains inadequately explored.

A study on behavioral adaptation examined the behavior of drivers with and without ACC experience when driving in ACC mode [26]. When utilizing the ACC system, drivers tend to adopt slightly lower driving speeds and larger time headways. Interestingly, drivers with ACC experience drive at faster speeds than regular drivers and maintain smaller headways in ACC mode. This behavior is attributed to an indirect behavioral adaptation or carryover effect from their experience of driving with the ACC system.

A crucial aspect in the design of dedicated lanes at interchanges involves addressing the carryover impacts of automated driving. As drivers exit these exclusive lanes, they must disengage automation and assume manual control based on either lane utilization policies or their personal choice regarding how to navigate regular lanes. Research indicates that behavioral adaptation persists during manual driving following automation exposure [27]. Chen et al. [28] conducted simulation experiments examining CACC effects on traffic in off-ramp freeway sections. Low CACC penetration degrades safety and operational efficiency, while near 1% penetration significantly improves traffic flow. Increasing the conservative mandatory lane change zone length enhances traffic speed, with optimum performance at 750 m. Excessive zone lengths show diminishing benefits. The findings offer valuable insights for alleviating road traffic congestion, suggesting that directing lane changes can be an effective strategy to improve overall traffic flow efficiency. In terms of selecting time headway in CACC mode, Nowakowski et al. [29] found that male drivers tend to maintain shorter time headways compared to females. Overall, drivers choose approximately 50% shorter time headways in CACC mode than in ACC mode. However, the authors expressed reservations, noting that events were often brief, with only half lasting two to three minutes. According to [30,31], further investigation is warranted to understand drivers' preferences in setting ACC or CACC parameters, considering factors like age, gender, and driving style in mixed traffic scenarios at road interchange facilities. In certain traffic scenarios, human drivers may choose to take control of the vehicle, either switching off the automation mode or activating it.

Based on the above, this study aims to fill scientific knowledge gaps in two key areas of transportation research. Firstly, there is a lack of studies on how human drivers adapt behavior when driving alongside autonomous vehicles and connected autonomous vehicles. Secondly, there is limited exploration of rehabilitation options for interchanges operating with mixed vehicles. Addressing these research areas is essential for improving road safety and optimizing infrastructure in the dynamic context of mixed-vehicle transportation systems. The present study aims to contribute scientific understanding and addressing social implications related to the interactions between CAVs and human-driven vehicles at road interchanges. Employing traffic microsimulation and surrogate safety measures allows for the examination of a variety of complex scenarios, enhancing the replicability of design solutions. Furthermore, the study may facilitate international meta-analysis, offering valuable insights for the development of globally applicable solutions in the intelligent transportation systems domain.

3. Materials and Methods

The first goal of this study is to conduct a thorough analysis and comparison of specific interchanges in Sicily, Italy, and Florida, USA. So far, correlations and distinctions that offer insights into factors pertaining to the design and safety of urban road infrastructure have been investigated.

The chosen case study locations comprise interchanges located on ring-roads within diverse urban settings, adhering to both Italian and American roadway design standards [15,16]. In Italy, the selected interchanges are within the road network of three Sicilian cities—Palermo, Catania, and Messina—while in the United States, the selected interchanges are situated in Florida, functioning as a reference for scalability. Each interchange facility is designated with a code for ease of identification and result interpretation as shown in Figure 1: Interchanges 1A, 2A, and 3A in the road network of Palermo City (Italy); Interchanges 4A and 5A in the road network of Catania City (Italy); Interchange 6A in Messina City (Italy), and Interchanges 1B, 2B, 3B, and 4B in the City of Miami, Florida (USA). Considering that it was not possible to conduct a traffic detection survey, traffic volume parameters, required to model the origin/destination (O/D) matrix, were calculated based on lane capacity, i.e., the conditions in which the maximum hourly volume occurs in a generic roadway section. Therefore, the roadway network was simulated with the most unfavorable conditions both in terms of the level of service and consequent greater interactions between vehicles. Vehicle routes were established using "dynamic assignment" to allow for the generation of dynamic routes between junction nodes rather than static ones. Itineraries were assigned based on the most significant accident projections to be observed. Specifically, Figure 1 shows the layout of each interchange, showcasing features of the urban context such as on and off ramps, allowing for the examination of various elevations. Meanwhile, Table 1 outlines the primary design and operational parameters relevant to each studied interchange.

Table 1. Design and operational features by interchange studied.

Parameter	Interchange [1]									
	1A	2A	3A	4A	5A	6A	1B	2B	3B	4B
Main roadway section width (m)	12.00	12.00	11.00	11.00	11.00	10.5	17.00	15.00	22.00	19.00
Main roadways length (km)	0.75	0.69	0.61	0.56	0.690	0.635	1.15	1.130	2.135	0.74
Main roadways lanes quantity	2.00	2.00	2.00	2.00	2.00	2.00	3.00	3.00	4.00	4.00
Entering ramps	2.00	3.00	2.00	2.00	2.00	2.00	2.00	2.00	6.00	4.00
Exit ramps	2.00	5.00	2.00	2.00	2.00	2.00	4.00	2.00	6.00	6.00
interchange land use (km^2)	0.25	0.12	0.14	0.16	0.26	0.27	0.52	0.27	0.55	0.53
Bridges	1.00	2.00	1.00	1.00	1.00	2.00	2.00	4.00	2.00	9.00

[1] the interchanges are codified as shown in Figure 1.

VISSIM Modeling

Building upon the information provided in the introduction section, a case study sample was chosen, comprising 10 freeway interchanges situated in urban areas. Section 4 will delve into the specifics of each interchange. The initial phase of this study involved integrating each freeway interchange model into VISSIM software (Version 10) [17]. Vehicle routes were configured using the "dynamic assignment" option, which was preferred for its capability to generate dynamic itineraries between junction nodes rather than static ones [32].

In order to incorporate CAVs into the simulation model, a new vehicle category corresponding to cars was configured. The calibration phase involved assigning driving behavior parameters specific to autonomous driving. These parameters encompass four types of behavior that CAVs can adopt, considering factors such as driving aggressiveness and the availability of roadway context data.

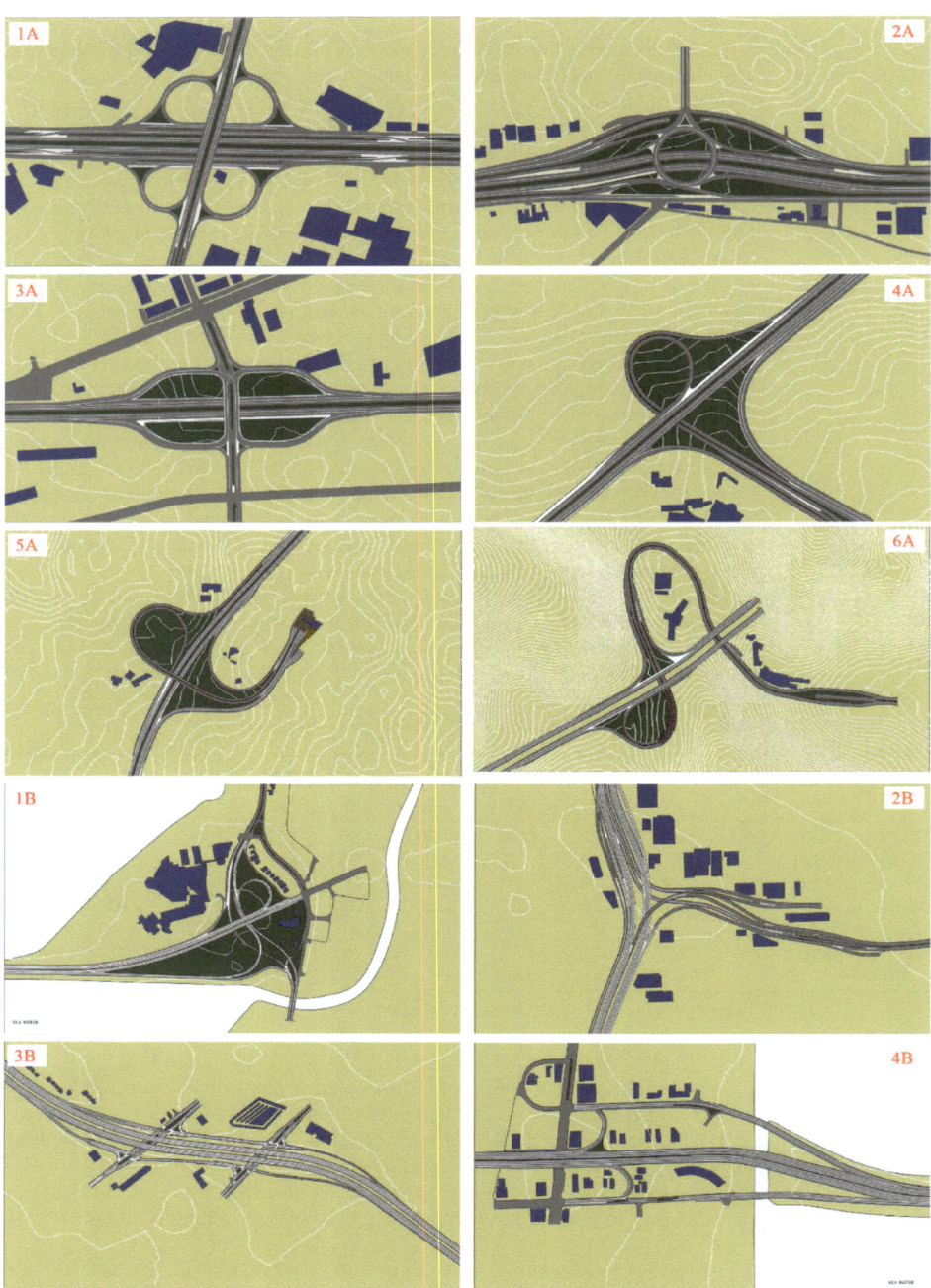

Figure 1. Sketches of the road interchanges within the investigated sample. A code is associated with each interchange facility as follows: Interchanges (**1A–3A**) in the road network of Palermo City (Italy); Interchanges (**4A,5A**) in the road network of Catania City (Italy); Interchange (**6A**) in Messina City (Italy); Interchanges (**1B–4B**) in the City of Miami, Florida (USA).

Table 2 displays the driving behavior parameters utilized concerning aggressive driving tendencies. In order to simulate a mobility context with highly efficient CAVs, the "all knowing" typology was selected.

Table 2. Driving behavior parameters in relation to aggressive driving way.

	Wiedemann's 99 Parameters	Real Safe	Cautious	Normal	All Knowing
CC0	(standstill distance) (m)	1.5	1.5	1.5	1
CC1	(mean headway time) (s)	1.5	1.5	0.9	0.6
CC2	(following variation) (m)	0	0	0	0
CC3	(threshold for entering following) (s)	-10	-10	-8	-6
CC4	(negative following threshold) (m/s)	-0.1	-0.1	-0.1	-0.1
CC5	(positive following threshold) (m/s)	0.1	0.1	0.1	0.1
CC6	(speed dependency of oscillation) (1/ms)	0	0	0	0
CC7	(oscillation acceleration) (m/s^2)	0.1	0.1	0.1	0.1
CC8	(standstill acceleration) (m/s^2)	2	3	3.5	4
CC9	(acceleration with 80 km/h) (m/s^2)	1.2	1.2	1.5	2

Consequently, the desired acceleration/deceleration and speed distributions were established, as depicted in Figure 2, where a behavior parameters setting is shown in terms of speed and acceleration/deceleration, basically due to the differences between CAV and HdV speed limit observations and acceleration/deceleration maneuver trends. HdVs do not respect speed limits as precisely as CAVs, and the latter operate with gradual speed increases or decreases, while HdVs are subjected to driver reactions that affect the way a desired speed or acceleration/deceleration is achieved. So, it can be observed that the spread is minimized, and the plots demonstrate how CAVs strictly adhere to the speed limit.

Operations at full capacity were simulated, as the introduction of CAVs into traffic is anticipated to encourage operating the entry mechanism at the highest possible level of utilization. According to [9], utilization is defined as the ratio of the number of entering vehicles (i.e., the throughput) to the maximum number of vehicles that each entry lane could accommodate (i.e., the capacity). Capacity calculations relied on capacity models and adjustment factors for connected and autonomous vehicles as suggested by the *Highway Capacity Manual* (HCM) [33]. In VISSIM, the integration of CAVs into the road interchange model was completed by configuring speed distribution functions and driving behavior parameters based on [34] and specific assumptions discussed by [35].

The values of the driving behavior parameters for both CAVs and human-driven vehicles were determined through the author's evaluation, incorporating insights from [34–36]. Also, each reported parameter indicates a specific value of following behavior belonging to the Wiedemann's 99 (W99) car-following model [37–39]; these are detailed in Table 3.

The traffic micro-simulator was coupled with the SSAM software to measure traffic conflicts [18]. Specifically, the VISSIM simulation process generates results that can be imported into the SSAM provided by [18]. The surrogate measures of safety, widely recognized for explaining the safety performance of road facilities through the vehicle trajectories provided by traffic micro-simulators [18,40], are integral to understanding safety dynamics. In this context, the SSAM reads trajectory files generated by VISSIM. By utilizing surrogate measures such as time to collision or post-encroachment time, the SSAM can evaluate the probability of conflict occurrence. Following the logic of SSAM, conflict events (i.e., conflicting vehicle pairs) are systematically listed, encompassing conflicts from preceding steps. For each interchange, eight trajectory (*.trj) output files were extracted from VISSIM and processed by SSAM, utilizing parameter thresholds to identify potential high severity conflicts and their specific locations within each sample interchange.

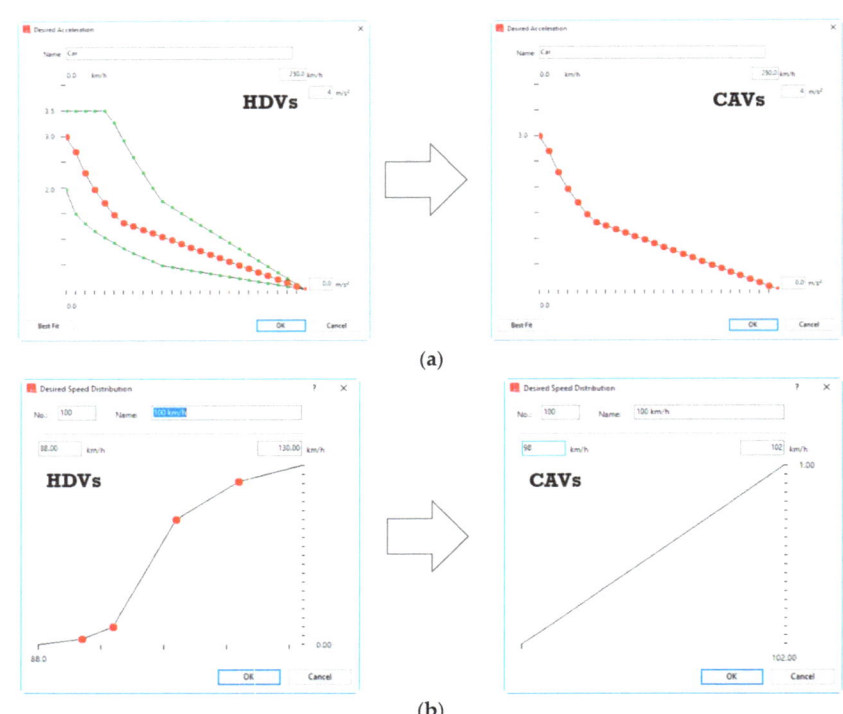

Figure 2. Examples of (**a**) desired acceleration settings with the range variation in green for HDVs; (**b**) speed distribution settings.

Table 3. Driving behavior parameters, defaults, and fine-tuned values for CAVs and human-driven vehicles.

	Wiedemann's 99 Parameters	Default Value	CAV Value	HdV Value
CC0	(standstill distance) (m)	1.50	1.00	1.50
CC1	(mean headway time) (s)	0.90	0.60	0.90
CC2	(following variation) (m)	4.00	0.00	4.00
CC3	(threshold for entering following) (s)	−8.00	−6.00	−8.00
CC4	(negative following threshold) (m/s)	−0.35	−0.10	−0.35
CC5	(positive following threshold) (m/s)	0.35	0.10	0.35
CC6	(speed dependency of oscillation) (1/ms)	11.44	0.00	11.44
CC7	(oscillation acceleration) (m/s^2)	0.25	0.10	0.25
CC8	(standstill acceleration) (m/s^2)	3.50	4.00	3.50
CC9	(acceleration with 80 km/h) (m/s^2)	1.50	2.00	1.50

It was determined that evaluation parameters significantly impacting potential conflicts among vehicular trajectories are the time-to-collision (TTC) and post-encroachment time (PET) [41]. It is observed that conflicts are more probable with smaller values of TTC and PET, with a TTC of zero indicating a collision. It is crucial, however, for the TTC to be shorter than the PET [18].

The upper limit for the time-to-collision (TTC) was set at 1.5 s, consistent with the default TTC value. Alternative threshold values below 1.5 s led to reduced overlap for the vehicle pair in the projection timeline, resulting in a revised maximum TTC threshold. It is important to highlight that the SSAM continuously updates the time-to-collision (TTC) values for each vehicle pair, ensuring that the projection timeline remains free of overlaps.

However, a collision occurs if the projection reaches zero with overlapping vehicles. In such cases, conflicts are considered resolved once the TTC value exceeds the threshold again [18]. Conversely, the threshold for the post-encroachment time (PET), representing the time interval between one vehicle exiting and another entering the conflict area, was adjusted to 2.50 s, in contrast to the default value of 5.00 s [18]. PET is associated with a conflict timestep, enabling the recording of the final PET value after a conflict concludes, even if the corresponding time-to-collision (TTC) value is below its threshold. Setting the minimum values for TTC at 0.3 s and PET at 0.50 s was necessary to address processing errors, as zero values were identified and removed [18,41]. The conflict type parameter allowed classification of conflicts based on the conflict angle, representing the hypothetical collision between the trajectories of conflicting vehicles: a rear-end conflict occurs if the absolute value of the conflict angle is less than 30 degrees and a crossing conflict occurs when the absolute value of the conflict angle exceeds 80 degrees, otherwise a lane-changing conflict occurs [42]. To clarify, a rear-end conflict occurs when two vehicles are in the same lane simultaneously, whereas lane changing involves two vehicles that have switched lanes. In the examination of each case study, trajectory files for each scenario underwent individual analysis using the SSAM. This method resulted in the initial identification of crucial areas at the interchanges, classifying potential conflicts into three main groups based on the conflict angle values between vehicular trajectories, as mentioned previously. To prevent unrealistic maneuvers, other surrogate safety measures related to driving behavior were kept at their default values.

Car following models are crucial for defining the driving behaviors of vehicles, and consequently, they were configured differently for human-driven vehicles and CAVs. This differentiation is made because it is anticipated that CAVs can follow the leading vehicle more closely and have shorter reaction times. The subsequent discussion in Section 5 is grounded in both the outcomes of traffic simulations and road safety analyses. This is attributed to the fact that potential traffic conflicts with significant severity can be linked to low values of TTC (time-to-collision) and PET (post-encroachment time) [18]. The selection of value thresholds aligns with the findings presented in [43]. To ensure a valid safety estimation, the time-to-collision (TTC) and post-encroachment time (PET) thresholds were adopted as evaluation parameters, as they are widely used in the freeway context and prevalent in the safety analysis literature [44].

The results of the potential traffic conflict analysis were utilized for road safety assessment, with special consideration given to CAVs based on insights provided by [45]. As is well known, potential traffic conflicts serve as the foundation for the application and modeling of the safety performance functions (SPFs), as elaborated in Section 5. The results revealed a significant increase in rear-end potential conflicts, exceeding 90%, compared to other conflict types. This is primarily attributed to the presence of on and off ramps, where capacity conditions and queuing situations often arise. In many documented cases, a higher concentration of CAVs resulted in elevated instances of rear-end conflicts, as previously discussed by [46], highlighting the heightened vulnerability of CAVs to such types of conflicts. To conduct safety analysis, trajectory output files from the VISSIM simulations were transferred into the SSAM software. Notably, the selected interchanges represented diverse design standards from two different countries, each characterized by varying traffic volumes. The findings were presented using a normalization factor, facilitating the comparison of potential conflicts for each interchange [40]. Figures 3 and 4 depict the quantity and types of conflicts for each interchange, with conflict quantities standardized to 1000 vehicles entering the bypass area. The CAV penetration rate was varied, and it was anticipated that rear-end conflicts would dominate, given that simulations considered the maximum vehicle capacity for each entry lane or ramp, operating in segments with low headway values.

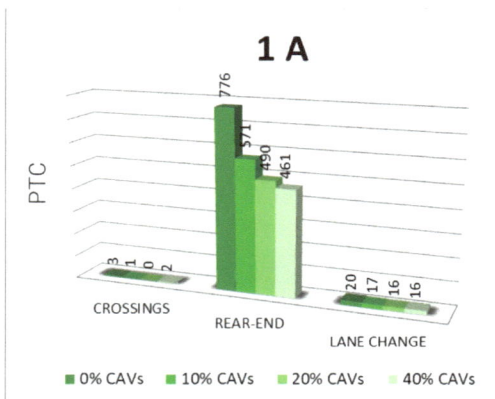
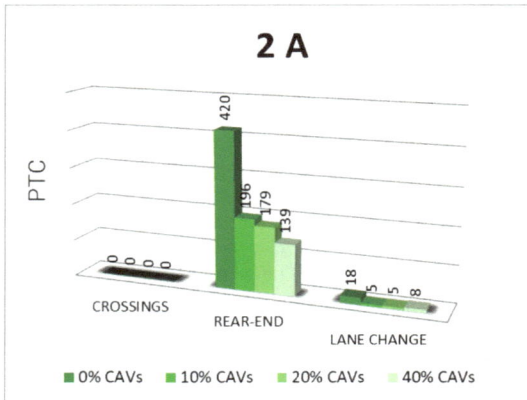
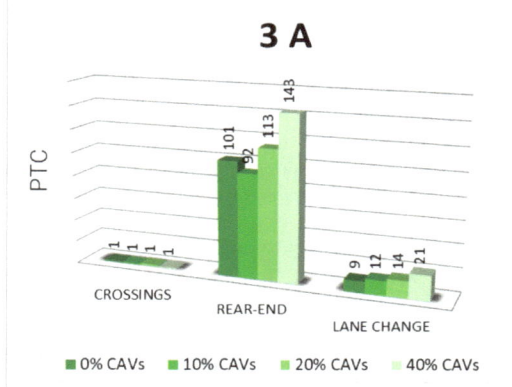
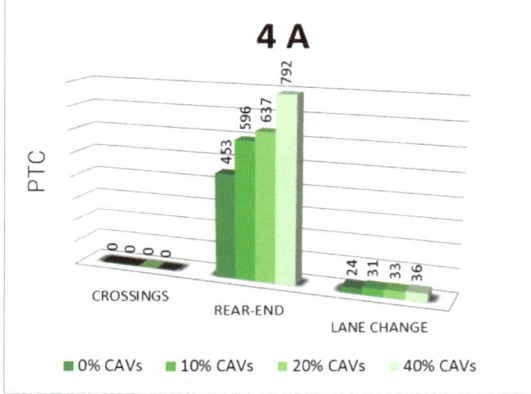
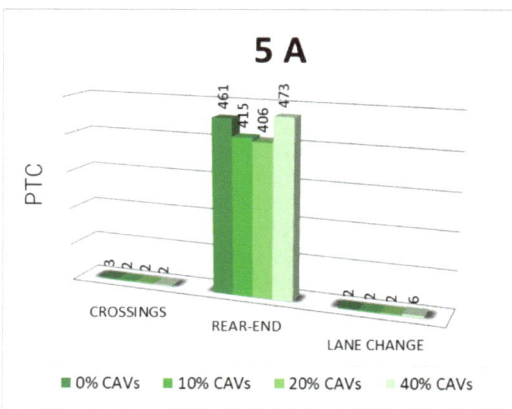
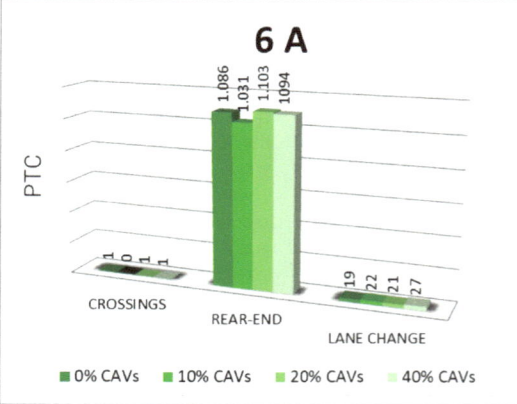

Figure 3. Potential traffic conflicts (PTCs) for 1000 entering vehicles at Italian interchanges under varying CAV rates. **Note:** A code is associated with each interchange facility in Figure 1 as follows: Interchanges (**1A–3A**) in the road network of Palermo City (Italy); Interchanges (**4A,5A**) in the road network of Catania City (Italy); and Interchange (**6A**) in Messina City (Italy).

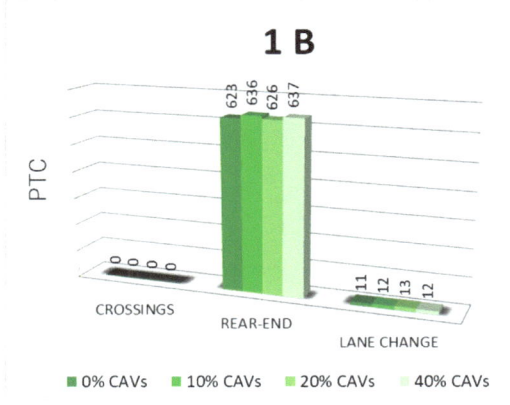
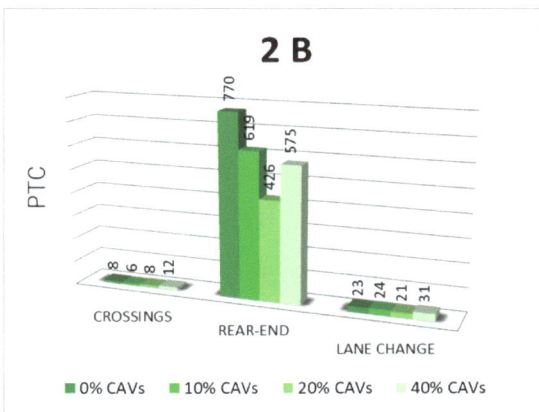
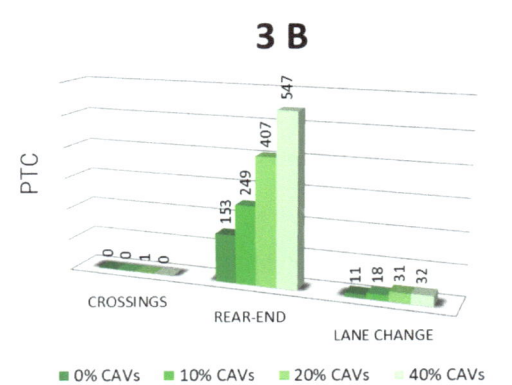
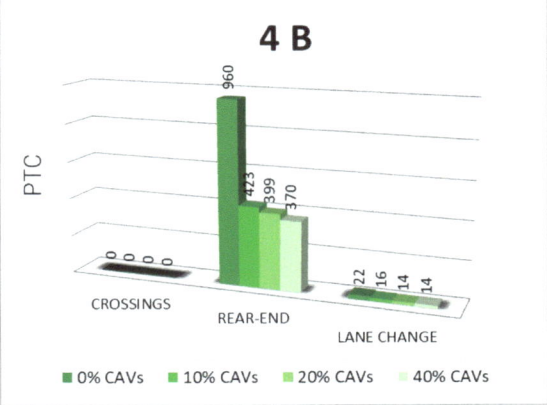

Figure 4. Potential traffic conflicts for 1000 entering vehicles at US interchanges under varying CAV rates. A code is associated with each interchange facility in Figure 1 as follows: Interchanges (**1B–4B**) in the City of Miami, Florida (USA).

Rear-end potential conflicts are notably elevated compared to other types, primarily due to the propensity for queuing along on- and off-ramps when traffic conditions approach capacity.

Interchanges 1A, 2A, and 4B show a gradual decrease in rear-end potential conflict, while 3A, 3B, and 4A are characterized by an increase. Rear-end potential conflicts results to be gradual for samples 3B–4A from 0% CAVs, while 3A shows a minimum initial decrease for 10% CAVs that does not affect the increasing trend. Samples 5A and 2B are characterized by the 40% CAV threshold that leads from a gradual rear-end decrease to a sudden increase, while 6A and 1B do not show significant outputs, even if the geometrical layouts are similar but are characterized by different operating schemes.

To provide a comprehensive safety assessment of interchange performance, with and without CAVs, the ex ante safety standards for the case studies were scrutinized. To achieve this, the ISATe tool [47] was applied to estimate crashes with a higher probability of resulting in fatalities.

Figure 5 displays potential crashes derived from the ISATe analysis tool applied to each interchange within the entire sample. As described in further detail in the following text, a safety performance function frequency has been estimated for 0% CAVs and different CAV penetration rates.

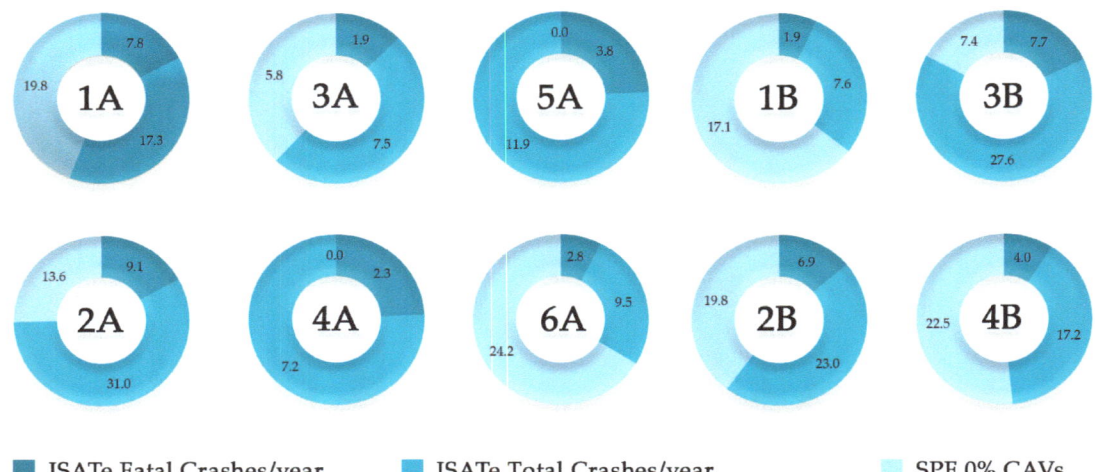

Figure 5. The crash estimates derived from the application of Interchange Safety Analysis Tool Enhanced (ISATe) for the entire sample with 0% CAVs. Note: (**1A–6A**) and (**1B–4B**) are the interchanges in Figure 1.

However, it must be emphasized that ISATe outputs are solely linked to geometric layout parameters and AADT values [48]. Consequently, the results exclusively pertain to the road network operations without CAVs in traffic.

Crashes calculated by ISATe exhibited significant variability owing to the characteristics of the sampled interchanges [49]. These findings hold true for both the Italian and North American groups of interchanges, as well as for the sample as a whole.

Furthermore, the results obtained concerning potentially fatal crashes are typical and are usually associated with the type of intersection under investigation. This circumstance affirms, on one hand, the feasibility of utilizing an "international" sample, and on the other hand, underscores the necessity to validate presumed characteristics such as the geometry of the junction, specific traffic conditions, and the susceptibility/adaptability to the presence of CAVs in traffic. These observations could be further enriched through an in-depth study of the urban context characteristics where each junction is installed.

To incorporate the penetration rate of connected and autonomous vehicles into the network models of the sampled interchanges, we sought a correlation between predicted collisions and predicted conflicts. For this purpose, the methodological framework provided by [50] was employed to forecast the number of conflicts and collision types. This framework involved a two-phase nested modeling process wherein a Poisson–gamma safety performance function (SPF) is utilized. This SPF uses traffic volume as an exposure parameter to predict conflicts, which are then employed in another Poisson–gamma SPF to predict collisions.

The traffic conflicts obtained from the SSAM were used for the application and modeling of SPFs. Following the methodology outlined in [50], the expected collision frequency as a function of average hourly conflicts was calculated using Equation (1):

$$E(Y) = a_0 AHC^{a_1} \qquad (1)$$

where $E(Y)$ is the expected collision frequency, AHC is the average hourly conflicts, $\ln(a_0)$ is equal to -1.1991, and a_1 is equal to 0.626. This equation is functional not only because this study was based on conflict prediction techniques but also because traffic conflicts represent a more appropriate predictor parameter for crashes, due to the fact that they are related to vehicle interactions. Then, it is possible to obtain expected collision frequency for a specific site based on specific site conflicts frequency and confidence level parameters (a_0, a_1).

Subsequently, the previously reported equations were applied for each scenario. The results of the expected yearly crash frequency at the examined intersections are depicted in Figure 6.

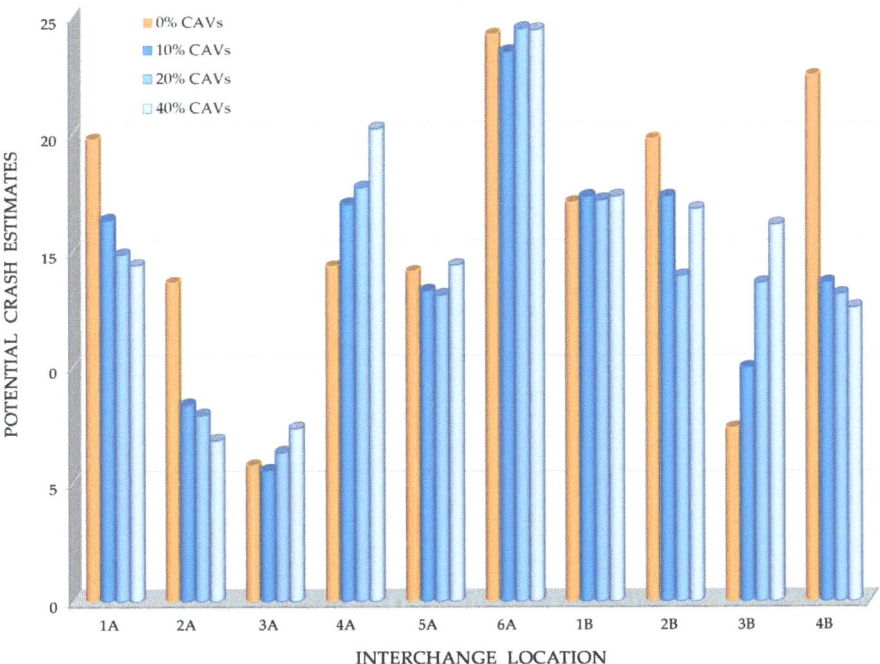

Figure 6. Yearly crash frequency estimates from Equation (1) related to CAV penetration rate at the sampled interchange.

4. Results

Anticipated disparities between the SPF outcomes and the ex ante modeling of ISATe arise from variations in road environment input data. SPF results are influenced by potential conflicts identified in microsimulation trajectory (trj) files, whereas ISATe predictions are derived from roadway geometrical and traffic volume data. Nonetheless, Figure 5's findings illustrate how distinct conditions at each interchange can result in varying sensitivity to the presence of CAVs, both in terms of absolute values (i.e., number of estimated collisions) and relative values (i.e., between different junction layouts), as discussed in the following paragraphs.

In Figure 7, presented below, an overarching view is depicted, illustrating the potential traffic conflict plot for all investigated interchanges. These plots, directly generated by the SSAM model, predominantly indicate the positions of potential traffic conflicts involving rear-end collisions and lane changes within the considered intersection. Notably, for the analyzed sample, the potential traffic conflict of crossing type remains relatively low. However, a noteworthy observation arises when comparing the traffic volumes of North

American junctions, which were higher than at Italian ones; despite this, the potential conflict rates appear like those found in Italian interchanges.

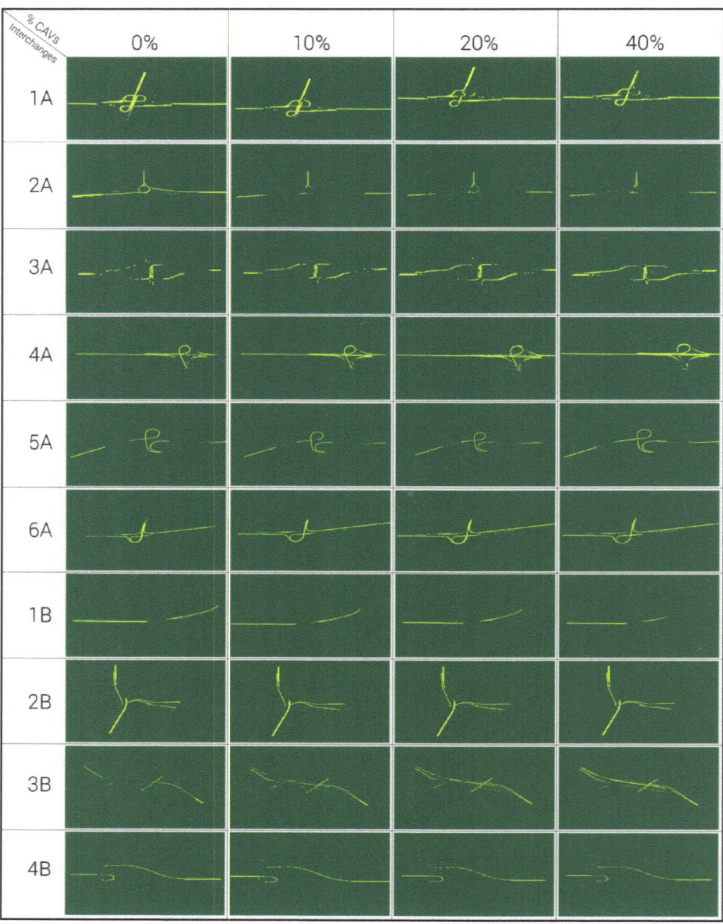

Figure 7. Location of potential rear-end traffic conflicts in interchanges with varying CAV rates.

The larger dimensions of North American interchanges, even with elevated traffic volumes, afford both human-driven vehicles and connected autonomous vehicles the ability to pre-determine the ramp for entry. This is facilitated by the option to maneuver across a greater number of lanes than the standard two lanes observed in Italian interchanges. Furthermore, despite the greater number of lanes in North American (average 3–4 lanes on main roadway) interchanges, potential lane-change conflicts also exhibit similar quantities.

In samples 1A, 2A, and 4B, characterized by a two- and three-lane main roadways, respectively, a gradual decrease in rear-end potential conflict can be attributed to long extensions on/off-ramps that lead to an interchange area that, compared to the entire interchange extension, is characterized by reduced speeds due to reduced radius curves. Therefore, CAVs are able to slow down traffic flow significantly in advance by exploiting ramp lengths, reaching the curved interchange area with reduced speed, and thus minimizing rear-ends potential conflicts. The greater presence of CAVs proportionally affects the impact on the overall slowdown harmonization of traffic flow close to the curved interchange areas.

As regards the increase in potential rear-end conflicts for samples 3A, 4A, and 3B, it is important to highlight that 3A and 3B belong to diamond interchanges category, where the prevalence of straight segments and the minimal presence of curves in interchange areas lead vehicles to increase their desired speed and are then subjected to a non-gradual slowdown in the interchange areas. For sample 4A, rear-end conflict increases are due to the queue formation in and close to ramps, due to high traffic volumes on interchange areas.

Ultimately, samples 5A and 2B are characterized by a 40% CAVs threshold that from a gradual rear-end decrease causes a sudden increase, and considering that interchanges layout are not comparable, this aspect needs to be further explored.

The increase in CAV penetration further leads to a rise in potential conflicts, as an overlap occurs between the decision-making capacity of CAVs (which choose their route well in advance) and human-driven vehicles that occasionally perform maneuvers due to sudden driver decisions. Consequently, in these areas, a high percentage of CAVs does not contribute to the reduction in lane-change conflicts.

Overall, Figure 8 shows the framework of the research model described above. Specifically, the methodological path should be explained using a real case study, as it should be possible to design the simulation environment and to then reach the appropriate safety analysis.

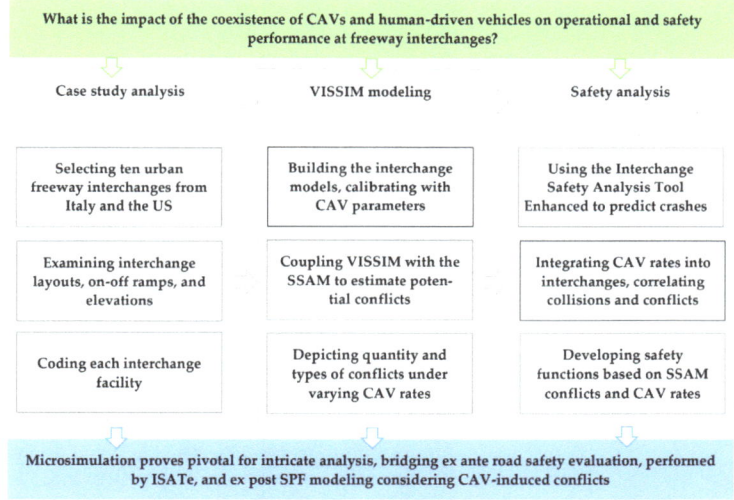

Figure 8. Methodological path of the study analysis.

5. Discussion

Results related to traffic safety estimation emphasize the analysis of conflict numbers and the application of the SSAM. The key performance indicators associated with varying penetration rates of CAVs have been shaped by the contextual analysis of the roadway network, spanning intersections and freeway roadways in prior studies. For instance, in the context of freeway traffic, a decrease in conflict numbers was noted [51], mirroring a similar trend observed with the implementation of a longer headway time [52]. Notably, achieving a 100% CAV penetration rate on freeways resulted in a remarkable 90% reduction in conflicts [53]. Intersection scenarios presented significant outcomes, particularly with the presence of 100% CAVs facilitating the mitigation of crossing conflicts [54].

Concerning other types of intersections, such as signalized intersections and roundabouts, a documented reduction of 65% in conflicts has also been observed [55]. However, it is crucial to highlight that in a roundabout environment, an increased percentage of CAVs may correspond to a heightened frequency of conflicts [56].

Although safety analysis using the SSAM is a consolidated methodology in road safety literature, it is necessary to highlight some critical issues due to the fact that this approach was developed before the implementation of CAVs, then avoiding their driving behavior and capabilities. Compared to human-driven vehicles, CAVs are able to share precise and complex data, such as close vehicle maneuvers, in real time with surrounding vehicles and infrastructure devices with much shorter reaction times. However, it is questionable whether they are enough accurate to carry out CAVs circulation assessment, which is why it would be necessary to develop dedicated SSAM parameters for CAV behavior, according to their automation/connectivity levels based on field data or driving simulator results. One of the aspects that should be studied in depth is related to the possibility of combining the planning of lateral positions with the optimization of the longitudinal trajectory, especially for maneuvers aimed at changing lanes or merging [57]. For example, when a lane change is simulated, rather than a convergence or divergence maneuver, considering the reduced gaps that CAVs can keep, in addition to the V2V or CACC systems, it is not certain that a crash risk would be established [58,59], but the SSAM will inevitably detect a potential rear-end or lane-change conflict.

6. Conclusions

The exploration of road interchanges and the integration of the latest generation of connected and autonomous vehicles (CAVs) is on the road. The significance of the CAV penetration rate is evident for both operational efficiency and safety features at specific road facilities. However, providing a general rule regarding the realistic consequences of CAVs on freeway bypasses remains challenging. The sample selected for this study, as detailed in Section 4, serves as a valuable starting point to elucidate the intricate relationship between interchange working conditions and safety standards.

The presence of ramps adjacent to the main transit corridor, featuring multiple lanes and additional parallel lanes external to the main ones, contributed to a smoother behavior for CAVs as enrollment increased. In certain scenarios, a reduction in potential rear-end conflicts has been observed. Additionally, intersections with auxiliary road arteries facilitated a reduction in dangerous interactions, primarily involving rear-end collisions with CAVs.

While this approach requires optimization, as methodologies in the literature were predominantly applied to smaller intersections with lower operating speeds and reduced vehicle numbers, this study emphasizes the necessity for safety preventive analyses in freeway and ring road interchanges due to the circulation of CAVs. The study suggests potential solutions that must be subjected to further analysis and trials, such as reducing lane width to increase the quantity and capacity of lanes and widening roadway sections. The former would benefit CAVs with precise lane-keeping systems but could lead to the increased potential for lane-change conflicts for traditional motor vehicles (MVs). The study also confirms the relationship between the trend of rear-end conflicts at 0% CAVs and 40% CAVs for Florida junctions.

CAVs exhibit consistent speed in zones near ramps, arriving with gradual slowdowns, unlike MVs that tend to accelerate in entrance and exit ramps, resulting in sudden slowdowns and evasive maneuvers. The configuration of an extended interchange layout significantly influences the number of potential conflicts, even with the presence of CAVs, and plays a crucial role in accommodating higher traffic volumes.

Considering the challenges associated with a 40% penetration of CAVs, each roadway interchange may have a specific threshold triggering cooperation difficulties with MVs.

Microsimulation emerges as the essential modeling tool for such analyses. Despite its complexity, this approach promises to link the ex ante evaluation of road safety, as seen in ISATe application, and the SPF modeling for ex post evaluation, which can consider potential traffic conflicts arising from CAV operations.

Our findings recommend implementing auxiliary lanes on arterial roads to reduce longitudinal traffic conflicts, particularly for CAVs in rear-end scenarios. This study also underscores the need for further tool development to efficiently analyze diverse urban morphologies and geometrical configurations, minimizing the potential conflicts in mixed traffic operations.

Author Contributions: Conceptualization, S.C., A.G. and T.G.; methodology, S.C., A.G. and T.G.; software, S.C.; validation, T.G. and A.G.; formal analysis, A.P., A.G. and T.G.; investigation, S.C. and A.P.; resources, A.G. and T.G.; data curation, S.C.; writing—original draft preparation, T.G. and A.G.; writing—review and editing, S.C., A.G., T.G. and A.P.; visualization, S.C., A.G., T.G. and A.P.; supervision, A.G. and T.G.; project administration, A.G. and T.G.; funding acquisition T.G. All authors have read and agreed to the published version of the manuscript.

Funding: This research received no external funding.

Data Availability Statement: Data are available after kind request to the corresponding authors.

Acknowledgments: The authors would like to thank the support of the Italian Ministry of University and Research, grant PRIN (Research Projects of National Relevance) number PRIN 2017 USR342. Authors thank the local office of Palermo, Catania and Messina City Council and all technicians that collaborated with the research team. The authors would also like to thank the support of the Sustainable Mobility Center (Centro Nazionale per la Mobilità Sostenibile—CNMS) under Grant CN00000023 CUP B73C22000760001.

Conflicts of Interest: The authors declare no conflicts of interest.

References

1. Serok, N.; Havlin, S.; Blumenfeld Lieberthal, E. Identification, cost evaluation, and prioritization of urban traffic congestions and their origin. *Sci. Rep.* **2022**, *12*, 13026. [CrossRef] [PubMed]
2. Shankar Iyer, L. AI enabled applications towards intelligent transportation. *Transp. Eng.* **2021**, *5*, 100083. [CrossRef]
3. Rana, M.M.; Hossain, K. Connected and Autonomous Vehicles and Infrastructures: A Literature Review. *Int. J. Pavement Res. Technol.* **2023**, *16*, 264–284. [CrossRef]
4. Othman, K. Exploring the implications of autonomous vehicles: A comprehensive review. *Innov. Infrastruct. Solut.* **2022**, *7*, 165. [CrossRef]
5. Like Jiang, L.; Chen, H.; Paschalidis, E. Diffusion of connected and autonomous vehicles concerning mode choice, policy interventions and sustainability impacts: A system dynamics modelling study. *Transp. Policy* **2023**, *141*, 274–290. [CrossRef]
6. Brost, M.; Deniz, Ö.; Österle, I.; Ulrich, C.; Senzeybek, M.; Hahn, R.; Schmid, S. Energy Consumption of Connected and Automated Vehicles. In *Book Electric, Hybrid, and Fuel Cell Vehicles, Encyclopedia of Sustainability Science and Technology*, 2nd ed.; Elgowainy, A., Ed.; Springer: New York, NY, USA, 2021; pp. 216–224.
7. Hamad, K.; Alozi, A.R. Shared vs. dedicated lanes for automated vehicle deployment: A simulation-based assessment. *Int. J. Transp. Sci. Technol.* **2022**, *11*, 205–215. [CrossRef]
8. Gora, P.; Katrakazas, C.; Drabicki, A.; Islam, F.; Ostaszewski, P. Microscopic traffic simulation models for connected and automated vehicles (CAVs)–state-of-the-art. *Procedia Comput. Sci.* **2020**, *170*, 474–481. [CrossRef]
9. Tumminello, M.L.; Macioszek, E.; Granà, A.; Giuffrè, T. Simulation-Based Analysis of "What-If" Scenarios with Connected and Automated Vehicles Navigating Roundabouts. *Sensors* **2022**, *22*, 6670. [CrossRef]
10. Chen, Y.; Kong, D.; Sun, L.; Zhang, T.; Song, Y. Fundamental diagram and stability analysis for heterogeneous traffic flow considering human-driven vehicle driver's acceptance of cooperative adaptive cruise control vehicles. *Phys. A Stat. Mech. Appl.* **2022**, *589*, 126647. [CrossRef]
11. Dong, J.; Luo, D.; Gao, Z.; Wang, J.; Chen, L. Benefit of connectivity on promoting stability and capacity of traffic flow in automation era: An analytical and numerical investigation. *Phys. A Stat. Mech. Appl.* **2023**, *629*, 129170. [CrossRef]
12. Hung, Y.-C.; Zhang, K. Impact of Cooperative Adaptive Cruise Control on Traffic Stability. *Transp. Res. Rec.* **2022**, *2676*, 226–241. [CrossRef]
13. Liu, H.; Kan, X.; Shladover, S.E.; Lu, X.-Y.; Ferlis, R.E. Modeling impacts of Cooperative Adaptive Cruise Control on mixed traffic flow in multi-lane freeway facilities. *Transp. Res. Part C Emerg. Technol.* **2018**, *95*, 261–279. [CrossRef]
14. Messer, C.J.; Bonneson, J.A.; Anderson, S.D.; McFarland, W.F. *Single Point Urban Interchange Design and Operations Analysis*; National Cooperative Highway Research Program (NCHRP) Report 345; Transportation Research Board (TRB): Washington, DC, USA.
15. Hancock, M.W.; Wright, B. *A Policy on Geometric Design of Highways and Streets*, 6th ed.; Green Book, Publication Code: GDHS-6; American Association of State Highway and Transportation Officials: Washington, DC, USA, 2013; p. 20001, ISBN 978-1-56051-508-1.

16. Functional and Geometric Standards for the Construction of Road Intersections. Ministry of Infrastructure and Transport, 19 April 2006. Available online: https://www.mit.gov.it/normativa/decreto-ministeriale-19042006 (accessed on 25 February 2024).
17. PTV Planung Transport Verkehr AG. PTV Verkehr. In *Städten–SIMulationsmodell (VISSIM), Version 10*; User Manual Karlsruhe: Karlsruhe, Germany, 2018.
18. Gettman, D.; Pu, L.; Sayed, T.; Shelby, S.G. *Surrogate Safety Assessment Model and Validation: Final Report*; Georgetown Pike (US) Report FHWA HRT 08–051; Federal Highway Administration: Washington, DC, USA, 2008.
19. Li, Q.; Chen, Z.; Li, X. A Review of Connected and Automated Vehicle Platoon Merging and Splitting Operations. *IEEE Trans. Intell. Transp. Syst.* **2022**, *23*, 22790–22806. [CrossRef]
20. Yang, Y.; Yuan, Z.; Meng, R. Exploring traffic crash occurrence mechanism towards cross-area freeways via an improved data mining approach. *J. Transp. Eng. Part A Syst.* **2022**, *148*, 04022052. [CrossRef]
21. Wu, Y.; Wang, D.Z.W.; Zhu, F. Influence of CAVs platooning on intersection capacity under mixed traffic. *Phys. A Stat. Mech. Appl.* **2022**, *593*, 126989. [CrossRef]
22. Lee, D.; Hess, D.J. Public concerns and connected and automated vehicles: Safety, privacy, and data security. *Humanit. Soc. Sci. Commun.* **2022**, *9*, 90. [CrossRef]
23. Hearne, R.; Siddiqui, A. Issues of dedicated lanes for an automated highway. In Proceedings of the Conference on Intelligent Transportation Systems, Boston, MA, USA, 12 November 1997; pp. 619–624. [CrossRef]
24. Liu, Z.; Song, Z. Strategic planning of dedicated autonomous vehicle lanes and autonomous vehicle/toll lanes in transportation networks. *Transp. Res. Part C Emerg. Technol.* **2019**, *106*, 381–403. [CrossRef]
25. Kulmala, R.; Rämä, P. Definition of behavioural adaptation. In *Behavioural Adaptation and Road Safety: Theory, Evidence and Action*, 1st ed.; Rudin-Brown, C., Jamson, S.L., Eds.; CRC Press (Taylor & Francis): Boca Raton, FL, USA, 2013; pp. 11–21.
26. Piccinini, G.F.B.; Rodrigues, C.M.; Leitão, M.; Simões, A. Driver's behavioral adaptation to Adaptive Cruise Control (ACC): The case of speed and time headway. *J. Saf. Res.* **2014**, *49*, 77–84. [CrossRef]
27. Miller, E.E.; Boyle, L. Adaptations in attention allocation: Implications for takeover in an automated vehicle. *Transp. Res. Part F Traffic Psychol. Behav.* **2019**, *66*, 101–110. [CrossRef]
28. Chen, X.; Wu, Z.; Liang, Y. Modeling Mixed Traffic Flow with Connected Autonomous Vehicles and Human-Driven Vehicles in Off-Ramp Diverging Areas. *Sustainability* **2023**, *15*, 5651. [CrossRef]
29. Nowakowski, C.; O'Connell, J.; Shladover, S.E.; Cody, D. Cooperative Adaptive Cruise Control: Driver Acceptance of Following Gap Settings Less than One Second. *Proc. Hum. Factors Ergon. Soc. Annu. Meet.* **2010**, *54*, 2033–2037. [CrossRef]
30. Tanveer, M.; Kashmiri, F.A.; Naeem, H.; Yan, H.; Qi, X.; Rizvi, S.M.A.; Wang, T.; Lu, H. An Assessment of Age and Gender Characteristics of Mixed Traffic with Autonomous and Manual Vehicles: A Cellular Automata Approach. *Sustainability* **2020**, *12*, 2922. [CrossRef]
31. Caruso, G.; Yousefi, M.K.; Mussone, L. From Human to Autonomous Driving: A Method to Identify and Draw Up the Driving Behaviour of Connected Autonomous Vehicles. *Vehicles* **2022**, *4*, 1430–1449. [CrossRef]
32. Sukennik, P. Micro-simulation guide for automated vehicles. *COEXIST* **2019**. Available online: https://www.rupprecht-consult.eu/project/coexist (accessed on 25 February 2024).
33. National Academies of Sciences, Engineering, and Medicine. *Highway Capacity Manual*, 7th ed.; A Guide for Multimodal Mobility Analysis; The National Academies Press: Washington, DC, USA, 2022.
34. Evanson, A. Connected autonomous vehicle (CAV) simulation using PTV Vissim. In Proceedings of the 2017 Winter Simulation Conference (WSC), Las Vegas, NV, USA, 3–6 December 2017; p. 4420. [CrossRef]
35. Abdelkader, G.; Elgazzar, K.; Khamis, A. Connected Vehicles: Technology Review, State of the Art, Challenges and Opportunities. *Sensors* **2021**, *21*, 7712. [CrossRef] [PubMed]
36. Gazder, U.; Alhalabi, K.; AlAzzawi, O. Calibration of autonomous vehicles in PTV VISSIM. In Proceedings of the 3rd Smart Cities Symposium (SCS 2020), Online Conference, 21–23 September 2020; pp. 39–42. [CrossRef]
37. Giuffrè, T.; Trubia, S.; Canale, A.; Persaud, B. Using Microsimulation to Evaluate Safety and Operational Implications of Newer Roundabout Layouts for European Road Networks. *Sustainability* **2017**, *9*, 2084. [CrossRef]
38. Weyland, C.M.; Baumann, M.V.; Buck, H.S.; Vortisch, P. Parameters Influencing Lane Flow Distribution on Multilane Freeways in PTV Vissim. *Procedia Comput. Sci.* **2021**, *184*, 453–460. [CrossRef]
39. Vortisch, P. Wiedemann-99 Source Code. Available online: https://www.researchgate.net/post/Where-I-can-find-the-mathematical-formulation-of-Wiedemann-99-car-following-model (accessed on 20 November 2023).
40. Giuffrè, T.; Curto, S.; Petralia, A. Freeway interchanges maintenance operations. Preliminary safety analysis of connected automated vehicles impact on road traffic. In Proceedings of the Roads and Airports Pavement Surface Characteristics-Proceedings of the 9th International Symposium on Pavement Surface Characteristics, SURF 2022, Milan, Italy, 12–14 September 2022. [CrossRef]
41. Saleem, T.; Persaud, B.; Shalaby, A.; Ariza, A. Can Microsimulation be used to Estimate Intersection Safety? *Transp. Res. Rec.* **2018**, *2432*, 142–148. [CrossRef]
42. Tumminello, M.L.; Macioszek, E.; Granà, A.; Giuffrè, T. A Methodological Framework to Assess Road Infrastructure Safety and Performance Efficiency in the Transition toward Cooperative Driving. *Sustainability* **2023**, *15*, 9345. [CrossRef]

43. Saulino, G.; Persaud, B.; Bassani, M. Calibration and application of crash prediction models for safety assessment of roundabouts based on simulated conflicts. In Proceedings of the 94th Transportation Research Board (TRB) Annual Meeting, Washington, DC, USA, 11–15 January 2015.
44. Sadid, H.; Antoniou, C. Modelling and simulation of (connected) autonomous vehicles longitudinal driving behavior: A state-of-the-art. *IET Intell. Transp. Syst.* **2023**, *17*, 1051–1071. [CrossRef]
45. Sharma, S.; Dabbiru, L.; Hannis, T.; Mason, G.; Carruth, D.W.; Doude, M.; Goodin, C.; Hudson, C.; Ozier, S.; Ball, J.E.; et al. CaT: CAVS Traversability Dataset for Off-Road Autonomous Driving. *IEEE Access* **2022**, *10*, 24759–24768. [CrossRef]
46. Schoettle, B.; Sivak, M. *A Preliminary Analysis of Real-World Crashes Involving Self-Driving Vehicles*; UMTRI-2015-34; The University of Michigan, Transportation Research Institute: Ann Arbor, MI, USA, 2015; pp. 1–24.
47. Interchange Safety Analysis Tool Enhanced (ISATe): User Manual; National Cooperative Highway Research Program Project 17-45 Enhanced Safety Prediction Methodology and Analysis Tool for Freeways and Interchanges. 2012. Available online: https://highways.dot.gov/research/safety/interactive-highway-safety-design-model/interactive-highway-safety-design-model-ihsdm-overview (accessed on 29 February 2024).
48. Saha, D.; Alluri, P.; Gan, A. Applicability of enhanced Interchange Safety Analysis Tool (ISATe) for local applications: A Florida case study. *Adv. Transp. Stud.* **2019**, *49*, 129–144.
49. Islam, M.B.; Perez-Bravo, D.; Silverman, K.K. *Freeway Safety Evaluation: A Quantitative Approach Using Highway Safety Manual to Informed Decisions*; CUTR Faculty Journal Publications, University of South Florida: Tampa, FL, USA, 2018; Volume 103. Available online: https://digitalcommons.usf.edu/cutr_facpub/103 (accessed on 29 February 2024).
50. Sacchi, E.; Sayed, T. Conflict-Based Safety Performance Functions for Predicting Traffic Collisions by Type. *Transp. Res. Rec. J. Transp. Res. Board* **2016**, *2583*, 50–55. [CrossRef]
51. Rahman, M.H.; Abdel-Aty, M.; Wu, Y. A multi-vehicle communication system to assess the safety and mobility of connected and automated vehicles. *Transp. Res. Part C Emerg. Technol.* **2021**, *124*, 102887. [CrossRef]
52. Li, Y.; Li, Z.; Wang, H.; Wang, W.; Xing, L. Evaluating the safety impact of adaptive cruise control in traffic oscillations on freeways. *Accid. Anal. Prev.* **2017**, *104*, 137–145. [CrossRef] [PubMed]
53. Papadoulis, A.; Quddus, M.; Imprialou, M. Evaluating the safety impact of connected and autonomous vehicles on motorways. *Accid. Anal. Prev.* **2017**, *124*, 12–22. [CrossRef] [PubMed]
54. Karbasi, A.; O'Hern, S. Investigating the impact of connected and automated vehicles on signalized and unsignalized intersections safety in mixed Traffic. *Transp. Plan. Technol.* **2022**, *2*, 24–40. [CrossRef]
55. Deluka Tibljaš, A.; Giuffrè, T.; Surdonja, S.; Trubia, S. Introduction of autonomous vehicles: Roundabouts design and safety performance evaluation. *Sustainability* **2018**, *10*, 1060. [CrossRef]
56. Morando, M.M.; Tian, Q.; Truong, L.T.; Vu, H.L. Studying the safety impact of autonomous vehicles using simulation-based surrogate safety measures. *J. Adv. Transp.* **2018**, *2018*, 1–11. [CrossRef]
57. Wang, C.; Xie, Y.; Huang, H.; Liu, P. A review of surrogate safety measures and their applications in connected and automated vehicles safety modeling. *Accid. Anal. Prev.* **2021**, *157*, 106157. [CrossRef]
58. Ren, T.; Xie, Y.; Jiang, L. New England merge: A novel cooperative merge control method for improving highway work zone mobility and safety. *J. Intell. Transp. Syst.* **2021**, *25*, 107–121. [CrossRef]
59. Xue, Y.; Zhang, X.; Cui, Z.; Yu, B.; Gao, K. A platoon-based cooperative optimal control for connected autonomous vehicles at highway on-ramps under heavy traffic. *Transp. Res. Part C Emerg. Technol.* **2023**, *150*, 104083. [CrossRef]

Disclaimer/Publisher's Note: The statements, opinions and data contained in all publications are solely those of the individual author(s) and contributor(s) and not of MDPI and/or the editor(s). MDPI and/or the editor(s) disclaim responsibility for any injury to people or property resulting from any ideas, methods, instructions or products referred to in the content.

Article

Enhancing Urban Intersection Efficiency: Utilizing Visible Light Communication and Learning-Driven Control for Improved Traffic Signal Performance

Manuela Vieira [1,2,3,*], **Manuel Augusto Vieira** [1,2], **Gonçalo Galvão** [1], **Paula Louro** [1,2], **Mário Véstias** [1,4] **and Pedro Vieira** [1,5]

1. DEETC-ISEL/IPL, R. Conselheiro Emídio Navarro, 1949-014 Lisboa, Portugal; mv@isel.pt (M.A.V.); a45903@alunos.isel.pt (G.G.); plouro@deetc.isel.ipl.pt (P.L.); mario.vestias@isel.pt (M.V.); pedro.vieira@isel.pt (P.V.)
2. UNINOVA-CTS and LASI, Quinta da Torre, Monte da Caparica, 2829-516 Caparica, Portugal
3. NOVA School of Science and Technology, Quinta da Torre, Monte da Caparica, 2829-516 Caparica, Portugal
4. Instituto de Telecomunicações, Instituto Superior Técnico, 1049-001 Lisboa, Portugal
5. INESC-ID, Instituto Superior Técnico, Universidade de Lisboa, 1000-029 Lisboa, Portugal
* Correspondence: mv@isel.ipl.pt

Citation: Vieira, M.; Vieira, M.A.; Galvão, G.; Louro, P.; Véstias, M.; Vieira, P. Enhancing Urban Intersection Efficiency: Utilizing Visible Light Communication and Learning-Driven Control for Improved Traffic Signal Performance. *Vehicles* **2024**, *6*, 666–692. https://doi.org/10.3390/vehicles6020031

Academic Editors: Deogratias Eustace, Bhaven Naik, Heng Wei, Parth Bhavsar and Mohammed Chadli

Received: 28 December 2023
Revised: 18 March 2024
Accepted: 1 April 2024
Published: 4 April 2024

Copyright: © 2024 by the authors. Licensee MDPI, Basel, Switzerland. This article is an open access article distributed under the terms and conditions of the Creative Commons Attribution (CC BY) license (https://creativecommons.org/licenses/by/4.0/).

Abstract: This paper introduces an approach to enhance the efficiency of urban intersections by integrating Visible Light Communication (VLC) into a multi-intersection traffic control system. The main objectives include the reduction in waiting times for vehicles and pedestrians, the improvement of overall traffic safety, and the accommodation of diverse traffic movements during multiple signal phases. The proposed system utilizes VLC to facilitate communication among interconnected vehicles and infrastructure. This is achieved by utilizing streetlights, headlamps, and traffic signals for transmitting information. By integrating VLC localization services with learning-driven traffic signal control, the multi-intersection traffic management system is established. A reinforcement learning scheme, based on VLC queuing/request/response behaviors, is utilized to schedule traffic signals effectively. Agents placed at each intersection control traffic lights by incorporating information from VLC-ready cars, including their positions, destinations, and intended routes. The agents devise optimal strategies to improve traffic flow and engage in communication to optimize the collective traffic performance. An assessment of the multi-intersection scenario through the SUMO urban mobility simulator reveals considerable benefits. The system successfully reduces both waiting and travel times. The reinforcement learning approach effectively schedules traffic signals, and the results highlight the decentralized and scalable nature of the proposed method, especially in multi-intersection scenarios. The discussion emphasizes the possibility of applying reinforcement learning in everyday traffic scenarios, showcasing the potential for the dynamic identification of control actions and improved traffic management.

Keywords: traffic management; intersection controlled by light; queue length; transmitters using white LEDs; silicon carbide light detectors; on–off keying (OOK) modulation method; density of pedestrians; model based on reinforcement learning

1. Introduction

Visible Light Communication (VLC) represents a cutting-edge technological paradigm, revolutionizing data communication through the innovative modulation of the intensity of the light produced by Light-Emitting Diodes (LEDs) [1,2]. This dynamic technology has a considerable impact on various applications, thanks to its straightforward design, operational efficiency, and wide geographic coverage. In the field of vehicular communications, VLC seamlessly integrates into the environment, as vehicles, streetlights, and traffic signals entirely adopt LEDs for illumination and signaling commitments [3]. This integration extends to the use of exterior automotive and infrastructure lighting, such as streetlamps, traffic signaling, and head and tail lamps, for both communication and illumination purposes [4,5]. While VLC does have directional constraints, it is possible that the use of VLC in combination with other communication technologies will overcome this limitation. For instance, VLC can be employed for high-speed, short-range communication, while other wireless technologies such as Wi-Fi or cellular networks can complement VLC for broader coverage and omnidirectional connectivity.

Traffic lights equipped with VLC transmitters can not only control traffic but also transmit data to vehicles and roadside sensors, maximizing the utility of existing infrastructure. Utilizing VLC technology to optimize traffic signal efficiency represents a novel approach to urban intersection management. VLC offers advantages such as high data rates, security, and interference-free communication, which can revolutionize traditional traffic signal systems. VLC systems can be designed to provide precise localization capabilities, allowing traffic control devices to accurately determine the position and movement of vehicles and pedestrians. This enables more precise control of traffic flow, including adaptive signal timing and dynamic lane control.

The advent of VLC localization paves the way for advancing security, efficiency, and scalability in multi-intersection traffic signal control, particularly within the context of mixed traffic flows [5]. To tackle the hurdles of coordination, scalability, and integration, our solution involves implementing a traffic signal control system based on distributed reinforcement learning, specifically designed for Vehicular Visible Light Communication (V-VLC). The model's concept is inherently versatile and can be applied to any outdoor pedestrian setting, provided there is access to street database and traffic data. A validation of the mobility model was undertaken using Lisbon's city center as a case study, affirming its efficacy [6]. Incorporating learning-based control algorithms introduces adaptability and intelligence into traffic signal optimization. By leveraging machine learning or artificial intelligence techniques, the system can continuously adapt and improve its performance based on actual traffic conditions and historical data, leading to enhanced efficiency and responsiveness.

The main goal of the paper is to help with the progress of Intelligent Transport Systems (ITS) technology, with a focus on optimizing traffic safety and efficiency. This endeavor involves leveraging enhanced situation awareness and reducing accidents through various communication modes, incorporating Vehicle-to-Vehicle (V2V), Vehicle-to-Infrastructure (V2I), and Infrastructure-to-Vehicle/Pedestrian (I2V/P) communication [7–9]. Recognizing the shortcomings of the conventional control of the traffic light cycle, marked by extended delays, our focus shifts towards dynamic adaptations driven by real-time traffic data. The final goal is to enhance safety and traffic flow at intersections by deploying cooperative drive strategies [10,11]. The combination of VLC technology and learning-based control represents a synergistic approach to urban intersection optimization. By integrating these two innovative technologies, the system can achieve greater efficiency, reliability, and adaptability than traditional traffic management systems, ultimately leading to improved traffic flow, reduced congestion, and enhanced safety in urban areas. The proposed approach may also offer scalability and sustainability benefits, as VLC infrastructure can be relatively easy to deploy and maintain, while learning-based control algorithms can adapt to varying traffic patterns and environmental conditions over time, contributing to long-term urban mobility solutions.

The structure of the paper is as follows: Following the introduction, Section 2 provides an in-depth examination of the V-VLC system, outlining its architecture, communication protocol, and coding/decoding techniques. Section 3 presents experimental results, system evaluations, and a Proof of Concept (PoC) through a phasing traffic flow diagram based on V-VLC. In Section 4, we delve into an agent-based dynamic traffic control simulation using SUMO, an urban mobility simulator tool. Finally, Section 5 summarizes the paper's findings and conclusions, highlighting the transformative potential of V-VLC in traffic signal control and intersection management.

2. Traffic Controlled Multi-Intersections

2.1. Multi-Intersection Complexity

The differences between vehicles and pedestrians, including disparities in speed, size, and movement patterns, introduce additional complexities. Interactions between pedestrians and vehicles can impede each other, leading to reduced traffic flow efficiency and potential safety hazards. Striking an optimal balance between these two components of traffic presents a significant challenge that necessitates thoughtful consideration [12,13].

In the context of multi-intersection scenarios, another obstacle arises. While a straightforward solution involves a single entity controlling traffic lights across all intersections, scalability issues hinder this approach. The rapid increase in state and action spaces makes it impractical for real-time control applications. While optimization schemes for single intersections exhibit scalability within their domain, extending their effectiveness to multi-intersection environments requires innovative solutions.

Researchers [14,15] have delved into collaborative mechanisms to tackle this challenge, incorporating elements like queue length in adjacent intersections and modeling the interdependencies among these intersections. These attempts aim to strike a balance between scalability and efficiency in multi-intersection scenarios, acknowledging the necessity for a new approach to optimize traffic control effectively. Our adaptive traffic control strategy aims to adapt to actual traffic demand by modeling current and expected future traffic flow data. In contrast to conventional ground coil detectors used in traffic settings, an adaptive traffic control system operating within a Vehicle-to-Everything (V2X) environment has the capability to collect comprehensive data, including precise vehicle positioning, speed, queue length, and stopping durations. While V2V connections are especially vital for safety features like pre-crash detection, Infrastructure-to-Vehicle/Pedestrian (I2V/P) links offer connected vehicles and pedestrians (Ps) access to a diverse array of information [16,17].

2.2. V-VLC Communication Link

The communication system shown in Figure 1a is designed to make it easy for different parts of the traffic control system to share and process data smoothly. At the heart of this system is a hybrid mesh cellular structure, which includes two types of controllers placed at street and traffic lights. This setup is crucial for improving the system's performance and scalability [18] The mesh controllers are placed at streetlights along roads at strategic intervals, acting as central nodes in the network. Their main role is to relay messages to nearby vehicles efficiently, thereby ensuring the timely distribution of information like geo-distribution, pose (q(x,y,t)), and traffic notifications. Positioned at intersections, the mesh/cellular hybrid controllers play a multifaceted role within the system. They serve as border routers facilitating edge computing (V2I), enabling seamless integration between mesh and cellular networks. Additionally, they serve as gateways for data exchange between edge devices and the central cloud infrastructure (I2IM), establishing robust communication pathways to ensure uninterrupted data flow. The system utilizes embedded computing platforms to enhance data processing capabilities at the network edge. These platforms enable tasks like real-time sensor data processing, the precise detection of traffic flow patterns, and geo-distribution. Through local data processing, the system decreases response times and alleviates the load on the central cloud infrastructure.

The V-VLC system consists of a transmitter emitting modulated light and a receiver detecting differences in the received light. Both are connected via a wireless channel. The LED light is modulated using ON–OFF keying amplitude modulation. The environment features a grid of square cells arranged orthogonally, with tetra chromatic white light (WLED) sources at corners. These WLED sources combine Red (R: 626 nm), Green (G: 530 nm), Blue (B: 470 nm), or Violet (V: 390 nm) chips to generate white light, facilitating various data channels along roads and intersections.

The modulation and conversion of information bits from digital to analog are achieved through signal processing techniques. Figure 1b depicts the mapping of the coverage of an intersection with four arms, highlighting nine distinct intersections (#1–#9) known as footprint regions, along the cardinal points; δ [19–22].

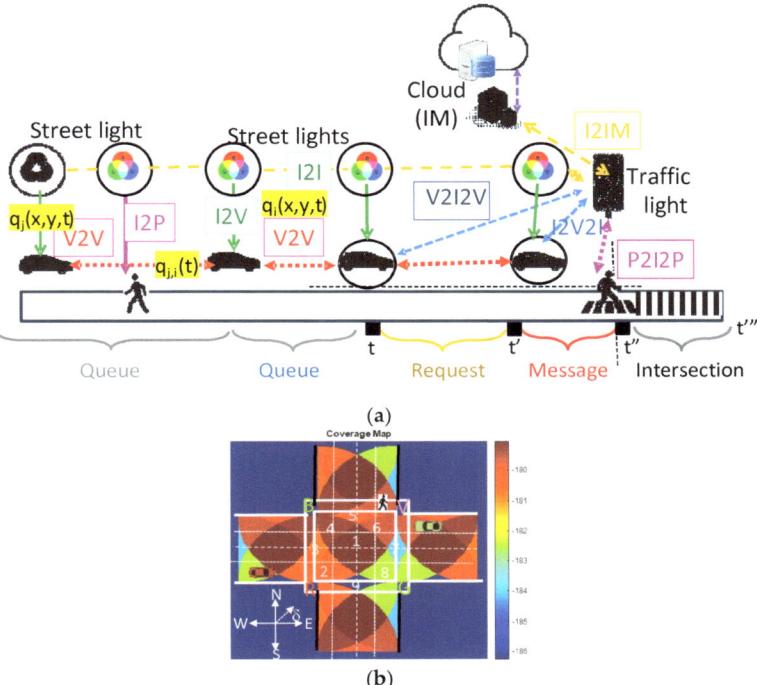

Figure 1. (**a**) Visual depiction in two dimensions of simultaneous localization relative to node density and transmission range. (**b**) Coverage map. Each region (footprint) is labeled from #1 to #9, and each region has a corresponding steering angle code ranging from 2 to 9 [22].

The system receives encoded signals from sources like road lamps and signal lights. Those indicators are meant for direct communication with identified vehicles (I2V) or indirect communication between vehicles using headlights (V2V). Each transmitter sends a message to vehicles (I2V) containing a unique identifier and traffic information. When a vehicle or pedestrian comes within range of the streetlight, upon receiving the light signal, the receiver reacts by allocating a unique identifier and the traffic message.

To control the flow of vehicles at intersections, methods such as queue/request/response and temporal/space relative pose concepts are used. PIN/PIN photodetectors with light filtering capabilities, integrated into mobile receivers, receive and decode the coded signals. The MUX receiver then combines various optical channels, performs different filtering processes (like amplification and switching), detects multiple signals, determines the centroid of received coordinates, and stores them as points of reference for the position. Nine reference points are identified for every unit cell, enabling the precise localization of pedestrians and vehicles within each cell. (See Figure 1b for illustration) [19].

2.3. Scenario, Environment, and Sequential Phases Used for the Simulation

The simulated scenario depicts a multi-intersection layout, as illustrated in Figure 2a, comprising a pair of four-way intersections (C1 and C2).

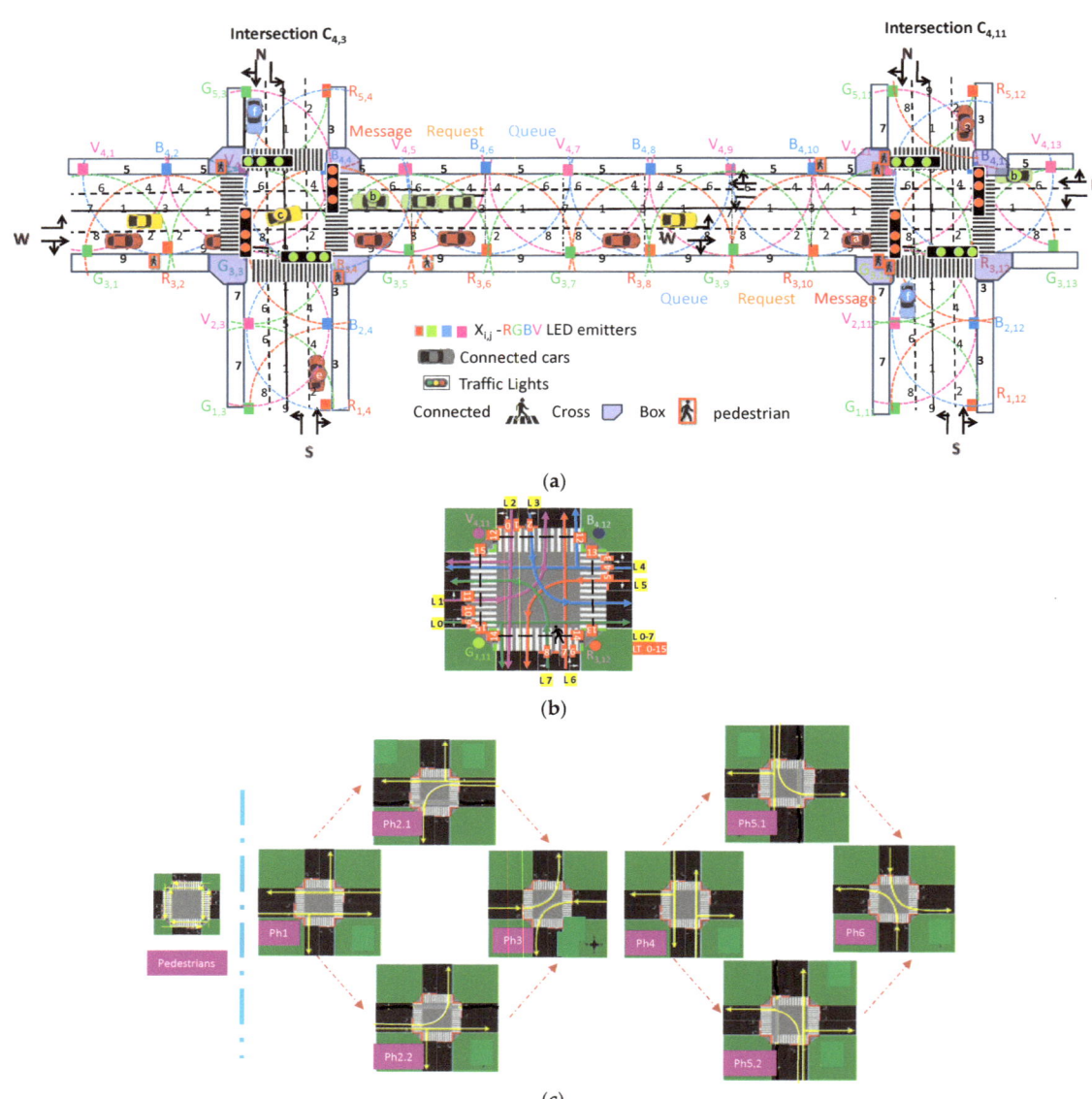

Figure 2. (**a**) Simulated scenario: depiction of an intersection with two sets of four arms and its surrounding environment featuring the optical infrastructure (X_{ij}), the resulting footprints (1–9), and the presence of connected cars and crossing pedestrians. (**b**) Identification of traffic lights (TL) and lanes (L), along with the illustration of possible trajectories for vehicles within an intersection. (**c**) Sequential progression of phases within the intersections, illustrating the evolution of operations over time [22].

Each intersection is equipped with two lanes on every arm, which approach from the cardinal points, leading to a configuration featuring two lanes on every arm. Each arm covers 100 m in length, with every lane measuring 3.5 m in width. Within each lane on every arm, specific directions for vehicle movement are delineated: the right lane accommodates vehicles turning right or proceeding straight ahead, while the left lane only permits left turns. Positioned at the intersection, a traffic system, overseen by the Intersection Manager (referring to the agent) handles the flow of incoming traffic. Emitters (streetlamps) are strategically positioned by the roadside with a spacing of 15 m between them. Each lane is subdivided into three distinct segments, each serving a specific purpose: the first segment is dedicated to accommodating vehicles in motion or queuing along the lane (queue distances); the second segment is reserved for vehicles requesting permission to cross the intersection (request distance); and finally, the third segment, known as the message distance, is where vehicles receive the requested permission to proceed with crossing.

In Figure 2b, a schematic of the intersection is presented, depicting potential trajectories for vehicles and pedestrians, coded lanes, and traffic signals. Meanwhile, Figure 2c provides a visual illustration, offering insight into the sequential evolution of phases within the intersections. This carefully arranged process follows a precisely organized cycle duration, comprising a dedicated pedestrian phase and eight separate vehicular phases arranged into two segments. The sequence of these phases depends on the ever-changing traffic patterns. Each phase is further divided into specific time intervals or states, creating a detailed temporal structure that regulates the intersection's functionality [20,21]. Throughout the pedestrian phase, all vehicular traffic comes to a halt.

The "environment" is based on clusters of unit cells, forming an orthogonal topology as shown in Figure 2a. Each transmitter, labeled as X subscript i,j, has its own color (Red, Green, Blue, or Violet) and horizontal and vertical position (i,j) in the network. In PoC, crossroads were assumed to be located at the intersections of columns 3 and 11 with line 4. Figure 2a illustrates four distinct traffic flows along the cardinal directions. A binary choice (turn left/go straight or turn right) is provided in the road request and response segments.

Each simulated car represents a percentage of the traffic flow. We have assumed a total influx of 2300 cars per hour approaching the intersections, with 80% originating from the east and west directions. Subsequently, 25% of these cars are expected to make either a left or right turn at the intersection, while the remaining 75% will continue straight. The pedestrian influx is approximately 11,200 per hour, generated from both vertical roads and crossing the intersection in all directions, with an average speed of 3 km/h.

To illustrate the diverse traffic flows within a cycle, let us examine the following scenario:

- Twenty-four vehicles, from the west (W), approach the intersection. Among these, twenty vehicles (category a_1) continue forward, depicted by the red flow, while four vehicles (category c_1) exclusively make left turns, represented by the yellow flow.
- Vehicles from the east (E) contribute to the green flow, with thirteen vehicles (category b_1) continuing straight, and two vehicles (category b_2) making left turns.
- The orange flow originates from the south (S) and consists of six vehicles (category e_1). Within these, two vehicles take a left-turn approach (category e_2), while the other four continue straight.
- Lastly, the blue flow comprises thirteen vehicles (category f_1) arriving from the north. Nine of them proceed straight ahead, while the others execute a left turn at the intersection.

This breakdown offers a glimpse into how traffic is distributed across each flow, outlining vehicle movements such as going straight or making left turns at the intersection. The top three requests are assumed to be a_1, b_1, and a_2, pursued, respectively, by b_2, a_3, and c_1 in the fourth, fifth, and sixth positions. In the seventh, eighth, and ninth request positions are, respectively, b_3, e_1, and a_4. The tenth position is taken by c_2, followed by a_5 in the second-to-last request and f_1 in the final one.

2.4. Communication Protocol

To encode information, we utilized an OOK modulation scheme with synchronous transmission employing a 64-bit data frame. Each infrastructure is outfitted with tetra chromatic LEDs (refer to Figure 1b), allowing the concurrent transmission of four signals.

Consequently, the PIN/PIN receiver must possess the capability to actively filter each channel, resulting in a quadruple increase in bandwidth.

The communication protocol, as outlined in Table 1, identifies the structure and regulations regulating information exchange. This protocol includes specifications for synchronization, identification, and payload sections within the transmitted frame.

Table 1. Simplified Protocol For Communication.

		COM	Position		ID (veic)			Time			Payload			
L2V	Sync	1	x	y	0 bits		END	Hour	Min	Sec				EOF
V2V	Sync	2	x	y	Lane (0–7)	Veic. (nr)	END	Hour	Min	Sec	Car IDx	Car IDy	nr behind	EOF
V2I	Sync	3	x	y	TL (0–15)	Veic. (nr).	END	Hour	Min	Sec	Car IDx	Car IDy	nr behind	EOF
I2V	Sync	4	x	y	TL (0–15)	ID Veic.	END	Hour	Min	Sec	Car IDx	Car IDy	nr behind	EOF
P2I	Sync	5	x	y	TL (0–15)	Direct.	END	Hour	Min	Sec				EOF
I2P	Sync	6	x	y	TL (0–15)	Phase	END	Hour	Min	Sec				EOF

Each frame within the communication protocol (designated as 1–6) adheres to a structured format, starting with a synchronization block, followed by identification blocks, and ending with an End-of-Frame (EoF) block. This organized framework ensures a systematic and standardized communication protocol for the Visible Light Communication (VLC) system.

The synchronization block initiates the frame with a 5-bit sequence, represented by the pattern [10101], which synchronizes receivers and marks the start of a new frame. Identification (ID) blocks are crucial as they encode information using binary representation for coded decimal numbers. This information includes the type of communication (1–6), the location of transmitters (x, y coordinates), and timeline details (END, Hour, Min, Sec). The time sub-block, identified by the pattern [111], informs the decoder that the following bit sequence (6 + 6 + 6) pertains to time identification rather than payload. Other ID blocks contain essential data such as the number and temporary identification of vehicles following the leader, details about the occupied lane (Lane 0–7), traffic signal requests (TL 0–15), cardinal direction, or active phase conveyed by the infrastructure in a "response" or "request" message at the intersection.

The traffic message, forming the core of the message, furnishes additional critical information. This encompasses vehicle details, x and y coordinates, and the order of cars behind the leader seeking or receiving permission to cross the intersection (Car IDx, Car IDy, number behind). The traffic information payload includes road conditions, average waiting time, and weather conditions. The frame concludes with a 4-bit EoF block, identified by the pattern [0000], indicating the end of the frame.

2.5. Transmitted and Decoded VLC Signals

Each RGBV signal transmitted carries a specific wavelength-calibrated amplitude, defining its unique characteristics. With four independent emitters in each VLC infrastructure, the optical signal received can have one to four excitations, resulting in 2^4 distinct combinations leading to 16 different photocurrent levels at the photodetector. A filtering operation is obtained through a double PIN/PIN demultiplexer, a critical component in the decoding process. With pre-established knowledge of calibrated amplitudes, the PIN/PIN demultiplexer precisely decodes the transmitted message.

Aiming to clarify both the communication protocol (see Table 1) and the decoding technique using calibrated signals (Figures 1 and 2), Figure 3a provides a visual representation. This illustration showcases the decoded optical signals (depicted at the topmost part of the figures) and the MUX signals received within a V2I (code 3) and a V2V (code 2) VLC scenario. In this scenario, at "10:25:46", the leader, a_o, positioned on lane L0 (direction E) at $R_{3,10}$, $G_{3,11}$, $B_{4,10}$, communicates with the IM (agent) at C2, asking permission to cross and informs the agent that, behind him, three additional vehicles (V_1, V_2, and V_3) positioned, respectively, at $R_{3,8}$, $G_{3,6}$, and $R_{3,4}$, are following him.

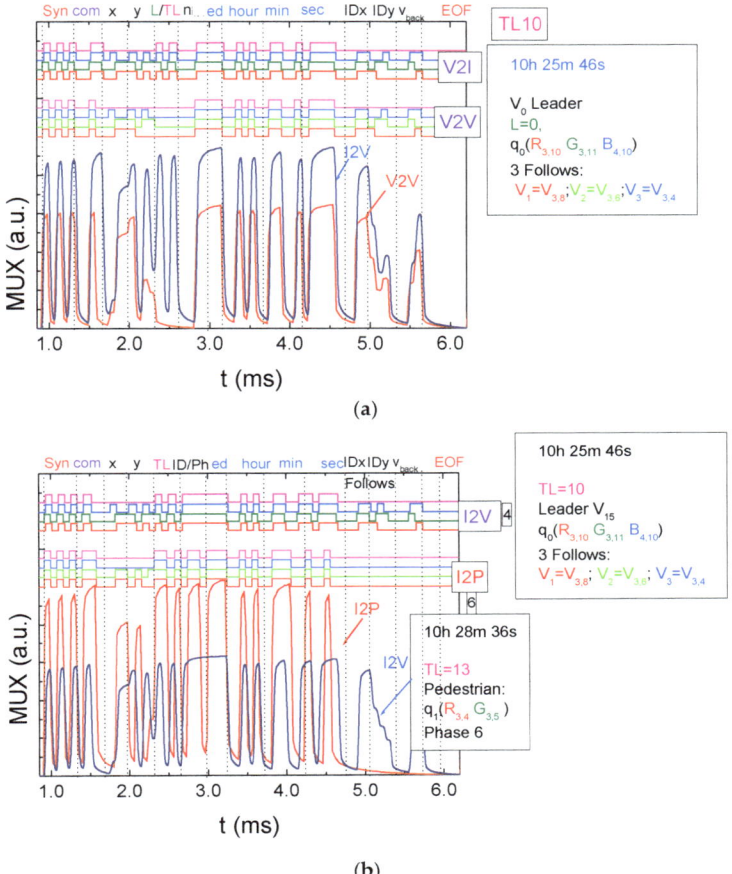

Figure 3. MUX signal requests (**a**) and responses (**b**) are categorized for different types of V-VLC communication. Deciphered messages are presented at the top of the figure [22].

Figure 3b demonstrates the infrastructure response, encompassing both I2V and I2P signals, issued by TL10 and TL13 traffic lights. These responses address the crossing requests initiated by the a_o vehicle and by the q_1 pedestrian positioned at the waiting corner "$R_{3,4}$, $G_{3,5}$" of C1. The response from TL10 was transmitted at "10:25:46", while the response from TL13 was sent at "10:28:66". To investigate pedestrian behavior, two variables are needed: average pedestrian speed and halting. The former evaluates how the cycle durations of vehicles affect pedestrian speed, while the latter enables the analysis of the number of inactive individuals in waiting corners at all intersections, offering insights into population density in the waiting zone over time.

Figure 4a depicts the MUX signal transmitted to the traffic lights (TLs) by two pedestrians at the corners ($P_{1,2}2I$) to cross C1 and C2, respectively. The top part of the figure exhibits the decoded messages, while the content of the message is outlined on the right-hand side. Similarly, in Figure 4b, the MUX signal sent by the traffic lights (I2P$_{1,2}$) is depicted. The upper section of the figure displays the decoded messages, while the right-hand side offers a summary of the message. This visual representation aids in understanding the communication between pedestrians waiting at corners and the corresponding traffic lights, shedding light on the signals exchanged for pedestrian crossings at both C1 and C2 intersections.

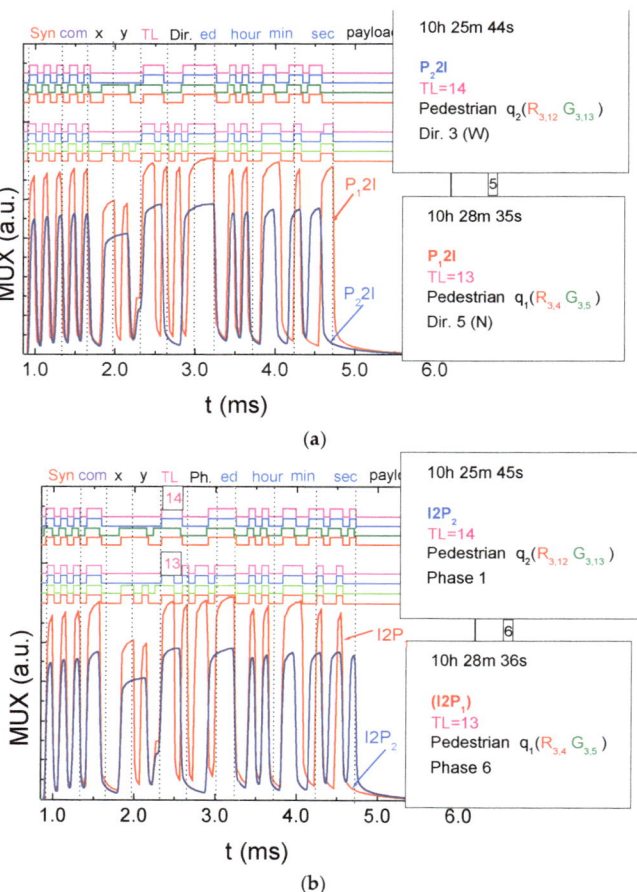

Figure 4. Requests and responses in normalized MUX signals and decoded signals (on the top) (**a**) transmitted by the waiting pedestrians ($P_{1,2}2I$) and (**b**) received by them ($I2P_{1,2}$) over different frame durations [22].

This illustration provides an understanding of the interaction between pedestrians and traffic lights across various intersections. The findings suggest that pedestrians initiate their crossing towards W, intending to traverse through TL14 waiting in positions $R_{3,12}$-$G_{3,13}$ before proceeding. At just "10:25:44", pedestrian 2 (P_22I), begins the communication with the TL14, and at "10:25:45", a response arrives ($I2P_2$). The pedestrian must wait until the pedestrian phase becomes active. With this information, it becomes evident that the current active phase is N-S (Phase 1) signifying that the pedestrian missed their designated phase (Phase 0). So, the pedestrian is required to wait for about 120 s before having the opportunity to cross.

3. Dynamic Traffic Flow Control: Simulation

This section introduces a dynamic control system model aimed at enhancing the efficient management of vehicular and pedestrian traffic at intersections. The model simulates expected outcomes resulting from implementing VLC technology for both vehicles and pedestrians. It utilizes information from V2V, V2I, and I2V communications to strategically make decisions regarding phase activation. This decision-making process prioritizes lanes with higher traffic, following a predetermined sequence of phases outlined in Figure 2b. Additionally, a comprehensive study analyzes the system's performance during high- and low-traffic cycles to estimate the number of vehicles efficiently managed within a one-hour timeframe.

3.1. SUMO Simulation: State Representation

The SUMO simulation environment, as shown in Figure 2, is constructed based on an existing Lisbon scenario. This scenario considers the impact of roads on traffic flow at two intersections. The traffic dynamics on the W–E arm, designated as the focal or "target" road, have a notable impact on traffic flow, with particular emphasis on this arm. The past influence on the target road of traffic conditions from other roads is restricted to a specific timeframe. The transmission of traffic flow and traffic waves quantifies the duration during which the traffic state of other roads affects the target road within the same timeframe. As vehicles continuously enter the system, the composition of traffic flow on the target road undergoes gradual changes over time, thereby influencing the cycle length at both intersections.

In order to improve traffic flow, adjustments were implemented to the originally suggested phases, as shown in Figure 2. These changes require a direct shift from the pedestrian phase (Ph0) to the N>S phase (Ph4), with subsequent phases proceeding as planned in both intersections. By reordering the phases and refining the traffic light control strategy based on simulation findings, enhancements in traffic flow, the alleviation of congestion, and overall intersection efficiency can be realized.

Regarding vehicle circulation, all vehicles are assumed to have an average speed of 10 m/s and a length of 4.5 m. However, as vehicles approach the traffic light at the beginning of the cycle, particularly during pedestrian evacuation, their speed is reduced to 5 m/s. Considering this adjusted speed, it is estimated that each vehicle requires approximately three seconds of green light to pass through the traffic signal. This represents the time needed for a vehicle traveling at 5 m/s to traverse a 15-m-long intersection. Therefore, considering the length of the cars, a minimum interval of 5 m between them is required to prevent collisions at this velocity. By incorporating this information into the incentive system, the agent is motivated to make decisions that optimize traffic flow, minimize delays, and ensure the efficient use of green light time, thereby enhancing overall intersection efficiency.

To accommodate pedestrians within the dynamic system, two scenarios were examined: the high- and low-traffic scenarios. In the high-traffic scenario, which lasts for 120 s, 76 cars are sent out, amounting to 2300 cars per hour. The low-traffic scenario, with a duration of 88 s, sends off 44 cars, equivalent to 1800 vehicles per hour. Each intersection experiences a pedestrian flow of 7200 at C1 and 4000 at C2. Pedestrians are introduced exclusively on the N and S roads, in both directions, at various distances from the intersection, mirroring real-life conditions where pedestrians originate from diverse starting points. All pedestrians are integrated into the SUMO simulator at a speed of approximately 1 m/s, which is equivalent to 3 km/h, a value closely resembling reality.

The IM, acting as the agent, strategically controls traffic signals to facilitate efficient and safe movement within the intersection. To achieve effective traffic optimization through learning, the state representation encompasses information about the environment, vehicle distribution obtained from V-VLC-received messages (refer to Table 1 and Figure 4b), and the proposed phasing diagram guiding agent actions (Figure 2b). The primary goal is to minimize the total accumulated waiting time in each intersection arm, a metric calculated based on vehicle speed and queue alerts [21].

The reward function evaluates the difference in accumulated waiting time between the current and previous steps in all lanes, with negative rewards indicating higher waiting times. The agent learns to optimize traffic by taking actions (dynamic phases; Figure 2b) based on the current state, with training involving stored data samples to enhance decision-making. These decisions are then communicated to drivers and pedestrians through VLC response messages (Figure 3b), where the vehicle ID is assigned.

The agent's state, denoted as s_t, serves as a representation of the environment's situation at a specific agent step t. Its effectiveness in facilitating the agent's learning to optimize traffic is contingent upon furnishing ample information about the car distribution on each road. Figure 5 illustrates the state representation of the target road at the intersections throughout a simulated timeframe [22]. This representation incorporates discrete sub-cells designated for "response," "request," and "queue" zones, enabling the detection of vehicle entry into incoming lanes. Preceding the stop line of the intersection, each lane is divided into five cells: 0 for messages, 1 for requests, and 2 to 5 for queues. Each lane is equipped with its own dedicated traffic light, resulting in a total of 40 state cells during simulation, with lanes denoted as L/0–7 and traffic lights as TL/0–15. The simulation monitors the physical positions of waiting vehicles across lanes (L; 0–7) at C1 and C2. Each lane is segmented into small cells from the intersection, with each cell capable of accommodating a single vehicle. Sub-states provide detailed information regarding the nearest cell's position to the intersection and the maximum queue length.

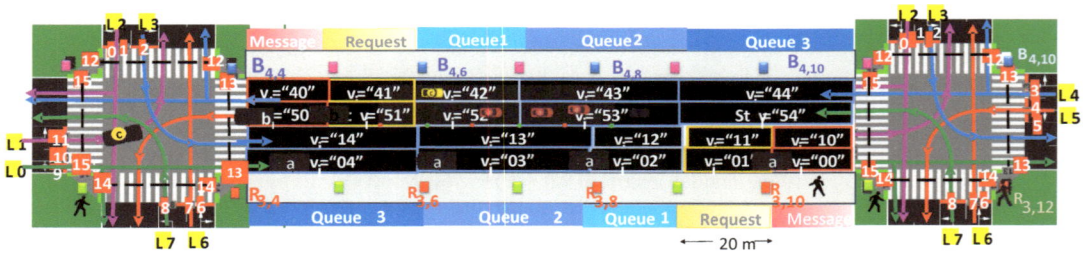

Figure 5. State representation (v_i) for the "target road" encompasses data on traffic lights (TL 0–15) and lanes (L 0–7), along with the visualization of vehicle and pedestrian trajectories [22].

The "complete state" refers to all the factors that contribute to the decision-making process. This could include various elements or aspects that are relevant to the environment or context in which an agent operates. Within this complete state, there are sub-states. These sub-states represent different facets or perspectives of the situation in a specific time step (t) as perceived by the agent. Each sub-state provides a unique representation of the environment at that particular moment. These representations help the agent understand and respond to the dynamic conditions of the environment at each step in its decision-making process. In the position state system at the intersection, a vehicle is referred to as "v_i", where "i" is the order of the request to cross, stated as a two-character sequence. The first character identifies the lane where the vehicle is located, while the second indicates its precise location within that lane. Referring to Figure 5, the states of the leader a_0 and subsequent vehicles are v_{15} = "00", v_{16} = "02", v_{17} = "03", and v_{18} = "04".

Each cell has the capability to measure the speed of a single vehicle. Vehicle speed is monitored during the simulation, representing the movement of vehicles among lanes (L; 0–7) segmented into small cells. Sub-states capture speeds ranging from "{0, 0.1, 0.2, ..., 0.9, 1}". A speed of "=1" denotes the maximum legal speed, such as 90 km/h, while "=0" indicates 0 km/h. As a result, the IM receives requests (V2I: illustrated in Figure 4a) from all leader vehicles and pedestrians seeking access to the intersection at different moments. The V2I data provide the IM with the precise location and speed details of all leading vehicles, as well as their followers' corresponding data, conveyed through V2V communication (Figure 3). Armed with this information, the IM can forecast the initial arrival times and speeds of vehicles at the different sections of the intersection.

In the queuing length system, the "queue length" denotes the count of stationary vehicles in a lane at the intersections. It fluctuates in response to incoming traffic and is influenced by departure rates. Vehicles at rest in the queue have a 0 km/h speed. The system's state is represented by the highest queue length across lanes (L; 0–7), and the number of possible states corresponds to this maximum among all lanes. For example, if the maximum queue length is 5, then the possible states could be "=0", "=1", "=2", "=3", "=4", and "=5". If there are three waiting vehicles in lane L5 at C 1, the queue states are indicated as"=1" for the three waiting vehicles and "=0" for the vehicles in motion. The queue length changes dynamically as vehicles arrive (increase queue length) and depart (decrease queue length). This representation allows for modeling and analyzing the traffic dynamics at the intersection based on the number of waiting vehicles in each lane. The goal is likely to optimize traffic flow and minimize congestion by understanding and managing the queuing system.

The traffic light state at each intersection changes between two states. When the signal is "Red Traffic Light (TL 5)," denoted as "=1", it indicates a red-light scenario. This state resets to "=0" when the light changes to green or yellow. Conversely, when the signal is "Green Traffic Light (TL 0)," represented as "=1", it signifies a green light situation. This state resets to "=0" when the light switches to red or yellow.

The traffic light phase state reflects the current traffic flow configuration at any given time "t". The simulation represents the current traffic phase at the intersection. For example, if "C 1 = (1, 0, 0, 0, 0, 0, 0)", it signifies that only traffic phase 1 is currently activated (Figure 2c).

The simulation considers the speed of pedestrians at pedestrian traffic light corners (TL; 12–15). The average pedestrian speed reflects the movement of pedestrians during the simulation. The term "halting pedestrian" refers to the count of pedestrians waiting at a corner of intersections C1 or C2. This count fluctuates due to pedestrian arrivals and is influenced by cross rates. Pedestrians at a standstill have a 0 km/h speed. The system's state is characterized by the maximum number of halting pedestrians across pedestrian traffic light corners (TL; 12–15), and the number of possible states corresponds to this maximum count. For example, if there is a certain number of waiting pedestrians at corner TL14 of C1, the states are expressed as "=n" for the pedestrians in waiting and "=0" for those in motion.

3.2. SUMO Simulation: Cycle and Phases Durations

The SUMO Application Programming Interface (API) allows for seamless interaction with external programs, enabling smooth integration with the simulation environment. SUMO offers an extensive array of statistics pertaining to overall traffic flow. Additionally, it produces a range of results, such as diagrams that visualize the duration of individual states or the traffic light colors observed throughout the simulation.

Utilizing the scenario illustrated in Figures 2 and 5, we constructed a state diagram for the peak traffic scenario, integrating both vehicles and pedestrians through the SUMO simulation. Figure 6a,c showcase the phase diagrams for the interconnected intersections, C1 and C2, spanning two cycles lasting 120 s each. Meanwhile, Figure 6b provides the SUMO environment characterized by high pedestrian and moderate vehicle traffic flows.

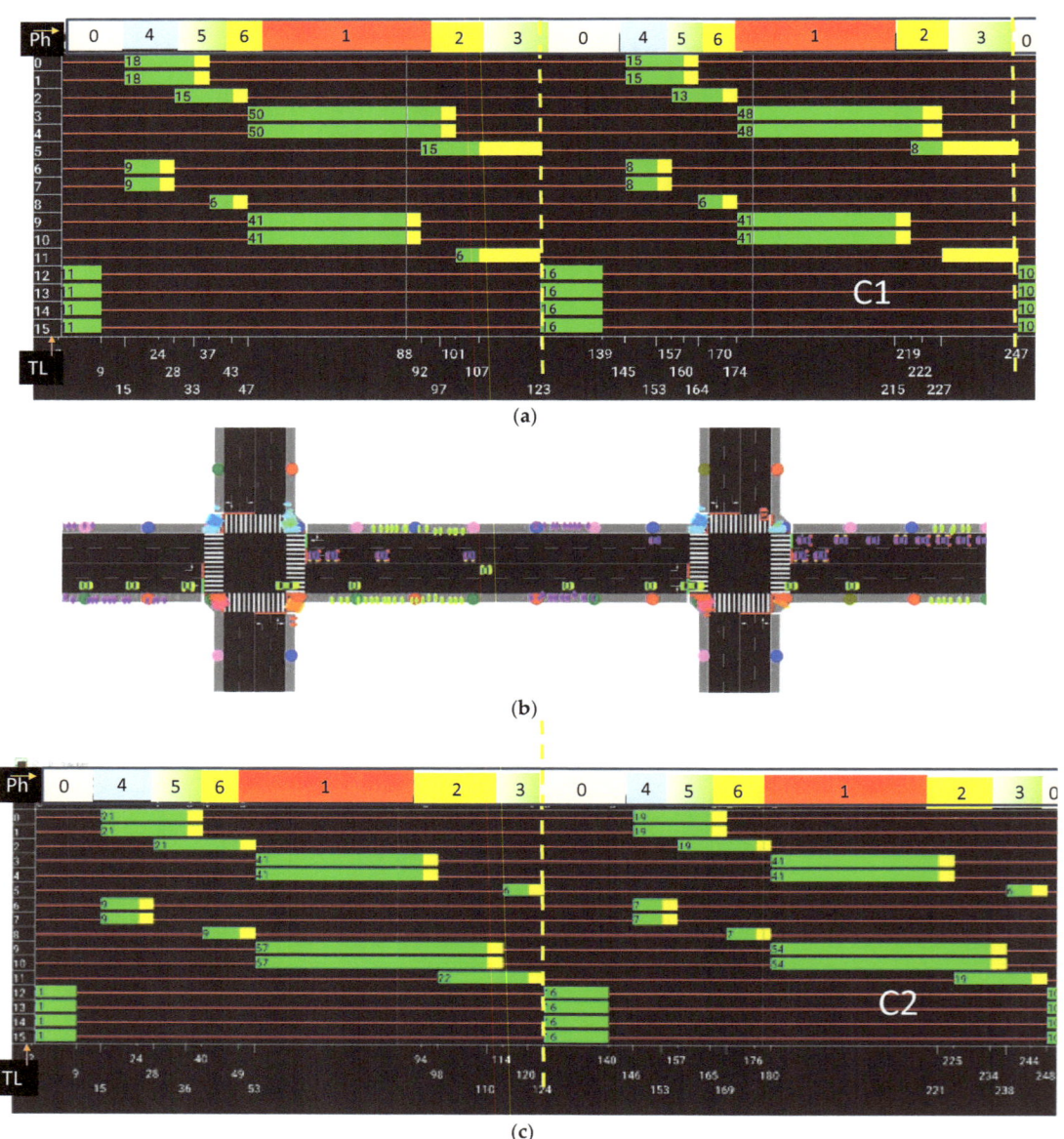

Figure 6. State phasing diagrams for two synchronized intersections are presented as follows: (**a**) Intersection C1; (**b**) the surrounding environment; and (**c**) Intersection C2. Phase numbers along the cycles are provided at the top of the state phase diagrams.

In Figure 6, we can discern the various cycles occurring during the simulation. It consistently kicks off with a pedestrian phase, allowing some individuals to cross the crosswalk, with the signal turning red for pedestrians after 11 s. Subsequently, phases dedicated to vehicles unfold until their conclusion at 123 s. Following this, the second cycle begins, marked by the reactivation of the pedestrian phase. This cycle repeats until 247 s, marking the conclusion of the second cycle and the commencement of a third cycle. These diagrams correlate with the analysis conducted for pedestrians that ensues.

3.3. Dynamic vs. Intelligent Traffic Management: Leveraging VLC and DRL

Dynamic traffic management systems involve real-time adjustments to signal timings and phases based on the actual traffic conditions. These systems rely on ground sensors, cameras, and other data sources to monitor traffic patterns continuously. Adjustments are made reactively in response to changes in traffic flow, aiming to optimize traffic flow and reduce congestion. While dynamic systems are effective in managing immediate traffic issues, they may lack foresight in anticipating future congestion or optimizing long-term traffic management strategies. The integration of VLC into dynamic traffic control systems has represented a novel approach to improving urban intersections [22].

Intelligent traffic management systems utilize advanced algorithms and artificial intelligence to optimize traffic management strategies proactively. These systems analyze large datasets from various sources, including VLC-enabled infrastructure, vehicles, and pedestrians, to predict traffic patterns and optimize traffic flow. By leveraging predictive modeling, machine learning, and optimization algorithms, intelligent traffic management systems can anticipate congestion before it occurs and implement preemptive measures to mitigate its impact. They continuously improve over time, adapting to changing traffic conditions and optimizing long-term traffic management strategies.

Some advantages of using VLC and DRL can be summarized as follows: VLC technology enables the collection of real-time data from various sources, providing valuable insights into traffic patterns and behavior. By combining VLC data with DRL algorithms, predictive modeling can anticipate traffic congestion and optimize traffic management strategies accordingly. Leveraging VLC and DRL allows traffic management systems to take a proactive approach by anticipating congestion before it occurs and implementing preemptive measures to alleviate traffic congestion and enhance traffic flow. Intelligent traffic management systems using VLC and DRL continuously learn from past experiences and adapt their strategies accordingly. This iterative learning process optimizes long-term traffic management strategies, resulting in improved traffic efficiency and reduced congestion over time. Integrating VLC and DRL enables efficient resource allocation, allowing traffic resource allocation systems such as traffic light durations and phases more effectively. This ensures optimal traffic flow while minimizing delays and congestion at intersections.

So, while dynamic traffic management systems focus on real-time adjustments to traffic conditions, intelligent traffic management systems using VLC and DRL take a proactive and data-driven approach. By leveraging advanced algorithms and predictive analytics, these systems can optimize traffic management strategies, anticipate congestion, and improve overall traffic efficiency.

4. Intelligent Traffic Flow Control Simulation

In traffic control problems, RL-based approaches consider traffic flow states at intersections as observable states (Figure 5). Signal timing plan changes are actions, with feedback on control performance being crucial. This section details building an urban traffic control system using reinforcement learning [23–25].

4.1. Reinforcement Learning and Deep Q-Learning

Reinforcement learning (RL) [26] represents a category within the machine learning (ML) framework, wherein an agent undergoes a learning process by actively engaging with an environment [27]. The RL algorithm is very suitable for automatic control [28] and, therefore, a promising approach to intelligent traffic light control. The primary objective for these agents is to attain a goal within an environment characterized by uncertainty and potential complexity. Feedback, in the form of rewards or punishments, serves as the guiding mechanism for the agent's learning process. The underlying concept involves the agent acquiring optimal behaviors or strategies through a series of trial-and-error experiences. The reward function assesses the disparity in accumulated waiting time across all lanes between the current and previous steps, with negative rewards denoting increased waiting times.

The reward function uses the accumulated total waiting time, $atwt_t$, as a metric which is defined in the following equation:

$$atwt_t = \sum_{veh=1}^{n} wt_{(veic,t)}$$

where $wt_{(veh,t)}$ denotes the duration in seconds during which a vehicle *veh* maintains a speed of less than 0.1 m/s at *agent step t* since its introduction into the environment, and *n* represents the total number of vehicles in the environment at agent *step t*. This metric ensures that when a vehicle exits without crossing the intersection, the $atwt_t$ value does not reset. The reward function, r_t, at agent *step t* is defined as follows:

$$r_t = atwt_{t-1} - atwt_t$$

with $atwt_t$ and $atwt_{t-1}$ denoting the accumulated total waiting time of all the vehicles in the intersection attained, respectively, at *agentstep t* and *agentstep t* − 1.

The agent optimizes traffic by taking actions (dynamic phases, as shown in Figure 2b) based on the current state, utilizing stored data samples during training to improve decision-making. These decisions are conveyed to drivers and pedestrians via VLC response messages (as depicted in Figure 3b), which include assigned vehicle IDs. At each discrete time step $t \in T$, the agent perceives its Markovian (or memoryless) decision-making factors (or state s_t) and obtains a state input based on the observed state of the environment and selects and performs an action (a_t) that transforms the observed state into a subsequent state (s_{t+1}). The reward (r_t) is then computed based on this action. Positive environmental rewards reinforce the likelihood of the agent reproducing the corresponding behavior, while negative rewards have the opposite effect. Following this action, the agent observes the subsequent state s_{t+1} and receives an immediate reward (or cost) $r_{t+1}(s_{t+1})$ which depends on the next state s_{t+1} for the state-action pair (s_t, a_t). The overarching objective is to maximize the cumulative discounted reward. Throughout this learning process, experiences in the form of (s_t, a_t, r_t, s_{t+1}) are stored in memory at each time step. Figure 7 provides a visual representation of the schematic for Deep Reinforcement Learning.

The replay memory comprises a dataset of an agent's experiences $D_t = (e_1, e_2, e_3...)$, accumulated as the agent interacts with the environment as time over time ($t = 1, 2, 3, ...$). In training, a batch of random samples is chosen to train the agent. This random selection of samples breaks the temporal correlation between consecutive samples. If the network learned only from consecutive samples of experiences as they occurred sequentially in the environment, the samples would be highly correlated and would therefore lead to inefficient learning. The neuronal network consists of a layered network, and the weight θ_k of the network is used to approximate its Q-values $Q(s, a; \theta_k)$ at iteration *k*.

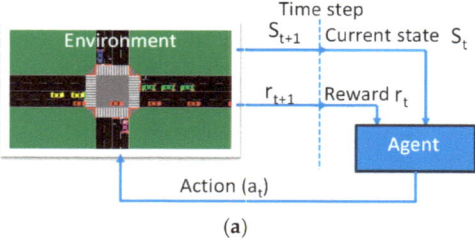

(a)

Figure 7. *Cont.*

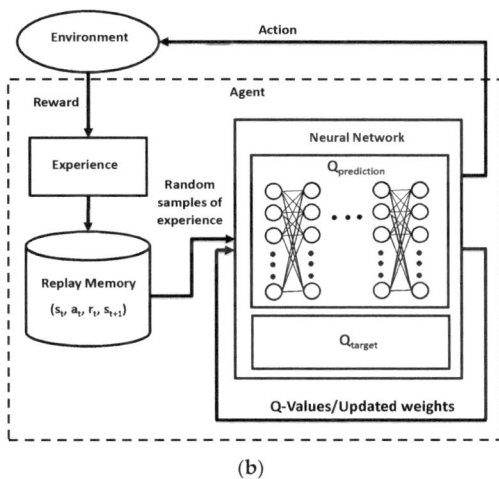

(b)

Figure 7. (a) The Deep Reinforcement Learning schematic. (b) Scheme of the deep neural network used.

To train the agent, the deep Q-Learning technique is employed, leveraging the Q-Learning algorithm. The Q-value (quality value) represents the expected cumulative reward of taking a particular action in a particular state and following the optimal policy thereafter. This algorithm introduces the Q-Function, an action-value function that estimates the value of selecting action a_t at state s_t. The Q-Function predicts the expected cumulative and discounted future reward. In traditional Q-Learning, the algorithm maintains a look-up table storing the Q-value coupled with each state-action pair, earning it the name "tabular Q-Learning". This method guarantees convergence to the optimal value with infinite visits to state-action pairs.

However, this tabular approach is effective only for problems with small-scale state and action spaces. Real-world challenges with continuous and large-scale state and action spaces led to the adoption of deep Q-Learning networks. In this approach, a neural network predicts Q-values, taking the state as input and outputting Q-values for each possible action. This contrasts with estimating Q-values for each state-action pair separately.

Each traffic lane approaching an intersection is represented by 10 discrete cells, each of which represents the presence of a vehicle, resulting in a representation of the state of the environment of 80 cells per intersection. The input layer of the neural network is then composed of 80 neurons representing the state of the environment. Following this, there are five hidden layers, each containing 400 neurons with rectified linear units (ReLUs). The network concludes with an output layer featuring eight neurons, displaying the Q-values for each potential action. To enhance Q-value predictions, a Mean Squared Error (MSE) function is employed. MSE quantifies the disparity between predicted Q-values and target Q-values, contributing to the refinement of the learning process.

$$\text{MSE}_{\text{Loss}} = \frac{1}{N} \sum_{i=1}^{N} \left(Q_{target} - Q_{pred} \right)^2$$

N is the number of samples stored in memory, and the target and predicted value, Q_{target} and Q_{pred}, respectively. After each episode of training, the target Q-values for action-state pairs are calculated based on the following equation:

$$Q_{target} = r_t + \gamma . max Q_{pred}(s_{t+1}, a')$$

where r_t is the reward obtained and γ is a discount factor applied to the $maxQ_{pred}$ value, lowering the importance of the future reward compared to the immediate reward.

The MSE_{Loss} function calculates the squared difference between each predicted and target value. During training, the objective is to minimize this loss, indicating that the model strives to make predictions as close as possible to the true target values. The process involves iteratively adjusting the weights, θ_k, of the neurons in the neural network to decrease the difference between the initial prediction and the target, influenced by the learning rate.

Through repeated updating iterations, the neural network refines its approximation of the Q-value, bringing it closer to the target Q-value. As the loss decreases, the quality of the prediction improves. Consequently, the agent becomes more adept at making decisions regarding actions based on the observed environment. The iterative adjustment of weights enables the model to learn and adapt, enhancing its ability to navigate the environment and make informed choices over time.

4.2. RL-Based Traffic Control Model with VLC Integration

In reinforcement learning scenarios, we operate under the assumption that an agent, such as traffic lights, engages with its environment across a series of discrete time steps with the aim of maximizing rewards [29,30].

The agent's state, st, captures a representation of the environment's condition at a specific time step t. In the RL framework, the objective is to optimize traffic lights at two intersections (Figure 2), each comprising four arms of different lengths ranging from 160 to 400 m. It is important to notice that through multi-V2V communication among follower vehicles and V2I communication from the leader to the infrastructure, we ensure uninterrupted transmission within lanes ranging from 160 to 400 m in length.

The state representation integrates data on vehicle distribution and velocities across each road. PIN/PIN sensors, deployed at traffic lights, monitor vehicles within request and response distances through V2I, and indirectly at queue distances via V2V. The state space is structured with 32 cells per intersection, delineating lanes (L/0–7) and traffic lights (TL/0–15), discretizing the continuous environment (as depicted in Figure 5). This design incorporates spatial information on vehicle presence, speed, and discretized cells. Figure 8 showcases the grid layout of the agent's state space (indicated by dotted lines), underscoring its pivotal role in enabling the RL agent to learn and optimize traffic control policies based on observed conditions.

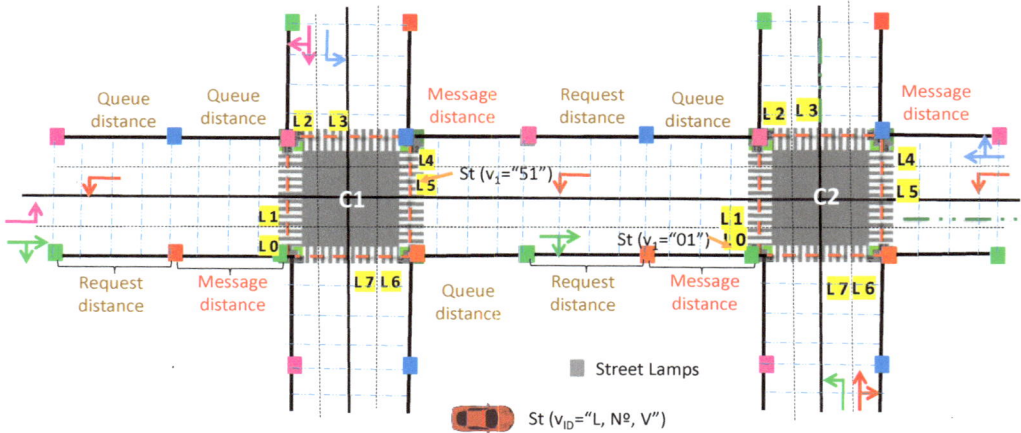

Figure 8. Agent state space grid representation with spatial information about vehicle presence and discretized cells.

The selection of the action space is a pivotal aspect of the RL model's effectiveness. In this scenario, a discrete action space is utilized, where the agent chooses a phase to execute at each time step t. The potential phases and their sequence for each intersection are predefined, as depicted in Figure 2.

The reward (r) signifies the environment's feedback to the agent's decision, serving as a measure of how beneficial or detrimental the agent's action was in terms of achieving specific objectives or optimizing performance metrics. This reward signal plays a crucial role in guiding reinforcement learning algorithms, shaping the agent's learning process, and enhancing its decision-making capabilities over time [31,32].

In this context, the total waiting time metric is employed, and a suboptimal action is defined by introducing more vehicles to queues in the current time step (t) compared to the previous time step ($t-1$). This results in an increased cumulative waiting time compared to the previous time step, leading to a negative reward. The degree of negativity in the reward (r_t) corresponds to the magnitude of additional vehicles introduced to queues at time step t, reflecting a more unfavorable evaluation of the agent's action. Conversely, positive rewards are associated with good actions, where minimizing waiting times contributes to an improved traffic flow. This positive feedback incentivizes the agent to make traffic-light control decisions that improve overall traffic conditions. The training process is divided into multiple episodes, with the total number of episodes determined by the user, where 300 episodes are utilized in this instance. Each episode acts as a training iteration. During an episode, actions are executed based on the activation of specific lanes by the traffic light system, following predetermined timings during the green phases as depicted in Figure 2. This iterative training approach allows the RL agent to gradually learn optimal traffic control policies across multiple episodes, refining its decision-making based on feedback from the environment, particularly concerning waiting times and traffic conditions. The duration of the yellow phase is standardized at four seconds, while the green phase persists for eight seconds.

When the action taken in the current agent step (t) matches the action from the previous step ($t-1$), no yellow phase is introduced, and the ongoing green phase is extended. On the contrary, when the action chosen differs from the previous one, a 4-s yellow phase is introduced between the two actions. This strategy ensures smoother transitions between distinct actions and allows vehicles ample time to adjust to evolving traffic signals. It is important to mention that in the SUMO simulation, each simulation step corresponds to one second, leading to eight simulation steps between two identical actions.

4.3. Implementing Symmetric Homogeneous Rewards in Training

In this study, two adjacent intersections within a (1 × 2) road network topology are examined, a setup previously used in dynamic system analyses. This configuration introduces nuanced considerations, particularly regarding the connecting roadways between the intersections. These roads serve as vital links for balancing traffic flow. Unlike scenarios involving a single intersection, traffic on these roads is influenced by the agent's decision to activate a phase allowing vehicle flow. However, a decision benefiting one intersection may detrimentally affect the other, potentially increasing pressure and wait times, and reducing overall traffic flow.

The observation made by the agent at each intersection is identical concerning the roadways and the occupancy of their cells. The distinction between the two intersections lies precisely in the decisions made by the agent. For instance, when the agent decides, at the first intersection (C1), to activate a green phase for the west direction in all directions, giving vehicles the possibility of going straight or turning right or left, this action will have a different impact on the environment when applied to the second intersection (C2), as can be seen in Figure 9a. At the first intersection when this phase is activated, the cars that do not go straight will leave the environment, while those that do go straight will take a critical lane, heading for the adjacent intersection. When this phase is active at the second intersection, regardless of which direction the cars are traveling, they will all leave the environment and will not return to it. This difference will cause problems when training the network, as the experiences observed at the first intersection will not be identical to those at the second. To address this issue, a phase relationship has been proposed between the first and second intersections, ensuring that both become entirely identical and homogeneous.

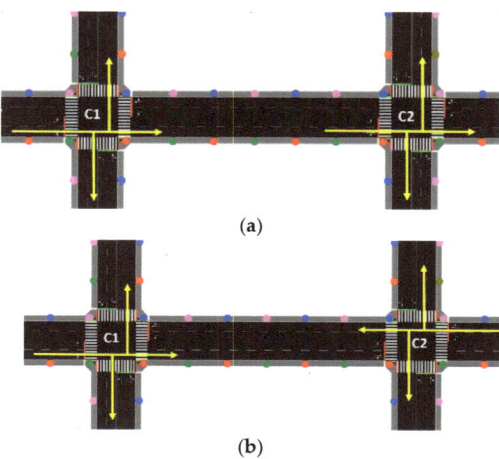

Figure 9. Agent's perception of (**a**) C1 and C2 intersections with north–south directions in both and of (**b**) C1 with a north–south direction and C2 with a south–north direction.

This approach allows for the attainment of an adjacent symmetric homogeneous reward, where actions taken at the first intersection have the same impact as those at the second, significantly contributing to reward improvement. The west all-direction action activated at the first intersection becomes equivalent to the east all-direction action at the second intersection (Figure 9b). The adjacent intersections with identical structures give rise to what is known as an adjacent symmetric homogeneous reward. This cooperative mechanism aids in the balancing of traffic flow between intersections and facilitates improved learning in both intersections, each with one agent.

Training typically involves multiple episodes (or epochs) to ensure effective learning from the data and convergence to an optimal solution. An "episode" refers to a single run or sequence of interactions that an agent undergoes with its environment from start to finish. The cumulative negative reward acts as a metric for evaluating the performance of the RL agent(s) in optimizing traffic control strategies throughout the training episodes.

Figure 10 displays cumulative negative rewards across successive episodes for intersections C1 and C2 in a 160 m (1×2) topology. States for training were obtained with either a single agent in C1 or C2, or with two agents, one in each intersection. This setup evaluates the RL model under different scenarios, including single-agent setups per intersection and the coordination of two agents, each managing one intersection (C1 or C2).

The results demonstrate that introducing a second agent accelerates the learning process with reduced oscillations towards the end of training. This behavior indicates the effective training of the network and validates the proposed solution's benefits for the traffic environment. Therefore, subsequent analyses and discussions assume the involvement of two agents in the learning process. This implies that collaborative efforts between agents in both intersections, C1 and C2, positively influence the learning dynamics, potentially leading to the more effective and efficient optimization of traffic control strategies in the multi-intersection environment.

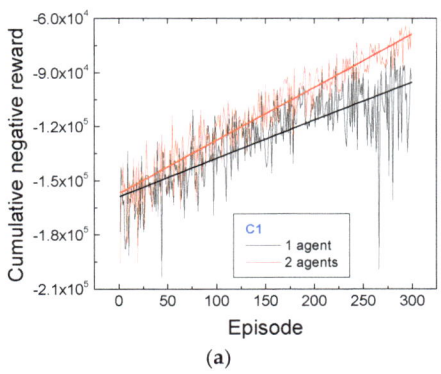
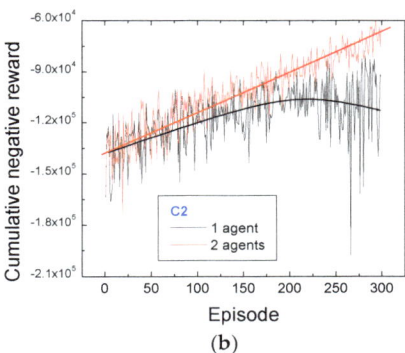

Figure 10. Cumulative negative rewards as a function of successive episodes acquired using a single agent or two separate agents at (**a**) intersection C1 and (**b**) intersection C2.

4.4. Analyzing the Performance of Neural Networks in High- and Low-Traffic Environments: A Study of a 160 m (1 × 2) Road Topology

Two scenarios were analyzed in a 160 m (1 × 2) topology: one with 2300 cars and the other with 1800 vehicles. The aim was to compare and contrast these scenarios with dynamic system findings, validating the feasibility of dispatching these car quantities within an hour. Neural networks for each scenario were trained over 300 episodes, each lasting 3600 s.

To characterize the scenarios, various traffic-related variables were utilized to assess the system's performance. These variables included queue sizes, with individual intersections in each scenario scrutinized to compare car flow. Additionally, the average queue size for each scenario was computed to gauge the impact of car numbers on the environment and the system's responsiveness in each instance. The average car speed was also considered, as it offers insights into traffic fluidity. Lastly, the number of cars halting (waiting) was analyzed to provide insights into the influence of vehicle volume on the environment.

Figure 11 depicts the queue length graph at both intersections (C1 and C2) for the scenario with 1800 and 2300 vehicles. It can be observed that until approximately 800 s, there is a significant increase in vehicles in the waiting queues, akin to a real-world rush hour scenario. There is a substantial influx of cars at both intersections, which gradually diminishes over time. During the neural network training, agents learn to make optimal decisions based on the observed environment. In testing, when agents are prompted to make these same decisions based on their observations, they respond accordingly, as evidenced by the decreasing number of cars in waiting queues over time. This results in clearing most of the vehicles from the intersections within the one-hour timeframe.

In the low-traffic scenario, with fewer vehicles in waiting queues, the intersections are less congested, aiding the agent in making better decisions and increasing the fluidity of vehicle movement throughout the environment. This translates to less time spent in waiting queues and more time in motion. Here, at around 3200 s, there were no longer any cars in the environment.

Figure 12a,b present a comparison that highlights the average speed and halting of vehicles in two distinct scenarios: one with 1800 vehicles per hour and the other with 2300 vehicles per hour.

By analyzing these factors, we aim to discern how varying vehicle volumes impact traffic dynamics and congestion levels.

As illustrated by the graphs, an evident peak in speed is noticeable during the initial phases of the halting simulations. This peak gradually diminishes over time as the simulation progresses. The initial flow in speed is attributed to the absence of vehicles at the intersections, allowing for smoother and faster movement. However, as the number of cars entering the intersections increases, there is a significant decline in average speed. Towards the end of the simulation, as cars start to clear out, the average speed experiences an upturn due to reduced congestion. This trend reflects the dynamic nature of traffic, where higher volumes of waiting cars lead to decreased speed, while lower volumes result in increased speed, in accordance with expected traffic patterns.

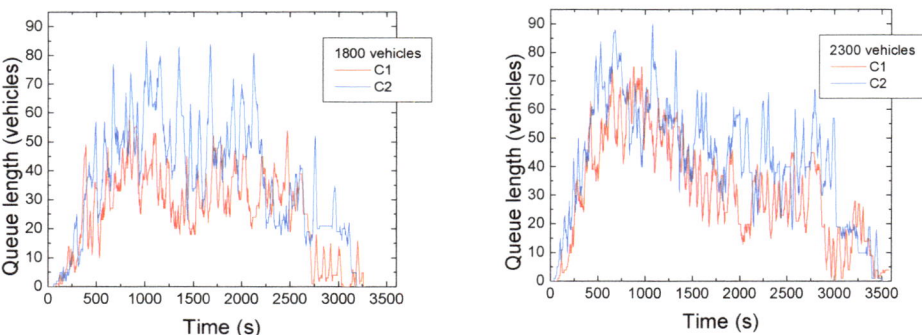

Figure 11. Queue length as a function of time in a scenario of 1800 (**left**) and 2300 vehicles (**right**) for both intersections during the training.

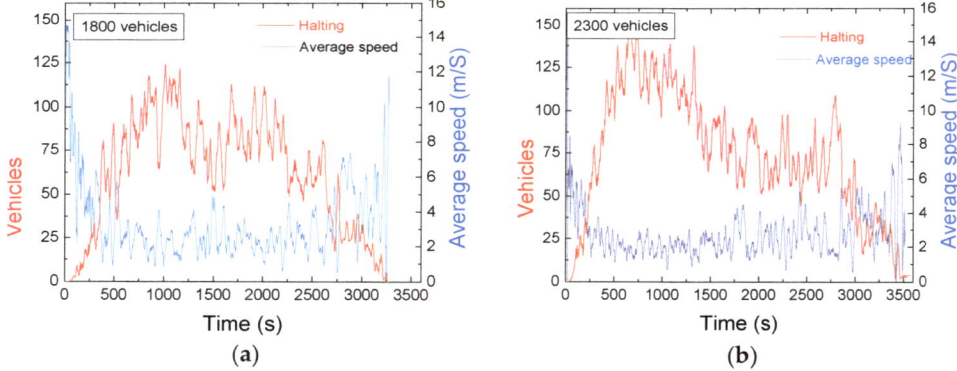

Figure 12. Comparative analysis of average speed and halting: (**a**) scenario with 1800 vehicles/hour and (**b**) scenario with 2300 vehicles/hour.

4.5. Inter-Intersection Roads: 160 m (1 × 2), 250 m (1 × 2), and 400 m (1 × 2) Road Network Topology

After examining the environmental impact of varying the number of vehicles, our focus shifts to investigating the size of critical lanes connecting two junctions. Each agent oversees its junction, monitoring lanes and car volumes through cell occupation. Following optimization in terms of intersection phase relationships, both intersections become homogeneous, rendering the experience identical. Despite this, inadequate communication among agents may elevate car volumes on critical roads. Agent decisions generate rewards based on vehicle wait times at respective junctions. When an action facilitates vehicle movement to target roads, the agent perceives it as beneficial locally, but this may adversely affect the adjacent intersection. Enhanced communication could manage actions based on neighboring intersection pressure. However, implementing this communication might escalate system complexity, potentially requiring a neural network for information exchange and facing scalability issues with more adjacent intersections.

Figure 13 illustrates the cumulative negative reward across successive episodes for the high-traffic scenario, where 2300 vehicles per hour are considered, across different target road lengths for both intersections. This depiction allows for an analysis of how varying road lengths impact the performance of the system in terms of negative rewards over time.

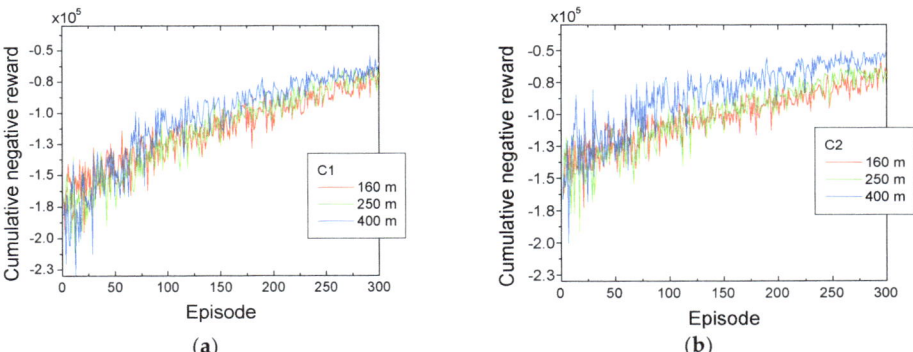

Figure 13. Cumulative negative reward through successive episodes in the high-traffic scenario (2300 vehicles/hour) and different target road lengths. (**a**) C1 intersection. (**b**) C2 intersection.

Here, the neural networks trained with different lane sizes exhibit expected reward behaviors. The findings indicate that with a higher target road length, waiting times decrease, leading to reduced queue sizes. This alleviates the pressure on the agent's junction and ensures sufficient space for vehicle circulation.

Figure 14a,b illustrate average queue sizes during network training episodes. The 400 m lane exhibits fewer queued cars than the other two, indicating minimal need for communication due to ample space for circulation. Conversely, for the 160 m and 250 m lanes, communication remains essential, as queue sizes are comparable to the 400 m lane, necessitating coordination to manage traffic effectively.

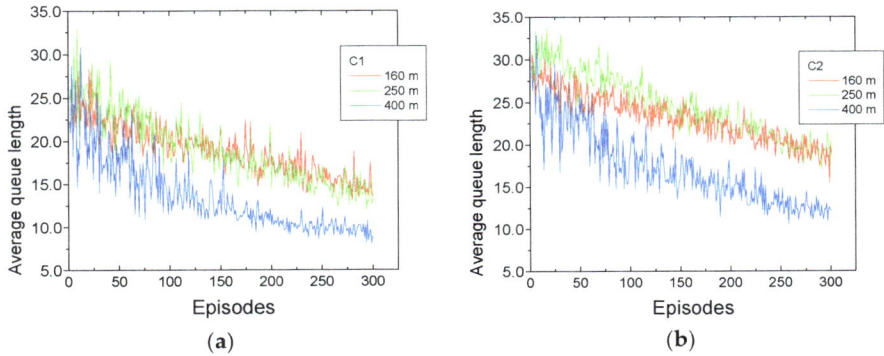

Figure 14. Average queue length (number of vehicles) across successive episodes in the high-traffic scenario (2300 vehicles/hour) and different target road lengths. (**a**) C1 intersection. (**b**) C2 intersection.

After completing the reinforcement learning (RL) training, we observed fluctuations in the learning curve, indicating challenges in achieving convergence. Nevertheless, the model demonstrated gradual improvement, reaching a moderate level of performance over the training period.

The results revealed consistent trends in both cumulative negative reward and average queue length at both intersections. Importantly, there was no significant separation between the cumulative rewards for the three types of road networks, highlighting the scalability of our distributed approach across road networks of varying sizes. The observed stability in these metrics, with a decreasing amplitude of oscillations as training progressed, suggests an enhancement in decision-making capabilities. Interestingly, in the shorter path, learning was faster initially but was later surpassed by longer paths as training advanced.

As anticipated, the average number of vehicles in the queue decreased at both intersections. Notably, the reduction in queue lengths was more pronounced and stable in the longer path at C1 compared to C2. This discrepancy can be attributed to the decreasing resistance of traffic flow with increasing path length, contributing to the observed effects.

In Figure 15, the average queue length across the time was tested for both intersections (C1 and C2) and different target road lengths.

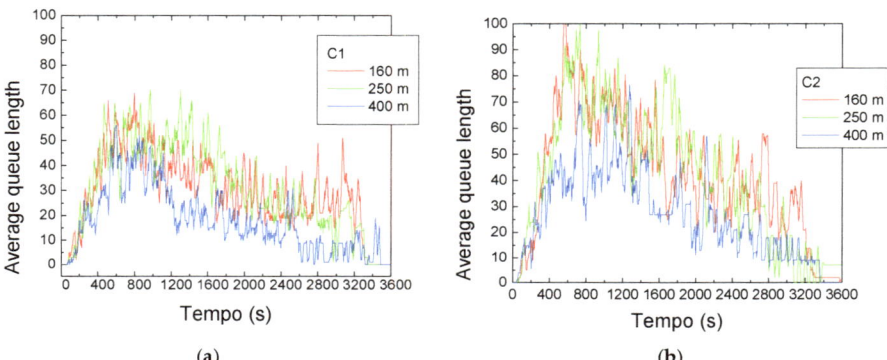

Figure 15. Average queue length (number of vehicles) test as a function of time in the high-traffic scenario (2300 vehicles/hour) and different target road lengths. (**a**) C1 intersection. (**b**) C2 intersection.

The observed average queue length can be explained by noting a notable surge in the number of vehicles in waiting queues until approximately 15 min, resembling a real-world rush hour scenario. Both intersections experience a substantial influx of cars, which gradually diminishes over time. Notably, as the road length increases, the queue length decreases at both intersections. Around the 45-min mark of training, at C1, there are no cars waiting, while at C2, the queue disappears only at the end. So, as the road length increases, fewer vehicles remain in waiting queues, resulting in less congestion at the intersections. This reduction in congestion aids the agent in making better decisions, ultimately enhancing the fluidity of vehicle movement throughout the environment. Consequently, less time is spent in waiting queues, allowing for more time in motion.

Throughout training, agents learn to make optimal decisions based on the observed environment. In testing, when agents are prompted to make these same decisions, they respond accordingly. This is evident in the decreasing number of cars in waiting queues over time, leading to the clearance of most vehicles from the intersections within almost half an hour.

Results show that reinforcement learning can optimize traffic flow by dynamically adjusting traffic signals, pedestrian crossing times, and other traffic management parameters. This adaptability helps reduce congestion, improve overall traffic efficiency, and minimize delays for both pedestrians and vehicles. Reinforcement learning is particularly effective in adaptive traffic signal control. Traffic signal timings can be dynamically adjusted based on current traffic conditions, reducing wait times and improving the overall throughput of intersections. In summary, reinforcement learning offers a flexible and adaptive approach to traffic management, providing the potential for significant improvements in efficiency, safety, and sustainability in both pedestrian and vehicle traffic scenarios. While RL offers several advantages, it is focal to consider potential challenges such as safety concerns, ethical considerations, and the need for careful validation and testing before deploying RL-based traffic control systems in real-world scenarios.

Comparison with previous works is hard to achieve since there is not a well-established benchmark or traffic scenario that allows a fair comparison between different solutions. Previous works consider different environments, different traffic conditions, different reward metrics, etc.

Considering some previous recent works, in [22], a formal analysis of the queue problem is addressed. The number of vehicles is very low and not comparable to the problem under study. The work proposed in [23] does not control traffic lights. It considers a crossroad and autonomous vehicles controlled by an intersection manager using 5G technology. Similarly, work [24] does not control traffic lights. It considers wireless communication to detect vehicles and take transfer information. The work from [25] has a similar solution but only focuses on the RL algorithm. There is no associated communication technology. In this case, there are four lanes per arm. The number of cars as well as the number of lanes doubles. Comparing both, our cumulative reward is lower.

5. Advancements in Urban Traffic Management through Integrated Technologies and Innovative Strategies

This study involves the integration of emerging technologies, the enhancement of intersection efficiency, the development of multi-intersection traffic control strategies, and the application of reinforcement learning algorithms. These advancements have the potential to significantly impact urban traffic management and contribute to the development of more efficient and sustainable transportation systems.

The integration of VLC into dynamic traffic control systems has represented a novel approach to improving urban intersections [22]. VLC technology offers advantages such as high data transmission rates, low latency, and immunity to electromagnetic interference. Incorporating this emerging technology has also contributed to advancing our study in the field of intelligent transportation systems. Also using VLC technology, here we have added a proposal for an intelligent traffic control system leveraging advanced algorithms and artificial intelligence to optimize traffic management strategies. This system, in the future, can analyze large datasets collected from various sources, including VLC-enabled infrastructure, vehicles, and pedestrians, to predict traffic patterns and optimize traffic flow proactively. Intelligent traffic control systems can anticipate traffic congestion before it occurs and implement preemptive measures to mitigate its impact. They may also incorporate features such as predictive modeling, machine learning, and optimization algorithms to continuously improve traffic management strategies over time. While dynamic traffic control focuses on real-time adjustments to optimize traffic flow, intelligent traffic control systems using VLC technology take a more proactive and data-driven approach, utilizing advanced algorithms and predictive analytics to optimize traffic management strategies and improve overall traffic efficiency.

The primary aim is to enhance the efficiency of urban intersections. Improving intersection efficiency can lead to shorter travel times, reduced congestion, and enhanced overall traffic flow, thereby benefiting both commuters and cities. By leveraging VLC for communication between vehicles and infrastructure, coupled with RL algorithms for traffic signal optimization, the research addresses a critical need in urban traffic management.

The development of a multi-intersection traffic control system is essential for managing complex urban traffic networks. By optimizing traffic signals across multiple intersections simultaneously, it addresses the challenges associated with urban traffic congestion and coordination. This approach demonstrates a holistic perspective on traffic management, contributing to the advancement of urban mobility solutions.

The utilization of a reinforcement learning scheme for traffic signal scheduling represents an innovative approach. RL enables the traffic control system to adapt and learn from real-time traffic conditions, leading to dynamic and adaptive signal control strategies. This adaptive nature enhances the system's responsiveness to changing traffic patterns, ultimately improving intersection efficiency and overall traffic management effectiveness.

By demonstrating the feasibility and efficacy of the integrated VLC-based traffic control system with reinforcement learning, the research provides evidence of its contribution. Real-world validation will enhance the credibility and applicability of the findings, showcasing the potential for practical implementation and impact.

6. Conclusions and Future Work

This paper sets the stage for future advancements in intelligent traffic management by emphasizing the potential of VLC technology in enhancing safety and efficiency at urban intersections through RL. The integration of VLC technology across pedestrians, vehicles, and surrounding infrastructure marks a significant breakthrough in optimizing traffic signals and vehicle trajectories. This integration facilitates the direct monitoring of critical factors such as queue formation, dissipation, relative speed thresholds, inter-vehicle spacing, and pedestrian corner density, ultimately leading to improved road safety.

Our dynamic control system model, designed to securely manage vehicular and pedestrian traffic at intersections, underwent detailed analysis under both high- (120 s) and low-traffic cycles (90 s) using the SUMO simulator. We introduced a SUMO extension for pedestrian modeling and made modifications to various tools within the SUMO package to facilitate the generation, simulation, and analysis of multi-modal traffic scenarios. The study aimed to assess the effective management of vehicles and pedestrians within a one-hour timeframe, taking into account various road network topologies.

In the realm of effective traffic optimization learning, our intelligent state representation incorporates environmental information, vehicle distribution from V-VLC messages, and a proposed phasing diagram guiding agent actions. A reinforcement learning model utilizing VLC technology to control traffic in dynamic scenarios was developed. Placing an agent at each intersection, the system optimizes traffic lights based on VLC-ready vehicle communication, calculating optimal strategies to enhance flow, and communicating with other agents to optimize overall traffic. The introduction of adjacent symmetric homogeneous rewards during training significantly improved the model's performance. Through training and testing, the reinforcement learning model showcased its ability to adapt to varying scenarios, emphasizing the importance of continuous learning in dynamic traffic environments. A comparative analysis of cumulative negative rewards across successive episodes and neural network tests for high and low vehicular scenarios using different road network topologies provided valuable insights into the model's efficiency and adaptability.

The improved results obtained with RL when compared to a traditional traffic control approach are traded-off by a higher computational cost since the RL requires the inference calculation of a neural network model. An optimized design is important to guarantee real-time computation in embedded systems near the sensors.

Future work will involve introducing the pedestrian phase, an aspect previously overlooked in the intelligent system. This addition aims to scrutinize agents' behavior, particularly regarding decision-making and environmental observations, with a focus on optimizing the activation timing of the pedestrian phase to ensure safety patterns for pedestrians. Relevant case studies will include analyzing the number of cars at intersections before initiating the pedestrian phase, pedestrian clearance time, and the number of individuals in waiting zones. Optimizing these factors will be crucial to ensuring an efficient system without a high concentration of people in designated areas.

Author Contributions: Conceptualization, M.A.V. and G.G.; validation: M.V. (Mário Véstias) and P.V.; formal analysis, P.L.; investigation, writing, and editing, M.V. (Manuela Vieira). All authors have read and agreed to the published version of the manuscript.

Funding: This research received support from FCT—Fundação para a Ciência e a Tecnologia, through the Research Unit CTS—Center of Technology and Systems, with references UIDB/00066/2020.

Data Availability Statement: No new data was created. The raw data supporting the conclusions of this article will be made available by the authors on request.

Acknowledgments: The authors acknowledge CTS-ISEL and IPL.

Conflicts of Interest: The authors declare no conflict of interest.

References

1. O'Brien, D.; Le Minh, H.; Zeng, L.; Faulkner, G.; Lee, K.; Jung, D.; Oh, Y.; Won, E.T. Indoor Visible Light Communications: Challenges and prospects. *Proc. SPIE* **2008**, *7091*, 60–68.
2. Parth, H.; Pathak, X.; Pengfei, H.; Prasant, M. Visible Light Communication, Networking and Sensing: Potential and Challenges. *IEEE Commun. Surv. Tutor.* **2015**, *17*, 2047–2077.
3. Memedi, A.; Dressler, F. Vehicular Visible Light Communications: A Survey. *IEEE Commun. Surv. Tutor.* **2021**, *23*, 161–181. [CrossRef]
4. Caputo, S.; Mucchi, L.; Cataliotti, F.; Seminara, M.; Nawaz, T.; Catani, J. Measurement-based VLC channel characterization for I2V communications in a real urban scenario. *Veh. Commun.* **2021**, *28*, 100305. [CrossRef]
5. Vieira, M.A.; Vieira, M.; Louro, P.; Vieira, P. Cooperative vehicular communication systems based on visible light communication. *Opt. Eng.* **2018**, *57*, 076101. [CrossRef]
6. Sousa, I.; Queluz, P.; Rodrigues, A.; Vieira, P. Realistic mobility modeling of pedestrian traffic in wireless networks. In Proceedings of the 2011 IEEE EUROCON-International Conference on Computer as a Tool, Lisbon, Portugal, 27–29 April 2011; IEEE: Piscataway, NJ, USA, 2011; pp. 1–4.
7. Elliott, D.; Keen, W.; Miao, L. Recent advances in connected and automated vehicles. *J. Traffic Transp. Eng.* **2019**, *6*, 109–131. [CrossRef]
8. Bajpai, J.N. Emerging vehicle technologies & the search for urban mobility solutions. *Urban Plan. Transp. Res.* **2016**, *4*, 83–100.
9. Wang, N.; Qiao, Y.; Wang, W.; Tang, S.; Shen, J. Visible Light Communication based Intelligent Traffic Light System: Designing and Implementation. In Proceedings of the 2018 Asia Communications and Photonics Conference (ACP), Hangzhou, China, 26–29 October 2018. [CrossRef]
10. Cheng, N.; Lyu, F.; Chen, J.; Xu, W.; Zhou, H.; Zhang, S.; Shen, X. Big data driven vehicular networks. *IEEE Netw.* **2018**, *32*, 160–167. [CrossRef]
11. Singh, P.; Singh, G.; Singh, A. Implementing Visible Light Communication in intelligent traffic management to resolve traffic logjams. *Int. J. Comput. Eng. Res.* **2015**, *5*, 1–5.
12. Oskarbski, J.; Guminska, L.; Miszewski, M.; Oskarbska, I. Analysis of Signalized Intersections in the Context of Pedestrian Traffic. *Transp. Res. Procedia* **2016**, *14*, 2138–2147. [CrossRef]
13. Han, G.; Zheng, Q.; Liao, L.; Tang, P.; Li, Z.; Zhu, Y. Deep Reinforcement Learning for Intersection Signal Control Considering Pedestrian Behavior. *Electronics* **2022**, *11*, 3519. [CrossRef]
14. Fruin, J.J. *Designing for Pedestrians a Level of Service Concept*; Polytechnic University: Kowloon, China, 1970.
15. Eskandarian, A.; Chaoxian, W.; Chuanyang, S. Research Advances and Challenges of Autonomous and Connected Ground Vehicles. *J. IEEE Trans. Intell. Transp. Syst.* **2021**, *22*, 683–711. [CrossRef]
16. Pribyl, O.; Pribyl, P.; Lom, M.; Svitek, M. Modeling of smart cities based on ITS architecture. *IEEE Intell. Transp. Syst. Mag.* **2019**, *11*, 28–36. [CrossRef]
17. Miucic, R. *Connected Vehicles: Intelligent Transportation Systems*; Springer: Cham, Switzerland, 2019.
18. Yousefpour, A.; Fung, C.; Nguyen, T.; Kadiyala, K.; Jalali, F.; Niakanlahiji, A.; Kong, J.; Jue, J.P. All one needs to know about fog computing and related edge computing paradigms: A complete survey. *J. Syst. Archit.* **2019**, *98*, 289–330. [CrossRef]
19. Galvão, G.; Vieira, M.; Louro, P.; Vieira, M.A.; Véstias, M.; Vieira, P. Visible Light Communication at Urban Intersections to Improve Traffic Signaling and Cooperative Trajectories. In Proceedings of the 2023 7th International Young Engineers Forum (YEF-ECE), Caparica/Lisbon, Portugal, 7 July 2023; pp. 60–65. [CrossRef]
20. Vieira, M.A.; Vieira, M.; Louro, P.; Vieira, P.; Fantoni, A. Vehicular Visible Light Communication for Intersection Management. *Spec. Issue Adv. Wirel. Sens. Netw. Signal Process. Signals* **2023**, *4*, 457–477. [CrossRef]
21. Zhang, J.; Wang, F.Y.; Wang, K.; Lin, W.H.; Xu, X.; Chen, C. Data-driven intelligent transportation systems: A survey. *IEEE Trans. Intell. Transp. Syst.* **2011**, *12*, 1624–1639. [CrossRef]
22. Vieira, M.A.; Galvão, G.; Vieira, M.; Louro, P.; Vestias, M.; Vieira, P. Enhancing Urban Intersection Efficiency: Visible Light Communication and Learning-Based Control for Traffic Signal Optimization and Vehicle Management. *Symmetry* **2024**, *16*, 240. [CrossRef]
23. Elbaum, Y.; Novoselsky, A.; Kagan, E. A Queueing Model for Traffic Flow Control in the Road Intersection. *Mathematics* **2022**, *10*, 3997. [CrossRef]
24. Antonio, G.-P.; Maria-Dolores, C. AIM5LA: A Latency-Aware Deep Reinforcement Learning-Based Autonomous Intersection Management System for 5G Communication Networks. *Sensors* **2022**, *22*, 2217. [CrossRef]
25. Shi, Y.; Liu, Y.; Qi, Y.; Han, Q. A Control Method with Reinforcement Learning for Urban Un-Signalized Intersection in Hybrid Traffic Environment. *Sensors* **2022**, *22*, 779. [CrossRef]
26. Kaelbling, L.P.; Littman, M.L.; Moore, A.W. Reinforcement learning: A survey. *J. Artif. Intell. Res.* **1996**, *4*, 237–285. [CrossRef]

27. Shokrolah Shirazi, M.; Chang, H.-F.; Tayeb, S. Turning Movement Count Data Integration Methods for Intersection Analysis and Traffic Signal Design. *Sensors* **2022**, *22*, 7111. [CrossRef] [PubMed]
28. Genders, W.; Razavi, S. Using a deep reinforcement learning agent for traffic signal control. *arXiv* **2016**, arXiv:1611.01142.
29. Vidali, A.; Crociani, L.; Vizzari, G.; Bandini, S. A Deep Reinforcement Learning Approach to Adaptive Traffic Lights Management. In Proceedings of the WOA 2019, the 20th Workshop "From Objects to Agents", Parma, Italy, 26–28 June 2019; pp. 42–50.
30. Kővári, B.; Tettamanti, T.; Bécsi, T. Deep Reinforcement Learning based approach for Traffic Signal Control. *Transp. Res. Procedia* **2022**, *62*, 278–285.
31. Lopez, P.A.; Behrisch, M.; Bieker-Walz, L.; Erdmann, J.; Flötteröd, Y.; Hilbrich, R.; Lücken, L.; Rummel, J.; Wagner, P.; Wiessner, E. Microscopic traffic simulation using sumo. In Proceedings of the 2018 21st International Conference on Intelligent Transportation Systems (ITSC), Maui, HI, USA, 4–7 November 2018; pp. 2575–2582. [CrossRef]
32. Touhbi, S.; Babram, M.A.; Nguyen-Huu, T.; Marilleau, N.; Hbid, M.L.; Cambier, C.; Stinckwich, S. Adaptive traffic signal control: Exploring reward definition for reinforcement learning. *Procedia Comput. Sci.* **2017**, *109*, 513–520. [CrossRef]

Disclaimer/Publisher's Note: The statements, opinions and data contained in all publications are solely those of the individual author(s) and contributor(s) and not of MDPI and/or the editor(s). MDPI and/or the editor(s) disclaim responsibility for any injury to people or property resulting from any ideas, methods, instructions or products referred to in the content.

Review

An Overview of the Efficiency of Roundabouts: Design Aspects and Contribution toward Safer Vehicle Movement

Konstantinos Gkyrtis * and Alexandros Kokkalis

Department of Civil Engineering, Democritus University of Thrace (D.U.Th.), 67100 Xanthi, Greece; akokkal@civil.duth.gr
* Correspondence: kgkyrtis@civil.duth.gr

Abstract: Transforming intersections into roundabouts has shown that a sufficient degree of road safety and traffic capacity can be achieved. Because of the lower speeds at the area of a roundabout, drivers tend to become more easily adaptive to any kind of conflict with the surrounding environment. Despite the contribution to safety, the design elements of roundabouts are not uniformly fixed on a worldwide scale because of different traffic volumes, vehicle dimensions, drivers' attitude, etc. The present study provides a brief overview of the contribution of roundabouts to road safety and the interactions between safety and the design elements of roundabouts. In addition, discussion points about current challenges and prospects are elaborated, including findings from the environmental assessment of roundabouts; their use and performance on the era of autonomous vehicles that will dominate in the near future; as well as the role and importance of simulation studies towards the improvement of the design and operation of roundabouts in favor of safer vehicle movement. The criticality of roundabouts, in terms of their geometric design as well as the provided road safety, lies upon the fact that roundabouts are currently used for the conventional vehicle fleet, which will be gradually replaced by new vehicle technologies. Such an action will directly impact the criteria for road network design and/or redesign, thereby continuously fostering new research initiatives.

Keywords: roundabouts; road design impact; traffic safety; capacity; pavement condition; environmental aspects; autonomous vehicles; simulation

Citation: Gkyrtis, K.; Kokkalis, A. An Overview of the Efficiency of Roundabouts: Design Aspects and Contribution toward Safer Vehicle Movement. *Vehicles* **2024**, *6*, 433–449. https://doi.org/10.3390/vehicles6010019

Academic Editors: Deogratias Eustace, Bhaven Naik, Heng Wei and Parth Bhavsar

Received: 23 January 2024
Revised: 22 February 2024
Accepted: 23 February 2024
Published: 25 February 2024

Copyright: © 2024 by the authors. Licensee MDPI, Basel, Switzerland. This article is an open access article distributed under the terms and conditions of the Creative Commons Attribution (CC BY) license (https://creativecommons.org/licenses/by/4.0/).

1. Introduction

Road crashes are considered to be amongst the eight top leading causes of deaths globally according to the World Health Organization [1]. The most critical locations and conflict points that are vulnerable to incidents and/or fatal accidents are at or near intersections. According to [2,3], almost one in every four fatal crashes occur at or near intersections.

Transforming intersections into roundabouts has shown that a sufficient degree of road safety and traffic capacity can be achieved without the need for traffic signals that induce traffic delays [4]. During the approach of a roundabout, drivers must reduce their speed, something that helps them move smoothly into, around, and out of a roundabout. Typical maximum, minimum, and mean speed profiles are shown in Figure 1. Lower speeds allow drivers to become adaptive to any kind of conflict with surrounding vehicles already in the circular pathway, such as pedestrians and bicyclists. Thus, converting junctions to roundabouts appears to be a commonly applied road safety measure in many countries [2,5–7].

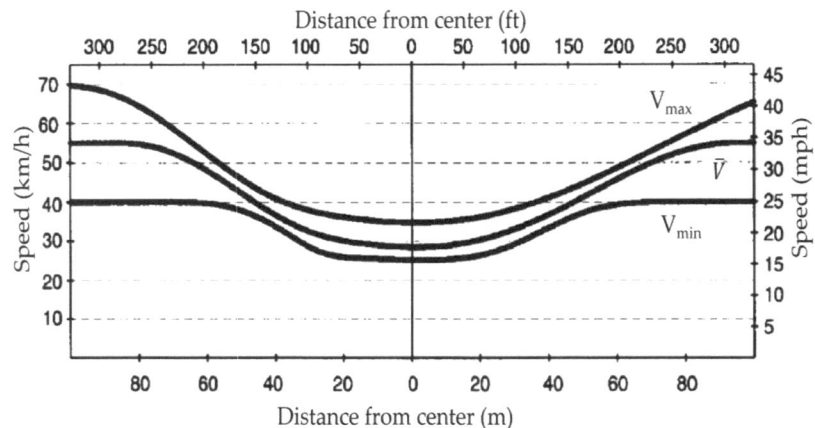

Figure 1. Typical speed profiles for vehicles travelling near a roundabout (adapted from [8]).

Despite these positive remarks, the design elements of roundabouts are not uniformly fixed on a worldwide scale because of the variety in the traffic volumes on the axes/legs of a roundabout, the available space at the area of a roundabout that could affect the number of the selected lanes, and the local traffic regulations or policies [5,9,10]. Most importantly, the trade-off of fulfilling safety and capacity criteria controls the design type and the efficiency of a roundabout [9]. The general rule is that the higher the number of lanes enabling parallel vehicle movement, the less safety levels of roundabouts because of the high-speed values that can be achieved [11]. On the contrary, single-lane roundabouts that force vehicles to drastically reduce their speeds can improve the level of the provided road safety. Moreover, due to lower speeds and fewer conflict points, roundabouts are considered to be a sustainable intersection type because of the safer travelling modes and the reduced vehicle emissions that limit the impact on air pollution [11,12].

Building upon these preliminary remarks, the aim of the present paper is to briefly overview the main design features of roundabouts, the contribution of roundabouts to road safety, and provide a collection of discussion points and thoughts on current challenges and future perspectives for that type of road element. First, the terminology related to roundabouts is recalled together with the types of roundabouts, their advantages, and disadvantages. Thereafter, aspects about the contribution of roundabouts to road safety and the interaction with the design elements are discussed, followed by current research findings on the use of roundabouts by autonomous vehicles (AVs) and challenges related to simulation analyses. Finally, the concluding remarks of this review are summarized. As such, the main contribution of this paper lies upon revealing that roundabouts are major contributors to a safer vehicle movement, provided that the importance of geometric design elements is well-understood for both the era of the current vehicle fleet, as well as for more modern vehicle technologies. The research's flowchart is given in Figure 2.

In respect to the survey methods, since more articles are covered in the Scopus database compared to other ones (e.g., Web of Science), it was decided to employ an advanced search in Scopus. Relevant articles mainly falling within the last decade (i.e., 2013 and thereafter) were selected to capture the most recent trends on roundabout design and safety interactions. Key indicators including road design impact, traffic safety and capacity, pavement condition, and environmental aspects were studied for both conventional and autonomous driving patterns. Both research and review papers were evaluated from multiple publishers, including Elsevier, MDPI, Springer, Taylor and Francis, etc. To a lesser extent, some conference papers were also overviewed.

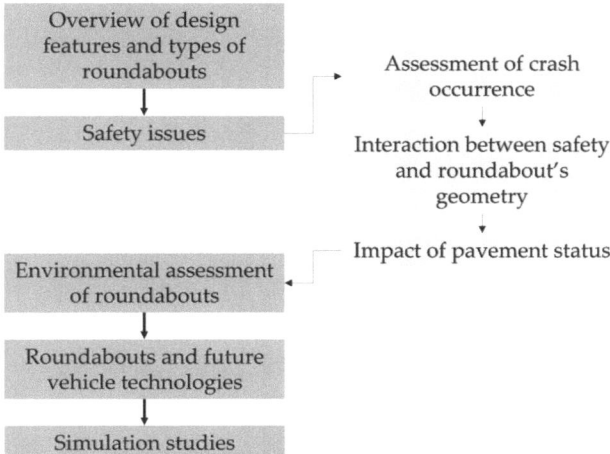

Figure 2. Research framework of this paper.

2. Characteristics of Roundabouts

2.1. Overview

Modern roundabouts were formally recognized in 1929 in the UK. Close cooperation between the Ministry of Transport and the Town Planning Institute led to the development of draft guidelines, according to which crossings of one or more major roads at the same conflict point required enough space, so that vehicle flow could be performed through a circulated traffic mode, or else a "roundabout" system [13].

It should be clarified that roundabouts differentiate from the conventional circular intersections. Vehicles moving in the circle yield to those entering the cyclic path. In these cases, drivers not experienced with circular intersections can indeed be confused by a poorly designed system and can eventually feel trapped when confronted by other vehicles in the circle. This behavioral pattern can result in travel delays, backed-up traffic. collisions, injuries, and even fatalities. On the contrary, a modern roundabout generally features a smaller footprint than a traditional traffic circle [14]. An important distinction between a modern roundabout and a traditional traffic circle is that the roundabout requires drivers who want to enter the circular intersection to yield to the vehicles already circling the roundabout, rather than completely stopping [15].

The level of maturity within the design and implementation processes for roundabouts is not unique. Several countries on a worldwide scale have adopted, to a variable extent, this type of road element for both urban and rural roads. The general trend is that roundabouts are mainly observed in Europe and Australia compared to America, where the term "rotary" is most commonly used in situations consisting of high radii [16]. Factors including variabilities in the traffic composition, the dimensions of design vehicles, driving habits, and culture explain the reason why little consensus exists about the optimal design of an "ideal" roundabout.

This fact justifies why research on roundabout features about optimal design, safety issues, crash patterns, traffic flow behavior, contribution to a sustainable traffic management, etc., continuously revive, so that design optimization and efficiency can be reached [14,17,18]. In addition, the transition era to the new types of AVs will definitively reveal new research capabilities for roundabouts [11].

2.2. Typical Structure

According to Figure 3, typical design elements in a roundabout include:

- The radii for the entry and exit curves; selecting small values for those radii ensure that drivers are easily guided into a transition area before and after the roundabout. As such, this component is most related to the aspect of safety.
- The flare length, which is the area of the approach that is widened. Usually, an additional lane is added at this length so that more vehicles can be accommodated. As such, traffic queues are reduced and better traffic flow is allowed [19]. This component is most related to the aspect of a roundabout's capacity.
- The central and splitter islands (if applicable) are usually concrete islands that are elevated compared to the pavement surface. They can improve both the deflection of vehicles, acting as a guide, and the pedestrian flow through the cross areas.

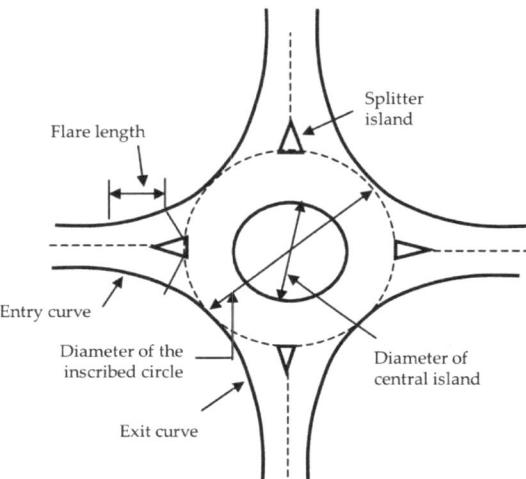

Figure 3. Core elements of a typical roundabout.

It is noted that both the diameter of the central island and the theoretical diameter of the inscribed circle have to also be defined. The latter is the largest circle that can be fitted into the junction outline.

2.3. Types of Roundabouts

Depending on their size and the number of lanes, roundabouts are divided into three categories (Figure 4): (a) mini roundabouts, (b) single-lane roundabouts, and (c) multi-lane roundabouts. The first type is suitable for urban areas and low-volume roads, where lower speeds are generally observed. The central island is of a relatively small diameter. A mini roundabout corresponds to a single-lane circulatory road path with a fully traversable central island, so that potential heavy vehicles can make use of the whole area available.

Single-lane roundabouts consist of a single lane for both entrance and exit at all legs and one circulatory lane. In those cases, higher diameters can be found for the central island, enabling higher operating speeds to be reached. In addition, the central island is non-traversable and it includes an apron.

Finally, multi-lane roundabouts are mainly applied in rural areas, or even suburban areas, where a higher number of vehicles is to be accommodated. In the circulatory path, vehicles travel side by side, and at least one entry has two or more lanes.

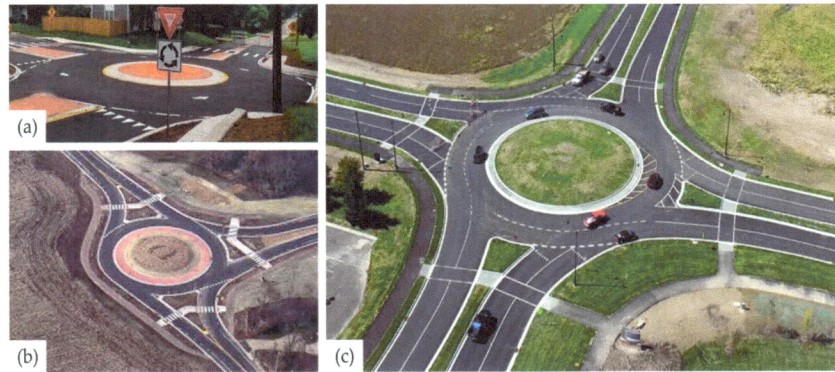

Figure 4. Basic types of roundabouts: (**a**) mini roundabout, (**b**) single-lane roundabout, and (**c**) multi-lane roundabout (with two lanes).

All of the aforementioned types belong to the category of modern roundabouts. Another classification for roundabouts takes into account the shape of the central island [20]. From this view, the following categories can be found (Figure 5): (a) modern (cyclic) roundabouts, (b) elliptical, and (c) turbo roundabouts. In the elliptical roundabout, the diameter ratio for the major and minor approaches is usually set to 2:1, which is consistent with most common design methodologies [21]. Comparative multi-parametric analysis has shown that elliptical roundabouts are more efficient for those cases where traffic congestions are expected [20]. Once avoided, higher speeds can be reached, thereby leading to increased crash severity at elliptical roundabouts, even though crash frequency is kept at low levels [20,22].

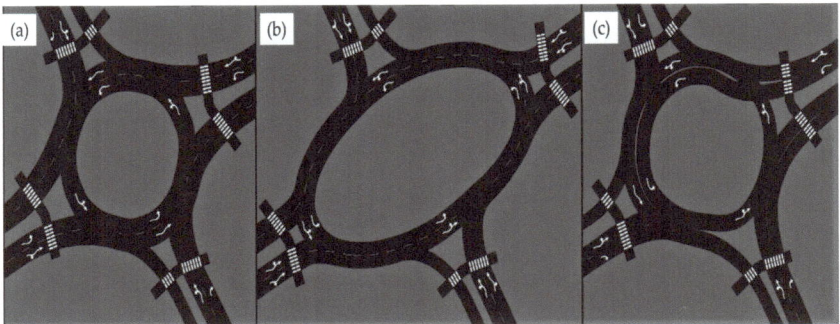

Figure 5. (**a**) Typical modern roundabout, (**b**) elliptical roundabout, and (**c**) turbo roundabout (adapted from [20]).

Before introducing the concept of a turbo roundabout, it is useful to clarify that a single-lane roundabout outperforms multi-lane ones in terms of safety because of the lower speeds. However, they fail to sustain higher traffic volumes (i.e., saturation). On the contrary, a multi-lane roundabout has a better traffic capacity, but may lack in traffic safety. Based on this contradiction, a turbo roundabout is a relatively new type, which provides a spiraling flow of traffic that forces drivers to choose their direction before entering the roundabout, thereby enhancing the levels of both safety and capacity [23,24].

The first attempt to construct a turbo roundabout was observed in the Netherlands in 2000, and it soon became so popular among other countries as well that it was followed by the development of design guidelines and recommendations during the early 2000s too [25,26]. Based on their concept, the deployment of turbo roundabouts has attracted increased research interests about the interaction between geometric design features and traffic aspects. For example, research by Dabiri et al. [23] based on microsimulation scenarios with three-legged and four-legged roundabouts, found that increasing the diameter of the central island can cause traffic congestions and delays, thereby reducing the provided level of service. On the contrary, the diameter increase was found to positively affect the performance of a turbo roundabout in terms of capacity. Table 1 provides characteristic values for typical configurations of four-legged circular roundabouts applied in the US according to [27].

Table 1. Characteristics of four-legged roundabouts (adapted from [27]).

Configuration Parameters	Mini Roundabout	Urban Single-Lane	Urban Double-Lane	Rural Single-Lane	Rural Double-Lane
Typical daily service volume (veh/day)	10,000	20,000	>100,000	20,000	>80,000
Typical inscribed circle diameter (m)	13–25	30–40	45–55	35–40	55–60
Recommended maximum entry design speed (km/h)	25	35	40	40	50
Maximum number of entering lanes	1	1	2	1	2
Splitter island configuration	Raised if possible, crosswalk cut if raised	Raised with crosswalk cut	Raised with crosswalk cut	Raised and extended with crosswalk cut	Raised and extended with crosswalk cut

Other researchers have investigated different design vehicles so that their swept paths are taken into consideration during the design of roundabouts [24]. Because of the different dimensions of vehicles that may use the roundabout and the necessity to select direction before entering the circulatory paths, the individual vehicle paths will be considered. Compared to the more conventional types of roundabouts, there is sufficient evidence on the necessity to (i) increase the width of the circulatory lanes in modern types, (ii) increase the radii of the entry and exit paths, and (iii) alter the positioning of the separator island [28,29]. Research on their design principles is still ongoing.

3. Road Safety at Roundabouts

Once properly designed and placed within a road network, roundabouts enclose many contributions compared to signalized intersections. The most critical component of a road network is to be able to sustain a certain amount of vehicle flows (i.e., capacity) and ensure safe travelling of all driving vehicles (i.e., safety). There is a general agreement on the international literature that roundabouts aim at enhancing both of the aforementioned parameters [5,12,19,30]. The reason is simple: conflict points are eliminated or at least altered, compared to conventional intersections, and drivers are forced to slow down, so it becomes much easier to control their potential to engage in an incident [19]. Of course, selecting a specific type of roundabout with proper values for its geometric elements aims at achieving a balance between capacity and safety. The latter is definitely affected by the geometric design elements, the drivers' perception of danger (related to their experience and driving performance), and the condition of pavement surface to some extent [31].

3.1. Overview of Crash Occurence at Roundabouts

At a roundabout's entry locations, cars should yield to the oncoming traffic rather than completely halt [15]. As a result, there may be fewer traffic waits and a smoother transition pattern at this kind of intersection. In fact, it has been documented that turning a signalized crossroad into a roundabout, results in an 89% decrease in traffic delays and a 56% decrease in vehicle stops [32]. With respect to the safety pillar of roundabouts, Burdett et al. [33] reported a 38% reduction in fatal injury and severe crashes because of the lower vehicle speeds. In the same context, De Brabander et al. [34] reported an average rate of reduction of 34%, 30%, and 38% for the total number of injury accidents, light injury accidents, and serious injury accidents, respectively. Reduction rates of 65% for the number of fatalities and 40% for injuries have also been mentioned elsewhere [5,35].

However, the effect of roundabouts on the number of non-injury crashes is yet to be clarified [35]. The international literature agrees on an increase in the total number of low-severity crashes. Polders et al. [36] confirm that despite the contribution of roundabouts to road safety, crashes still occur. Indeed, an even conservative increase of 12% has been reported [33], but the increase rate exhibits some variability. Zubaidi et al. [2] jointly studied the impact of roundabouts on the road safety level and reported that despite the advantages of roundabouts, road crashes still occur. Noticeably, crash frequency at roundabouts is higher in the US compared to Europe and Australia.

On the one hand, severe fatalities and injuries appear to be limited, probably because of the reduced speeds of vehicles at roundabouts, but the evolutionary trend of injury crashes or damage-related (with no injury) crashes is unambiguous. The number of available lanes is critical. Mamlouk and Souliman [37] indicated that single-lane roundabouts decreased the overall rate of accidents by 18%, while double-lane roundabouts increased the accident rate by 62%. The damage rate increased by 2% and 60% for single-lane and double-lane roundabouts, respectively. Most recently, Johnson [38] also observed a significant increase in property-damage-only (PDO) crashes for multi-lane roundabouts. Therefore, the higher the number of lanes, the higher potential for light non-injury crashes.

Moving forward, studies focusing on the crash patterns at roundabouts have been performed over the past decades. In particular, Daniels et al. [39] looked at the severity of crashes at roundabouts to see what elements were most important. Data from 1491 crashes at 148 roundabouts in Flanders, Belgium, were gathered by the researchers. To evaluate the data, they employed hierarchical binomial logistic regression and logistic regression approaches. The findings indicated that a higher frequency and severity of accidents were caused by the presence of vulnerable users. Furthermore, this effect was exacerbated throughout the night by inadequate street illumination [39,40]. Polders et al. [36] investigated four dominant crash types with data from urban roundabouts in Belgium too. These include rear-end crashes, collisions with vulnerable road users, entering–circulating crashes, and single-vehicle collisions with the central island. It was found that about 80% of the crashes occurred on the entry lanes (i.e., roundabout approach area). Road users who were found to be susceptible to the risk of being involved in a serious injury crash were the cyclists and moped riders.

No matter the cause of crashes, the increase in those less serious incidents can lead to a negative public perception about roundabout benefits [33]. This is expected to affect (i) younger drivers because of lack of driving experience and reactions in a complex environment, (ii) older drivers, and (iii) pedestrians in the case of urban areas. As per the older drivers, their vulnerability lies upon the fact that an increasingly complex road network raises the demand for their adaptability. In other words, older drivers experience difficulties in regulating their operational level of driving behavior [41].

With regards to pedestrians, a random crossing at the roundabout definitively limits its capability in terms of both vehicle capacity management as well as pedestrian safety [42]. Vignali et al. [43] recognized that research about roundabout safety usually focuses on drivers and vehicle movement and, unfortunately, overlooks the importance of safety for the vulnerable users including pedestrians and bicyclists. A solution to this issue could be the improvement of infrastructure conditions, like proper pavement marking (Figure 6). Indeed, in a recent study, the potential of moving the pedestrian crossings before the entrance to the roundabout has been commented as a contributor to road safety in the case of urban roundabouts [43].

Figure 6. Entrance to a roundabout and markings for pedestrians.

The importance of proper pavement markings and improved guide signs (Figure 7) at a roundabout are also critical for better directional management of the vehicle position within the circulatory path, especially for younger and older drivers [33]. This aspect becomes even more pronounced for cases of larger roundabouts that could mimic the concept of traffic circles. Herein, the problem of limited directional information for drivers can occur. Wan et al. [44] claim that in those cases, drivers take more time to identify the exit they want to follow, thereby influencing the intersection capacity and safety.

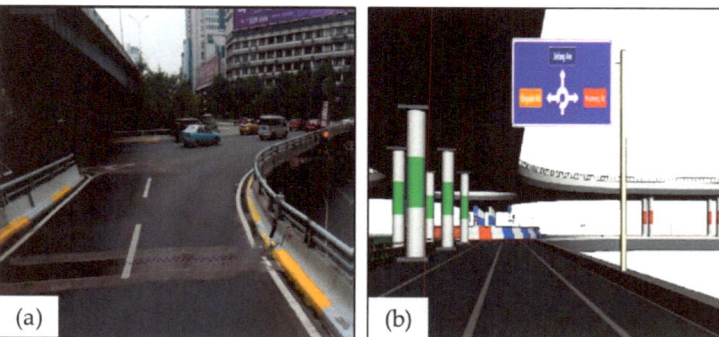

Figure 7. (**a**) View of the approach area of a roundabout, and (**b**) theoretical location for guide sign positioning in favor of improving road safety.

Following the example of Figure 7, it can be expected that properly selected positions for pedestrian crossings, together with improved guide signs, can help drivers in urban areas to accurately identify the exit they want to follow, thereby reducing the travel time needed to drive in a roundabout [44].

3.2. Interaction with Geometric Design Elements

So far, many studies on roundabouts have shown that, despite the high level of safety recognized for this type of intersection, there are several factors influencing the drivers' behavior [31]. The geometry of the roundabout is the major contributor to drivers' behavior. Design elements including entry and exit width, circulatory roadway width, entry radius, deflection angle, etc., can definitively have an impact on the way a driver adjusts its speed and driving performance during three critical types of maneuvers: (i) at the entry, (ii) at the circulatory path, and (iii) at the exit. This implies that the path of the turning vehicles is a matter in need of research in order to continuously improve the design of roundabouts. Table 2 provides a collection of roundabout geometry elements related to traffic conflict events.

Table 2. Dependency of traffic incidents on geometric design elements (adapted from [45]).

Characteristic	Type of Conflict	Contributing Factor
Radius of entry and exit approach	Run-off-road/entering–circulating/exiting–circulating	Vehicle speed or deflection angle
Inscribed circle diameter	Entering–circulating/exiting–circulating/rear-end/sideswipe	Length of weaving section/interactions between circulating and entering vehicles
Number of legs	Rear-end/entering–circulating	Increase in conflict points
Number of lanes and lane width of an approach	Exiting–circulating/rear-end/sideswipe	Increase in conflict points/distance between parallel vehicles

Furthermore, a methodology to assess and generate the variations in the vehicle paths as a result of geometric elements at circle intersections was developed in [46]. It was experimentally observed that the paths of right-turning vehicles are more sensitive to the vehicle speed and turning angle, whereas those of left-turning vehicles are additionally sensitive to the intersection corner radius.

As already known, the most frequent case of using roundabouts is within urban roads [43,47]. Anjana and Anjaneyulu [48] identified the crash causes and assessed safety performance measures for Indian urban roundabouts with the consideration of geometric design elements. They found that increasing the circulatory roadway width, exit angle, angle to the next leg, and splitter island width is associated with reduced crash rates at the roundabout approaches. Kim and Choi [49] coordinated field surveys in order to investigate the real movement of vehicles at several urban roundabouts. Their aim was to correlate the speed of vehicles for a given geometric design of a roundabout and the crash likelihood.

In another study, the importance of geometric design was also emphasized as being a crash contributing factor at urban roundabouts [50]. Factors related to the improper design of roundabouts, thus not related to the drivers' attitude and vehicle condition, were identified. The radius of deflection and the deviation angle were considered to be the most critical ones. Low entry angles force drivers into merging positions, where they must either look over their shoulder to their left or attempt a true merge using their mirrors. In the latter case, sight issues appear, as the drivers could disregard the give-way line and reach high entry speeds that contradict the road safety benefits of roundabouts. On the other hand, low values of the deflection angle contribute to failures to give way, increased pass-through speeds, and underestimations of these speeds by other vehicles being positioned in conflict points, like the subsequent approach on the right [50].

Based on an in-depth statistical analysis about the users' perception of road geometric elements, it seems that drivers prefer simple roundabout configurations, and in particular, single-lane circulatory pathways [31]. Furthermore, because of the interaction between markings, signs, and geometric design, it can be confidently stated that improving markings, i.e., complete vertical and horizontal signs, can significantly improve the road safety levels at sites where geometric design deficiencies are indeed contributors of crashes and incidents [44,50]. In other words, clear guidance can alert the drivers of the potential black spots of roundabout geometry.

Thereafter, for a given geometry of roundabout and a given set of available markings, vehicles tend to reach certain speeds. Relevant studies have demonstrated a strong relationship between the number of lanes, entry width, and exit leg speed [51,52]. Larger entry width and multi-lane roundabouts make the drivers increase their average vehicle speed at the entry legs [52]. In addition, a positive correlation has also been reported between the speed and the diameters of both the inscribed circle and the central island (recall Figure 3). Davidovic et al. [52] developed a regression model for the prediction of vehicle speed based on the radius of the circulatory lane. Of course, the speed of a vehicle may come as an additional result of the driver's perception, experience, and age, as well as the vehicle's status, like its age, the service performance, brake condition, etc. However, the purpose of the paper was mainly to investigate the interaction between speed and geometric design features.

Estimating speed as precisely as possible is an iterative process of roundabout design [51]. Once the characteristic speeds at a roundabout are known, either through measurements or model-based estimations, an improvement on the design or redesign of a roundabout can be made, as well as more accurate simulation models can be used to assess how these roundabouts affect traffic conditions and road safety levels in a road network. This guarantees that the street network can be managed more sustainably, making it possible to evaluate how each component affects travel times and traffic conditions. This is very important for a trustworthy determination of design parameters during the road planning phase and when choosing a long-term strategy for improving road safety and traffic flow management [52].

3.3. Pavement Condition

In terms of the pavement surface, the most impactful parameter is the skid resistance, or the frictional force that develops in the tire–pavement area. Provided that adequate construction quality has been achieved for pavement layers and materials [53,54], the focus is usually being put on the functionality of the pavement [55]. Less vehicle stops correspond to non-zero speeds, thereby rutting, shoving, or other severe distresses, typically observed at simple intersections, tend to be absent provided that shear-resistant asphalt mixtures are properly designed.

On the other hand, there is sufficient literature evidence that surface texture and skid resistance are considered contributing factors to traffic incidents, as they can interact with the skidding event of vehicles that affects road users' safety [56,57]. The peculiarity of roundabouts is that because of the circulatory paths, increased demand for lateral friction is required to ensure vehicle stability. However, this can be counterbalanced by low vehicle speeds occurring, especially at single-lane roundabouts. The impact of weather conditions has to also be highlighted; rainy or icy surfaces tend to reduce the provided skid resistance levels. Moreover, adverse weather conditions are known to be highly interrelated to increased accident rates that can hinder road safety.

In addition, even for a dry surface, the presence of oils or other contaminants on the pavement has been reported to cause traffic crashes at roundabouts [50]. Therefore, frequent visual inspections and/or friction measures could help preserve the condition status of the roundabout's pavement at acceptable levels. Other types of pavement-related contributing factors include the presence of surface defects, like potholes [50], that may limit the operational capabilities of the travelling vehicles.

4. Current Challenges and Prospects for Roundabouts

4.1. Environmental Implications

The environmental impact of traffic is well-known and has been growing during the recent decades, posing challenges for both the vehicle industry as well as traffic and road engineers too. Vehicular emissions are dependent on the total amount of traffic, intersection control type (e.g., signalized, roundabout, etc.), driving patterns, vehicle age, and vehicle condition [58].

The design of modern roundabouts has become dominant across many European countries in the 1980s [59]. Frequent construction activities have been observed in Europe over the last 30 years. Based on the "yield-to-entry" rule, complete vehicle stops that corresponds to abrupt decelerations and re-accelerations are limited. The longer the time of the stop, the more fuel is consumed. Thus, the required fuel is reduced during the entry to a roundabout with additional improvement in the air quality, apart from the contribution to road safety.

From this perspective, roundabouts help achieve the goals of sustainable transportation modes, according to which the environment is protected and resources are conserved by considering societal needs, benefits, and costs [60]. Ahac et al. [61] explain that the fulfillment of sustainability goals in road network planning, design, and management can be ensured through the incorporation of roundabouts in the road design network. Modern roundabouts have been commented to outperform traditional signalized intersections in terms of environmental sustainability, since a reduction is observed in the idling time as well as the rates of acceleration and deceleration that definitively contribute to a positive trend in the level of pollutant emissions and fuel consumption rates [62,63]. The level of noise pollution is also known to be reduced at the vicinity of roundabouts [61]. Reported average reduction rates of approximately 16–60% for the emissions of carbon monoxide and dioxide and a reduction of 1–4 dB in noise emission argue in favor of the sustainable potential of roundabouts [61,63,64].

However, careful environmental considerations have to be made before the decision on the type of new roundabout during the feasibility study of a new project. For example, detailed field investigations from pollutant emission measures at urban turbo roundabouts have yielded no considerable environmental improvement compared to the conventional ones [12]. Therefore, a balance between all of the individual aspects could lead to an optimized design and functionality of roundabouts. Considering environmental implications of roundabouts is definitely an open issue subject to additional research.

Finally, the aesthetic contribution of roundabouts should not be overlooked. Roundabouts, among others, are located in critical city places (i.e., with or no monuments); thus, they can also serve as a landmark in the city [65]. They can also be constructed at the boundary of two roads of different classification or areas with different functions, so that drivers are properly alerted to adjust their speed. In this context, roundabouts are considered to constitute an organizational landscape feature. Hence, beyond its basic functions, a roundabout with the appropriate central island arrangement is an aesthetic and easily identifiable place that characterizes the architecture of the local area [66].

4.2. Autonomous Vehicles and Roundabouts

The relationship between roundabouts and the autonomous driving mode, which is expected to become increasingly prevalent in the near future, is another noteworthy observation. It is important to mention that the majority of communities across the globe are currently grappling with the transition to an autonomous driving future, whereby new mobility patterns are anticipated. Truck platooning, connected autonomous vehicles (CAVs), and autonomous vehicles (AVs) are terms that both scholars and practitioners are starting to use more frequently. The scientific community, industry, and automation technologies are collaborating to improve the efficiency of the movement of people and products. The deployment of AVs has led to the development of new research studies examining modifications to road markings, lane width, roadway capacity, and pavement design elements [67–69].

Investigating the role of AVs on the status of current road infrastructure provides a unique opportunity for the transportation engineering community. Among others, the contribution of AVs on the roundabout capacity and safety have attracted a lot of research interests [11]. Autonomous driving in roundabouts requires the understanding of complex relationships between road design features, traffic rules, and the performed maneuvers of various road users [70]. According to Figure 8, in a fully autonomous vehicle driving environment, the Internet of Things will be responsible for any kind of decision maneuvers, where the driving behavior and drivers' perception will have no impact.

For the theoretical case of full AV dominance, it is currently impossible to evaluate the real performance of a roundabout against AVs in terms of safety and capacity; therefore, field tests and observations can be replaced by microscopic traffic simulations and driving simulator tests in order to gain further insights into this area. The research so far does not produce consistent remarks. Double-lane roundabouts were assessed in a study [71] through microsimulation with the Vissim software [72]. Different penetration rates of CAVs into the routine traffic flow were assumed, and it was found that for higher rates, significant benefits occur for the maximum queue length, travel times, and delays. For a fully CAV-based traffic scenario, it was claimed that the roundabout performance of the road network worsens [71]. On the other hand, Friedrich [73] reported a disproportional increase in the capacity of the road network as the share of AVs increases. Nevertheless, reaching the maximum possible speeds will become feasible, once AVs appear at a rate of 100% within traffic composition [74].

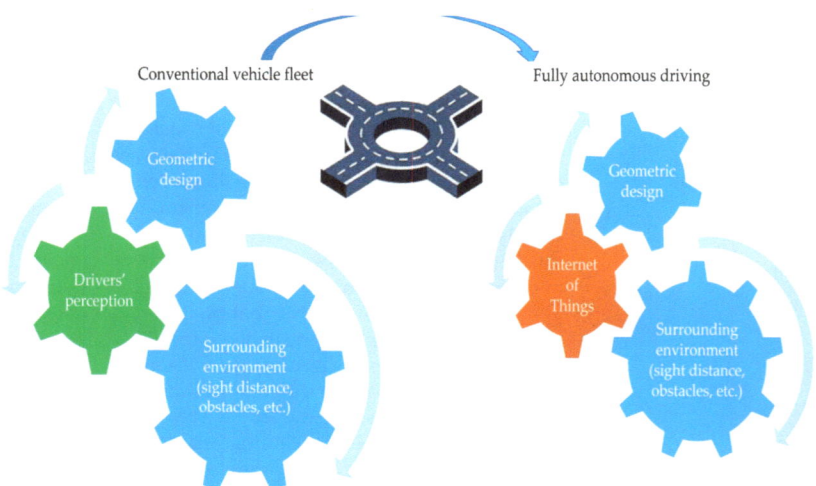

Figure 8. Pillars affecting vehicle movement at roundabouts for a conventional fleet and a fully autonomous fleet.

A path planning strategy for autonomous vehicle driving was developed by Gonzalez et al. [75] to better comprehend the patterns of AVs traveling at a roundabout. They proposed a system that generates continuous paths, dividing the driving process into three stages as follows: (a) entrance maneuver, (b) driving within the roundabout, and (c) exit maneuver. In their parametric study, they considered double-lane roundabouts, but with different exit scenarios. Their contribution was to allow for rational real-time planning, easily adjustable to any AV architecture [75].

Overall, the international literature agrees that a roundabout is safer than a traffic signalized intersection for AVs [74,76], since the progress in vehicles' sensors will help them better manage merges in different lanes of traffic. Further to this, the aspect of connectivity enables a better operational management of lateral distances, time gaps, etc., thereby increasing the traffic capacity and the quality of traffic flow at a roundabout [77]. Nevertheless, the major challenges until the full absorption of AVs into a typical traffic composition is achieved include the joint consideration of conventional vehicles with autonomous vehicles at different rates [74].

4.3. The Role of Simulation

Real-scale measurements on the vehicle performance at roundabouts does not offer the opportunity of design optimization; rather, they offer a reactive potential instead of a proactive one. At the same time, it is not feasible to strictly overview the impact of multiple features of road design (e.g., the width and radius of entry lanes, the diameter of the inscribed circle, sight distance, etc.) and assess how the drivers' response to changes in the roundabout's geometric design is affected [78]. In order to investigate how the safety and operational characteristics of the traffic will change when AVs are added, traffic micro-simulation appears to be a valuable approach. At the same time, it is necessary to explore data mining and artificial intelligence techniques to find an effective method of understanding traffic behavior at roundabouts [79,80]. This is even more important in the case of AVs, where despite the general trend of safer vehicle movements at roundabouts, there is a risk of some self-driving cars to enter a roundabout without a complete realization of the driving environment.

Using driving simulation technology appears to be an effective means to evaluate driving behavior, without taking risks as per the real driving activities [81]. Thanks to their use, it becomes feasible to evaluate the liaison between drivers and geometric design principles. This is important in order to achieve a balance between capacity and safety and maximize the performance of roundabouts. Besides, the joint effect of geometric elements is more important than their individual impacts [2]. Related research should be directed towards the accuracy improvement of safety performance models through the consideration of geometric parameters of the road design process, the use of automated video analysis for the description of traffic incidents, and the use of reliable simulators. Of course, it has to be acknowledged that the problems of considering physical limitations or obstacles, the lack of realism, the case of drivers' fatigue, as well as the validity challenges are among the major shortcomings of driving simulators [78,82].

In the same context, Alozi et al. [20] highlighted the role of simulation for a balanced design of roundabouts by jointly considering three pillars: (i) the separation of particular movements, (ii) the achievement of desirable speed profiles, and (iii) the satisfaction of geometry constraints. Neglecting traffic design elements implies that any enhancement in road safety will not necessarily be accompanied by better mobility and vice versa [20]. Hence, the authors developed a novel multi-criteria approach to simultaneously incorporate the different evaluation criteria in a meaningful way. Micro-simulation enabled them to conduct analyses in a controlled environment and assume multiple scenarios with different volumes for traffic and pedestrians. They considered modern, elliptical, and turbo roundabout design. They concluded that a turbo roundabout excels for low to medium traffic congestion as well as for the total vehicle emissions. Elliptical roundabouts were found to be more prone to incidents and safety was better only for cases of higher congestion rates.

Thanks to simulation studies, one can obtain useful implications about the speed profiles at roundabouts. In addition, maintaining suitable speeds for all vehicles while travelling in a roundabout is the most crucial design goal. However, because of the non-common consensus on the roundabout design, it is rather difficult to quantitively evaluate the effect of alternative safety measures on the resulted speed and the related control parameters. Of course, speed surveys can prove beneficial, since speed provides a link between roundabout safety and geometry [6]. However, direct observation and geometrical parameter measurements that may lead to the collection of other variables related to driving behavior do not necessarily guarantee consistent and solid remarks. Therefore, the joint analysis of using simulation analysis and real-scale supportive measures would enable an optimized assessment. To this end, robust research efforts should be targeted to ameliorate the design standards and guidelines of roads and roundabouts towards the optimization of the design parameters that have conflicting effects.

5. Conclusions

Roundabouts have been advocated by many transportation professionals as an effective alternative to conventional intersection designs. They provide a convenient solution by reducing vehicle delays and enhancing safety among other presumed benefits. The most predominant safety benefits are usually attributed to the geometry and priority rules of roundabouts, which force approaching vehicles to reduce their speeds and, subsequently, face a lower risk of collisions.

Roundabout implementation, integrated design, and proper evaluation are a necessity to achieve beneficial results. Despite this fact, limited literature exists focusing on roundabout quality evaluation (level of service versus quality of service); this is something that could be rather useful for transportation engineers and policymakers during the design stage, maintenance, or while deciding on the construction of a new roundabout [10].

The criticality of roundabouts in terms of their geometric design as well as the provided road safety lies upon the fact that roundabouts are currently used for the conventional vehicle fleet, which will be gradually replaced by new vehicle technologies. Such an action will directly impact the criteria for road network design and/or redesign, thereby continuously fostering new research initiatives. Towards this direction, the role of microsimulation studies was highlighted. Related research is ongoing aiming at shedding light on the optimized geometric design of roundabouts with an efficient traffic flow, enabling both "safety" and "capacity" potentials to become maximized, thereby offering sustainable traffic management at roundabouts.

Author Contributions: Conceptualization, K.G. and A.K.; methodology, K.G. and A.K.; literature review, K.G. and A.K.; writing—original draft preparation, K.G. and A.K.; writing—review and editing, K.G. and A.K. All authors have read and agreed to the published version of the manuscript.

Funding: This research received no external funding.

Data Availability Statement: Not applicable.

Conflicts of Interest: The authors declare no conflicts of interest.

References

1. Wolrd Health Organization (WHO). *Global Status Report on Road Safety*; Wolrd Health Organization (WHO): Geneva, Switzerland, 2018. Available online: https://www.who.int/publications/i/item/9789241565684 (accessed on 3 January 2024).
2. Zubaidi, H.A.; Jason, C.A.; Salvador, H. Understanding roundabout safety through the application of advanced econometric techniques. *Int. J. Transp. Sci. Technol.* **2020**, *9*, 309–321. [CrossRef]
3. Haleem, K.; Abdel-Aty, M. Examining traffic crash injury severity at unsignalized intersections. *J. Saf. Res.* **2010**, *41*, 347–357. [CrossRef] [PubMed]
4. Jamal, A.; Tauhidur Rahman, M.; Al-Ahmadi, H.M.; Ullah, I.; Zahid, M. Intelligent Intersection Control for Delay Optimization: Using Meta-Heuristic Search Algorithms. *Sustainability* **2020**, *12*, 1896. [CrossRef]
5. Evlik, R. Road safety effects of roundabouts: A meta-analysis. *Accid. Anal. Prev.* **2017**, *99*, 364–371.
6. Ambros, J.; Novák, J.; Borsos, A.; Hóz, E.; Kiec, M.; Machciník, S.; Ondrejka, R. Central European comparative study of traffic safety on roundabouts. *Transp. Res. Procedia* **2016**, *14*, 4200–4208. [CrossRef]
7. Han, I. Safety analysis of roundabouts and avoidance of conflicts for intersection-advanced driver assistance systems. *Cogent Eng.* **2022**, *9*, 2112813. [CrossRef]
8. National Cooperative Highway Research Program (NCHRP). *Roundabouts: An Information Guide*, 2nd ed.; Chapter 4 Operation, Chapter 6 Geometric Design; US Department of Transportation: Washington, DC, USA, 2010.
9. Nikiforiadis, A.; Nikiforiadis, A.; Mitropoulos, L.; Basbas, S.; Campisi, T. International Design Practices for Roundabouts. In *Computational Science and Its Applications–ICCSA 2023 Workshops. ICCSA 2023*; Gervasi, O., Murgante, B., Rocha, A.M.A.C., Garau, C., Scorza, F., Karaca, Y., Torre, C.M., Eds.; Lecture Notes in Computer Science; Springer: Cham, Switzerland, 2023; Volume 14111, pp. 308–326. [CrossRef]
10. Damaskou, E.; Kehagia, F.; Karagiotas, I.; Anagnostopoulos, A.; Pitsiava-Latinopoulou, M. Driver's Perceived Satisfaction at Urban Roundabouts—A Structural Equation-Modelig Approach. *Future Transp.* **2022**, *2*, 675–687. [CrossRef]
11. Boualam, O.; Borsos, A.; Koren, C.; Nagy, V. Impact of Autonomous Vehicles on Roundabout Capacity. *Sustainability* **2022**, *14*, 2203. [CrossRef]
12. Fernandes, P.; Salamati, K.; Rouphail, N.M.; Coelho, M.C. Identification of emission hotspots in roundabouts corridors. *Transp. Res. Part D Transp. Environ.* **2015**, *37*, 48–64. [CrossRef]
13. Tollazzi, T. *Alternative Types of Roundabouts. An Information Guide*; Springer Tracts on Transportation and Traffic; Roess, R.P., Ed.; Springer: New York, NY, USA, 2015; Volume 6, pp. 40–47.
14. Giuffrè, T.; Trubia, S.; Canale, A.; Persaud, B. Using Microsimulation to Evaluate Safety and Operational Implications of Newer Roundabout Layouts for European Road Networks. *Sustainability* **2017**, *9*, 2084. [CrossRef]
15. Wang, C.; Wang, Y.; Peeta, S. Cooperative Roundabout Control Strategy for Connected and Autonomous Vehicles. *Appl. Sci.* **2022**, *12*, 12678. [CrossRef]
16. Wang, W.; Yang, X. Research on Capacity of Roundabouts in Beijing. *Procedia-Soc. Behav. Sci.* **2012**, *43*, 157–168. [CrossRef]
17. Vichovaa, K.; Heinzovaa, R.; Dvoraceka, R.; Tomastika, M. Optimization of Traffic Situation Using Roundabouts. *Transp. Res. Procedia* **2021**, *55*, 1244–1250. [CrossRef]

18. Macioszek, E. Roundabout Entry Capacity Calculation—A Case Study Based on Roundabouts in Tokyo, Japan, and Tokyo Surroundings. *Sustainability* **2020**, *12*, 1533. [CrossRef]
19. Damaskou, E.; Kehagia, F. Quality of service (QOS) of Urban roundabouts: A literature review. *Int. J. Transp. Syst.* **2017**, *2*, 37–45.
20. Alozi, A.R.; Hussein, M. Multi-criteria comparative assessment of unconventional roundabout designs. *Int. J. Transp. Sci. Technol.* **2022**, *11*, 158–173. [CrossRef]
21. Mohamed, A.I.Z.; Ci, Y.; Tan, Y. A novel methodology for estimating the capacity and level of service for the new mega elliptical roundabout intersection. *J. Adv. Transp.* **2020**, *2020*, 8467152. [CrossRef]
22. Ahac, S.; Ahac, M.; Majstorović, I.; Bašić, S. Speed Reduction Capabilities of Two-Geometry Roundabouts. *Appl. Sci.* **2023**, *13*, 11816. [CrossRef]
23. Dabiri, A.R.; Aghayan, I.; Hadadi, F. A comparative analysis of the performance of turbo roundabouts based on geometric characteristics and traffic scenarios. *Transp. Lett. Int. J. Transp. Res.* **2021**, *13*, 674–685. [CrossRef]
24. Sołowczuk, A.B.; Benedysiuk, W. Design of Turbo-Roundabouts Based on the Rules of Vehicle Movement Geometry on Curvilinear Approaches. *Sustainability* **2023**, *15*, 13882. [CrossRef]
25. CROW. *Turborotondes*; Publication No. 257; CROW: The Hague, The Netherlands, 2008. (In Dutch)
26. Overkamp, D.P.; van der Wijk, W. *Roundabouts—Application and Design—A Practical Manual*; Royal Haskoning DHV, Dutch Ministry of Transport, Public Works and Water Management, Partners for Roads: Hague, The Netherlands, 2009.
27. Robinson, B.W.; Rodegerdts, L.; Scarborough, W.; Kittelson, W.; Troutbeck, R.; Brilon, W.; Bondzio, L.; Courage, K.; Kyte, M.; Mason, J.; et al. *Roundabouts: An informational guide*. Federal Highway Administration; FHWA-RD-00-67; Turner-Fairbank Highway Research Center: McLean, VA, USA, 2000.
28. Chan, S.; Livingston, R. Design vehicle's influence to the geometric design of turbo-roundabouts. In Proceedings of the International Roundabout Conference, Seattle, WA, USA, 7–10 April 2014; pp. 1–17.
29. Guerrieri, M.; Ticali, D.; Corriere, F. Turbo roundabouts: Geometric design parameters and performance analysis. *GSTF J. Comput.* **2012**, *2*, 227–232.
30. Mathew, S.; Dhamaniya, A.; Arkatkar, S.S.; Joshi, G. Roundabout Capacity in Heterogeneous Traffic Condition: Modification of HCM Equation and Calibration. *Transp. Res. Procedia* **2017**, *27*, 985–992. [CrossRef]
31. Distefano, N.; Leonardi, S.; Pulvirenti, G. Factors with the greatest influence on drivers' judgment of roundabouts safety. An analysis based on web survey in Italy. *IATSS Res.* **2018**, *42*, 265–273. [CrossRef]
32. Retting, R.A.; Mandavilli, S.; McCartt, A.T.; Russell, E.R. Roundabouts, Traffic Flow and Public Opinion. *Traffic Eng. Control* **2006**, *47*, 268–272.
33. Burdett, B.; Alsghan, I.; Chiu, L.H.; Bill, A.R.; Noyce, D.A. Analysis of Rear-End Collisions at Roundabout Approaches. *Transp. Res. Rec.* **2016**, *2585*, 29–38. [CrossRef]
34. De Brabander, B.; Nuyts, E.; Vereeck, L. Road safety effects of roundabouts in Flanders. *J. Saf. Res.* **2005**, *36*, 289–296. [CrossRef]
35. Leich, A.; Fuchs, J.; Srinivas, G.; Niemeijer, J.; Wagner, P. Traffic Safety at German Roundabouts—A Replication Study. *Safety* **2022**, *8*, 50. [CrossRef]
36. Polders, E.; Daniels, S.; Casters, W.; Brijs, T. Identifying Crash Patterns on Roundabouts. *Traffic Inj. Prev.* **2015**, *16*, 202–207. [CrossRef] [PubMed]
37. Mamlouk, M.; Souliman, B. Effect of traffic roundabouts on accident rate and severity in Arizona. *J. Transp. Saf. Secur.* **2019**, *11*, 430–442. [CrossRef]
38. Johnson, M.T. Effects of Phi and View Angle Geometric Principles on Safety of Multi-Lane Roundabouts. *Transp. Res. Rec.* **2023**, *2677*, 362–371. [CrossRef]
39. Daniels, S.; Brijs, T.; Nuyts, E.; Wets, G. Externality of risk and crash severity at roundabouts. *Accid. Anal. Prev.* **2010**, *42*, 1966–1973. [CrossRef] [PubMed]
40. Daniels, S.T.; Brijs, T.; Nuyts, E.; Wets, G. Explaining variation in safety performance of roundabouts. *Accid. Anal. Prev.* **2010**, *42*, 393–402. [CrossRef]
41. Sun, Q.; Xia, J.; Foster, J.; Falkmer, T.; Lee, H. Unpacking older drivers' mobility at roundabouts: Their visual-motor coordination through driver–vehicle–environment interactions. *Int. J. Sustain. Transp.* **2019**, *13*, 627–638.
42. Vijayawargiya, V.; Rokade, S. Identification of factors affecting pedestrian level of service of crosswalks at roundabouts. *Int. Res. J. Eng. Technol.* **2017**, *4*, 342–346.
43. Vignali, V.; Pazzini, M.; Ghasemi, N.; Lantieri, C.; Simone, A.; Dondi, G. The safety and conspicuity of pedestrian crossing at roundabouts: The effect of median refuge island and zebra markings. *Transp. Res. Part F* **2020**, *68*, 94–104. [CrossRef]
44. Wan, H.; Chen, X.; Du, Z. Improving Safety and Efficiency of Roundabouts Through an Integrated System of Guide Signs. *Sustainability* **2019**, *11*, 5202. [CrossRef]
45. Li, L.; Zhang, Z.; Xu, Z.-G.; Yang, W.-C.; Lu, Q.-C. The role of traffic conflicts in roundabout safety evaluation: A review. *Accid. Anal. Prev.* **2024**, *196*, 107430. [CrossRef]
46. Alhajyaseen, W.K.M.; Asano, M.; Nakamura, H.; Minh Tan, D. Stochastic approach for modeling the effects of intersection geometry on turning vehicle paths. *Transp. Res. Part C Emerg. Technol.* **2013**, *32*, 179–192. [CrossRef]
47. Hydén, C.; Várhelyi, A. The effects on safety, time consumption and environment of large-scale use of roundabouts in an urban area: A case study. *Accid. Anal. Prev.* **2000**, *32*, 11–23. [CrossRef]

48. Anjana, S.; Anjaneyulu, M.V.L.R. Development of safety performance measures for urban roundabouts in India. *J. Transp. Eng.* **2015**, *141*, 1. [CrossRef]
49. Kim, S.; Choi, J. Safety analysis of roundabout designs based on geometric and speed characteristics. *KSCE J. Civ. Eng.* **2013**, *17*, 1446–1454. [CrossRef]
50. Montella, A. Identifying crash contributory factors at urban roundabouts and using association rules to explore their relationships to different crash types. *Accid. Anal. Prev.* **2011**, *43*, 1451–1463. [CrossRef]
51. Surdonja, S.; Dragcevic, V.; Deluka Tibljaš, A. Analyses of maximum-speed path definition at single-lane roundabouts. *J. Traffic Transp. Eng.* **2018**, *5*, 83–95. [CrossRef]
52. Davidović, S.; Bogdanović, V.; Garunović, N.; Papić, Z.; Pamučar, D. Research on Speeds at Roundabouts for the Needs of Sustainable Traffic Management. *Sustainability* **2021**, *13*, 399. [CrossRef]
53. Loizos, A.; Spiliopoulos, K.; Cliatt, B.; Gkyrtis, K. Structural pavement responses using nonlinear finite element analysis of unbound materials. In Proceedings of the 10th International Conference on Bearing Capacity of Roads, Railways and Airfields (BCRRA), Athens, Greece, 28–30 June 2017; pp. 1343–1350.
54. Gkyrtis, K. Pavement Analysis with the Consideration of Unbound Granular Material Nonlinearity. *Designs* **2023**, *7*, 142. [CrossRef]
55. Plati, C.; Gkyrtis, K.; Loizos, A. A Practice-Based Approach to Diagnose Pavement Roughness Problems. *Int. J. Civ. Eng.* **2024**, *22*, 453–465. [CrossRef]
56. Pomoni, M.; Plati, C. Skid Resistance Performance of Asphalt Mixtures Containing Recycled Pavement Materials under Simulated Weather Conditions. *Recycling* **2022**, *7*, 47. [CrossRef]
57. Plati, C.; Pomoni, M.; Stergiou, T. From Mean Texture Depth to Mean Profile Depth: Exploring possibilities. In Proceedings of the 7th International Conference on Bituminous Mixtures and Pavements (ICONFBMP), Thessaloniki, Greece, 12–14 June 2019; pp. 639–644. [CrossRef]
58. Mandavilli, S.; Rys, M.J.; Russell, E.R. Environmental impact of modern roundabouts. *Int. J. Ind. Ergon.* **2008**, *38*, 135–142. [CrossRef]
59. Ahac, S.; Dragčević, V. Geometric Design of Suburban Roundabouts. *Encyclopedia* **2021**, *1*, 720–743. [CrossRef]
60. Ariniello, A.; Przybyl, B. Roundabouts and Sustainable Design. In Proceedings of the Green Streets and Highways 2010: An Interactive Conference on the State of the Art and How to Achieve Sustainable Outcomes, Denver, CO, USA, 14–17 November 2010; Weinstein, N., Ed.; American Society of Civil Engineers (ASCE): Reston, VA, USA, 2010.
61. Ahac, S.; Ahac, M.; Domitrović, J.; Dragčević, V. Modeling the Influence of Roundabout Deflection on Its Efficiency as a Noise Abatement Measure. *Sustainability* **2021**, *13*, 5407. [CrossRef]
62. Granà, A.; Giuffrè, T.; Guerrieri, M. Exploring Effects of Area-Wide Traffic Calming Measures on Urban Road Sustainable Safety. *J. Sustain. Dev.* **2010**, *3*, 38–49. [CrossRef]
63. Guerrieri, M.; Corriere, F.; Casto, B.L.; Rizzo, G. A model for evaluating the environmental and functional benefits of "innovative" roundabouts. *Transp. Res. Part D* **2015**, *39*, 1–16. [CrossRef]
64. Distefano, N.; Leonardi, S. Experimental investigation of the effect of roundabouts on noise emission level from motor vehicles. *Noise Control Eng. J.* **2019**, *67*, 282–294. [CrossRef]
65. Macioszek, E. Roundabouts as aesthetic road solutions for organizing landscapes. *Sci. J. Silesian Univ. Technol. Ser. Transport.* **2022**, *115*, 53–62. [CrossRef]
66. Tumminello, M.L.; Macioszek, E.; Granà, A.; Giuffrè, T. Evaluating Traffic-Calming-Based Urban Road Design Solutions Featuring Cooperative Driving Technologies in Energy Efficiency Transition for Smart Cities. *Energies* **2023**, *16*, 7325. [CrossRef]
67. Rana, M.M.; Hossain, K. Impact of autonomous truck implementation: Rutting and highway safety perspectives. *Road Mater. Pavement Des.* **2022**, *23*, 2205–2226. [CrossRef]
68. Machiani, S.G.; Ahmadi, M.; Musial, W.; Katthe, A.; Melendez, B.; Jahangiri, A. Implications of a Narrow Automated Vehicle-Exclusive Lane on Interstate 15 Express Lanes. *J. Adv. Transp.* **2021**, *2021*, 6617205.
69. Okte, E.; Al-Qadi, I.L. Impact of Autonomous and Human-Driven Trucks on Flexible Pavement Design. *Transp. Res. Record.* **2022**, *2676*, 144–160. [CrossRef]
70. Keler, A.; Malcolm, P.; Grigoropoulos, G.; Hosseini, S.A.; Kaths, H.; Busch, F.; Bogenberger, K. Data-Driven Scenario Specification for AV-VRU Interactions at Urban Roundabouts. *Sustainability* **2021**, *13*, 8281. [CrossRef]
71. Anagnostopoulos, A.; Kehagia, F. CAVs and roundabouts: Research on traffic impacts and design elements. *Transp. Res. Procedia* **2020**, *49*, 83–94. [CrossRef]
72. Giuffre, O.; Grana, A.; Tumminello, M.L.; Giuffre, T.; Trubia, S. Surrogate Measures of Safety at Roundabouts in AIMSUN and VISSIM Environment. In *Roundabouts as Safe and Modern Solutions in Transport Networks and Systems. TSTP 2018*; Macioszek, E., Akcelik, R., Sierpinski, G., Eds.; Lecture Notes in Networks and Systems; Springer: Cham, Switzerland, 2019; Volume 52, pp. 53–64.
73. Friedrich, B. The Effect of Autonomous Vehicles on Traffic. In *Autonomous Driving; Technical, Legal and Social Aspects*; Springer: Berlin, Germany, 2016; pp. 317–334.
74. Deluka Tibljaš, A.; Giuffrè, T.; Surdonja, S.; Trubia, S. Introduction of Autonomous Vehicles: Roundabouts Design and Safety Performance Evaluation. *Sustainability* **2018**, *10*, 1060. [CrossRef]

75. González, D.; Pérez, J.; Milanés, V. Parametric-based path generation for automated vehicles at roundabouts. *Expert Syst. Appl.* **2017**, *71*, 332–341. [CrossRef]
76. Gill, V.; Kirk, B.; Godsmark, P.; Flemming, B. *Automated Vehicles: The Coming of the Next Disruptive Technology*; The Conference Board of Canada: Ottawa, ON, Canada, 2015.
77. Johnson, C. *Readiness of the Road Network for Connected and Autonomous Vehicles*; RAC Foundation: London, UK, 2017; pp. 16–17.
78. Montella, A.; Aria, M.; D'Ambrosio, A.; Galante, F.; Mauriello, F.; Pernetti, M. Simulator evaluation of drivers' speed, deceleration and lateral position at rural intersections in relation to different perceptual cues. *Accid. Anal. Prev.* **2011**, *43*, 2072–2084. [CrossRef] [PubMed]
79. Wei, S.; Shen, X.; Shao, M.; Sun, L. Applying Data Mining Approaches for Analyzing Hazardous Materials Transportation Accidents on Different Types of Roads. *Sustainability* **2021**, *13*, 12773. [CrossRef]
80. García Cuenca, L.; Sanchez-Soriano, J.; Puertas, E.; Fernandez Andrés, J.; Aliane, N. Machine Learning Techniques for Undertaking Roundabouts in Autonomous Driving. *Sensors* **2019**, *19*, 2386. [CrossRef]
81. Ibanez, G.; Meuser, T.; Lopez-Carmona, M.A.; Lopez-Pajares, D. Synchronous Roundabouts with Rotating Priority Sectors (SYROPS): High Capacity and Safety for Conventional and Autonomous Vehicles. *Electronics* **2020**, *9*, 1726. [CrossRef]
82. Montella, A.; Aria, M.; D'Ambrosio, A.; Galante, F.; Mauriello, F.; Pernetti, M. Perceptual measures to influence operating speeds and reduce crashes at rural intersections: Driving simulator experiment. *Transp. Res. Rec.* **2010**, *2149*, 11–20. [CrossRef]

Disclaimer/Publisher's Note: The statements, opinions and data contained in all publications are solely those of the individual author(s) and contributor(s) and not of MDPI and/or the editor(s). MDPI and/or the editor(s) disclaim responsibility for any injury to people or property resulting from any ideas, methods, instructions or products referred to in the content.

Article

Driving Standardization in Infrastructure Monitoring: A Role for Connected Vehicles

Raj Bridgelall

Transportation, Logistics, & Finance, College of Business, North Dakota State University, P.O. Box 6050, Fargo, ND 58108-6050, USA; raj@bridgelall.com

Abstract: This study tackles the urgent need for efficient condition monitoring of road and rail infrastructure, which is integral to a nation's economic vitality. Traditional methods proved both costly and inadequate, resulting in network gaps and accelerated infrastructure decay. Employing connected vehicles with integrated sensors and cloud computing capabilities can provide a cost-effective, sustainable solution for comprehensive infrastructure monitoring. In advocating for international standardization, this study furnishes compelling evidence—encompassing trends in transportation, economics, and patent landscapes—that underscores the necessity and advantages of such standards. The analysis confirmed that trucks and rail will remain dominant in freight transport as infrastructure limitations intensify. A noteworthy finding is the absence of patented solutions in this domain, which simplifies the path toward global standardization. By integrating data from diverse sources, agencies can optimize maintenance triggers and allocate funds more strategically, thus preserving vital transportation networks. These insights not only offer an effective alternative to current practices but also have the potential to influence policymaking and industry standards for infrastructure monitoring.

Keywords: cloud-based computing; international standardization; maintenance optimization; patent analysis; sensor fusion; transportation economics

Citation: Bridgelall, R. Driving Standardization in Infrastructure Monitoring: A Role for Connected Vehicles. *Vehicles* **2023**, *5*, 1878–1891. https://doi.org/10.3390/vehicles5040101

Academic Editors: Deogratias Eustace, Bhaven Naik, Heng Wei and Parth Bhavsar

Received: 5 October 2023
Revised: 12 December 2023
Accepted: 16 December 2023
Published: 18 December 2023

Copyright: © 2023 by the author. Licensee MDPI, Basel, Switzerland. This article is an open access article distributed under the terms and conditions of the Creative Commons Attribution (CC BY) license (https://creativecommons.org/licenses/by/4.0/).

1. Introduction

A nation's economic health hinges on the ability of its transportation systems to support the movement of people and goods in a safe, reliable, and timely manner. The transportation system, however, presents complexity due to its vast and open nature. The U.S. multimodal system, for example, comprises at least four million miles of public road, at least 600,000 bridges, more than 92 thousand miles of rail, pipelines spanning more than two million miles, at least 25 thousand miles of navigable waterways, 185 container ports, and almost 20,000 airports [1]. Such characteristics present significant challenges in monitoring the condition and preserving the health of the infrastructure. This paper aims to address this critical issue.

Current monitoring methods suffer from manual operation, inconsistency, bias, and safety risks because they require human inspectors in the field. These methods thus demand extensive time and labor. Particularly in rural and tribal areas, skilled labor remains scarce. Critical issues include worker safety and data consistency. Manual surveys produce assessments that contain human bias.

The goal of this study is to advocate for the widespread use of connected vehicles (CVs) to automate road and rail condition monitoring. The author undertook comprehensive research and writing, including a doctoral dissertation on the topic in 2015 [2]. Despite this, industry adoption of the approach lags and there have been no standards developed to prescribe its use in CVs. Instead, a plethora of academic studies exist that evaluated the use of smartphones, all reporting limited success due to high variability and uncontrollable scenarios of both the devices and vehicles [3]. The merit of employing CVs lies in leveraging the standardized sensors, computing capacity, and wireless networks to exchange data in micro clouds and to enable remote processing by artificial intelligence and other advanced data analytics. A standard CV-based approach will expand spatial coverage, reduce inspection costs, and limit the exposure of practitioners to risky situations. International standardization will promote the widespread implementation of automation in preserving the surface transportation infrastructure while enhancing safety, reducing costs, and offsetting labor shortages. In advocating for the development of international standards, the author provided evidence of the need and potential benefits by highlighting trends in transportation, economic, and patenting activity.

The organization of the rest of this paper is as follows: Section 2 conducts a literature review of the motivations, current methods, and methods using smartphones and connected vehicles. Section 3 presents the methodology to mine transportation, economic, and patent data to provide evidence of the need and potential benefits. Section 4 presents the results of mining the various datasets. Section 5 discusses use cases, their potential benefits, and limitations of the study. Section 5 concludes the research and suggests future extensions of the research.

2. Literature Review

The literature addressing the impacts of poor transport infrastructure condition is extremely broad. Figure 1 illustrates the author's perspective on how characteristics of transportation relate to infrastructure supply and the broader impacts in achieving national objectives.

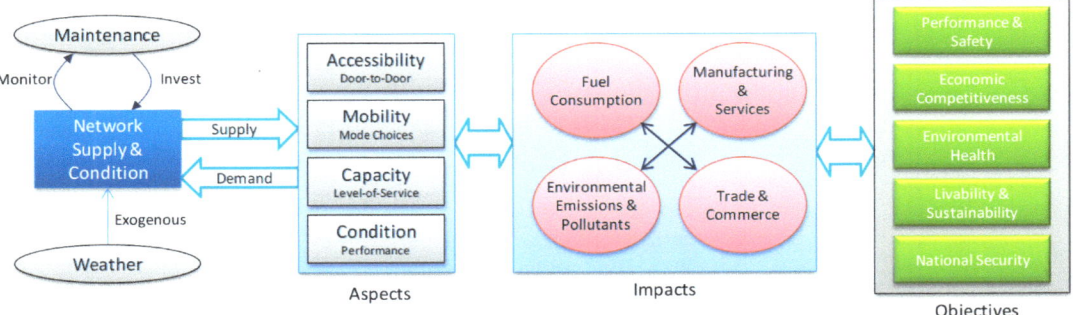

Figure 1. Broader impacts of poor transport infrastructure condition.

The figure illustrates that weather influences surface transportation condition whereas demand increases loading. While there are differences in infrastructure capabilities and utilization around the world, both loading and weather are common factors that cause deterioration, requiring discovery for remediation to maintain a stable supply. Important measures of transportation performance include accessibility, mobility, capacity, and condition, all of which depend on supply and can induce demand. These transportation aspects affect the environment, fuel consumption, manufacturing and services, and trade. Such impacts have implications to achieving national objectives such as economic competitiveness, sustainability, safety, and security. The next three subsections explore and report on the narrower recent literature covering motivations, current methods, and emerging methods of road and rail condition monitoring.

2.1. Motivations

With the expansion of the economy leading to increased traffic load density, the resulting surge in heavy vehicle traffic will intensify the strain on infrastructure, hastening its deterioration. Autonomous trucks emerged to enhance supply chain performance by reducing the impacts of driver cost and labor shortages [4]. Such developments have the potential to induce the mode shift to trucks as both cost and transport time decrease, consequently increasing the load on pavements and bridges [5].

Studies revealed that timely preventative maintenance extends the lifecycle of pavement [6]. Initial studies have indicated that a successful preservation program, capable of extending the lifespan of pavement by two-fold, can lead to cost savings for states, thrice exceeding the expense of reconstruction [7]. In a similar vein, a well-optimized track maintenance strategy can significantly lower the likelihood of train derailments [8]. However, the efficacy of such preventative maintenance programs is inherently tied to both the volume and the caliber of the data gathered for monitoring the condition of the infrastructure.

The American Society of Civil Engineers (ASCE) has consistently given the condition rating of roads in the United States ranging from D− to D+ [9]. Even when traffic is light, deteriorated road surfaces reduce the overall capacity, as vehicles are unable to travel safely at designated speed limits. A recent study has calculated that the cost incurred by U.S. drivers due to travelling on roads requiring repair amounts of an average of $621 per year in additional vehicle maintenance and operation expenses [10]. Furthermore, rough road surfaces can result in loss of vehicle control, potentially leading to accidents [11]. The consequences extend beyond just the loss of lives, encompassing damage to freight due to excessive vibrations, revenue losses from delayed deliveries to manufacturing sites, and damages to vehicles themselves.

2.2. Current Methods

Globally, transportation agencies allocate trillions of dollars for infrastructure maintenance, traditionally prioritizing the most deteriorated areas first. However, research indicates that adopting a preservation-first strategy could cut the cost of maintaining assets by at least threefold, potentially saving hundreds of billions of dollars [6]. This approach necessitates more regular monitoring to accurately determine the best times for maintenance.

Monitoring every segment of railroad networks is uncommon due to the high costs associated with manual and vehicle-based inspection methods [12]. The United States federal government require Class I railroads to conduct visual inspections of most tracks at least bi-weekly to adhere to federal safety standards. Yet, as traffic increases, so does the rate of defect formation [13], and with a trend towards reduced staffing in railroads [14], the disparity between the occurrence of defects and the ability to detect them in time to prevent accidents, delays, and financial losses is growing.

To enhance the effectiveness of visual inspections, railroads incorporate automated inspection vehicles that help identify both emerging and established defects. This method, however, requires repair teams to schedule track access and keep pace with defect detection, dependent on weather conditions [15]. Moreover, manual on-foot inspections remain essential as several types of defects are not detectable using vehicles. Consequently, railroad tracks become less available for commercial use during these inspection periods to ensure safety.

Agencies classified road pavement distress types caused by repeated traffic loading and temperature change into more than ten categories, including several types of cracks, potholes, plastic movement deformations, and raveling due to aggregate separation [16]. This complexity increases the difficulty and cost of frequent monitoring using specially instrumented vehicles and extensive visual inspections. Moreover, the shortage of skilled personnel and the requirement for fast computational resources constrain the efficiency and precision of these methods. Relying on visual inspections carried out by trained professionals also poses challenges, as it tends to be laborious, variable, and sporadic in nature.

2.3. Smartphone Methods

In 2014, the Michigan Department of Transportation anticipated that in 3 to 5 years agencies can use connected vehicle data to improve transportation asset management [17]. However, the prediction did not materialize as of 2023. Nevertheless, since then there has been a growing interest in utilizing smartphones and more recently to incorporate artificial intelligence (AI) to develop more efficient methods of road condition monitoring. Ranyal et al. (2022) reviewed the literature on AI-assisted smartphone-based road condition monitoring and found that those methods can be more accurate and faster than traditional methods [16]. The study also emphasized the challenges of obtaining and labeling large datasets to encourage widespread utilization. In a similar review of 130 articles that focused on response-based methods, Nguyen et al. (2019) concluded that the dependence on data limits data-driven methods including machine learning techniques as compared with profile reconstruction and estimation methods [18].

Yu et al. (2022) systematically reviewed 192 scholarly academic publications in the field and found that variations in data collection speed, vehicle type, smartphone specifications, and mounting configuration affected the accuracy and robustness of these methods [19]. Jeong and Jo (2023) similarly concluded that in lieu of a precisely calibrated setup, the unknown mechanical characteristics of vehicles, variable driving speed, and sensor location require expensive signal processing to address those uncertainties [3]. Yang et al. (2022) demonstrated how differences in the sensitivity of smartphone inertial sensors and their sample rate can result in measurement inconsistencies [20]. Janani et al. (2022) further demonstrated that vehicle speed variations significantly affected the accuracy of smartphone-based measurements of ride quality [21].

2.4. Connected Vehicles

Mahlberg et al. (2022) argued that despite the lack of standards for using connected vehicles to monitor pavement quality, agencies can begin to benefit from its potential for network level monitoring [22]. The authors suggested that even non-connected modern vehicles have extensive instrumentation that agencies can leverage to monitor infrastructure condition including pavement markings, signs, and pavement smoothness. In a more recent investigation, Samie et al. (2023) concluded that using CVs in a crowd-sourced manner can result in a cost-effective approach to collect pavement data for evaluation [23]. Similarly, Ruseruka et al. (2023) recently investigated the potential to leverage the built-in cameras and GPS receivers of vehicles to monitor road condition. The study found that applying You Only Look Once, Version 5 (YOLOv5) deep learning to the captured and labeled images resulted in up to 85% precision and 95% recall scores in classifying pavement distress conditions [24].

Hijji et al. (2023) proposed a federated learning framework to exploit developments in vehicle cellular communications and Convolution Neural Networks (CNN) to detect potholes [25]. The authors reported that the method achieved comparable performance with existing approaches, but it can be more computationally efficient to deploy. Hu et al. (2023) similarly utilized onboard cameras and an artificial Recurrent Neural Network (RNN) to detect and map slippery road conditions in real time [26]. The authors reported that the method achieved more than 98% prediction accuracy for icy pavement.

Drones also emerged as an alternative type of connected vehicle to implement systemwide multimodal infrastructure condition monitoring. Askarzadeh et al. (2023) recently conducted an extensive systematic literature review of drone utility in railway condition monitoring [27]. The research found that key motivations for using drones are to reduce costs, improve safety, save time, improve mobility, increase flexibility, and enhance reliability. In related work, Afsharnia and Ghavami (2023) compared smartphone-based and drone-based approaches for estimating the international roughness index (IRI), which is a standard measure of ride quality. The study found that although both methods provided comparable accuracy in IRI estimates, the smartphone method was more cost and time effective [28]. The research reviewed makes it evident that agencies have many new opportunities to use a combination of emerging techniques to enhance performance and reduce costs. However, the lack of standards can hamper adoption.

Predictive modeling approaches that use digital twin technologies have also emerged [29]. These methods create a virtual model of a physical object such as a road or rail section and link that model to its real-world counterpart in real-time. Sensors and other data sources collect real-time data about the physical object, which allows for continuous monitoring and predictive analysis of the object's state. While digital twins can provide valuable insights to help improve decision-making, developing and maintaining such models for extensive road and rail networks can be expensive due to the need for massive networks of advanced sensors and their maintenance. Furthermore, digital twins rely heavily on the quality of the data collection in the field, and any breach of these data sources could potentially compromise the security of the digital twin and the physical system it represents. Although CV technology is also susceptible to cyber-breaches, an unattended and stationary sensor in the field would be more vulnerable to an attack.

In summary, extensive research exists in the domains of infrastructure condition monitoring and CV technology. While there is significant literature on the use of CVs for various applications, there is a noticeable gap in their application specifically for efficient infrastructure condition monitoring. This research addresses that gap in the literature by compiling data evidence of the need and offering implications for policymaking, industry standards, and the future of infrastructure maintenance and safety.

3. Methodology

The methodology focuses on advocating for international standardization, supported by patterns in data from transportation, economics, and patents. The analysis of transportation and economic data highlighted gaps in the current and future demand for freight movements relative to infrastructure capacity. The patent analysis identified gaps in the practical or commercial application of connected vehicles to monitor the transportation infrastructure. Subsequent subsections detail the data mining, patent mining, and the datasets utilized.

3.1. Data Mining

Four objectives guided the data analysis. First, the study ranked the importance of freight movements by the trends in the mode share. Second, it correlated the dominant freight movement trends with economic growth. Third, it emphasized the importance of preserving and maintaining the infrastructure by identifying gaps between the demand for the dominant freight movements and their infrastructure capacity. Lastly, the study estimated the future demand for infrastructure capacity by forecasting the demand for the dominant freight movements. Figure 2 illustrates the developed data mining workflow to achieve these objectives.

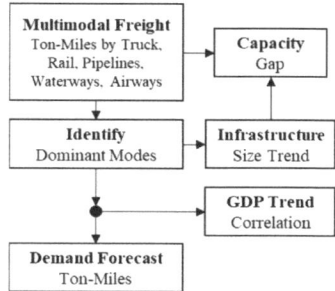

Figure 2. Data mining workflow.

3.2. Patent Mining

Figure 3 illustrates the workflow developed to analyze trends in patenting activity related to road or rail condition monitoring using connected vehicles. Three main procedures constituted the patent analysis workflow: data filtering, relevance filtering, and topic identification. Data filtering sub-procedures cleaned the text data by removing non-standard characters (non-UTF8) that downstream procedures could not recognize. Normalizing the text by lower-casing all characters increased the accuracy of automating the search for key phrases such as "connected vehicle" and "road" or "rail" in the patent summary. Patent summaries do not include all details such as patent claims. Different patents consequently displayed similar or identical summaries, which the procedure removed.

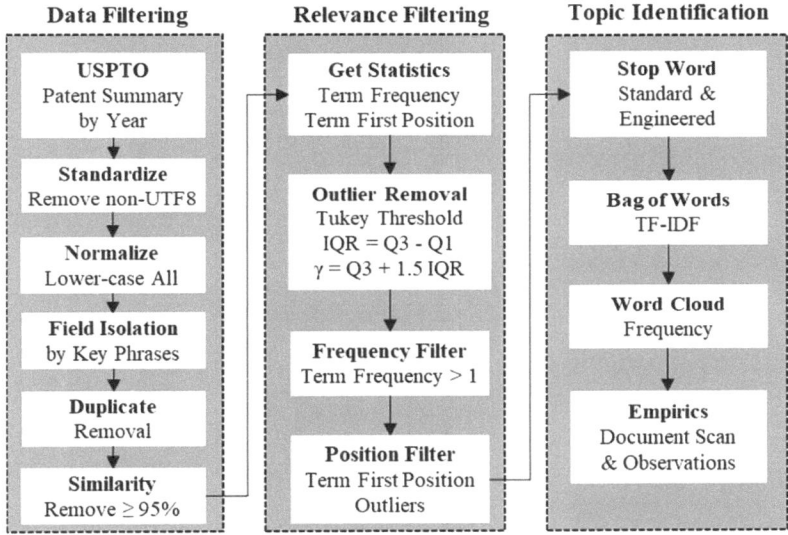

Figure 3. Workflow for patent mining.

Relevance filtering discerned and removed patent summaries that did not focus on the topic of connected vehicles. The strategy was to determine the distribution of both the frequency and first word position of a key phrase, and then remove documents that were outliers. For instance, validation confirmed that patent summaries that mentioned "connected vehicle" only once were only mildly related to the topic, so the frequency filter removed those. Similarly, patent summaries that mentioned "connected vehicle" extremely late in the description (an outlier in word position) were also only mildly related to the topic, so the position filter removed those. The Tukey threshold γ of outlier was based on the interquartile range (IQR) given by the equation shown in the figure [30].

Topic identification relied on an empirical observation of the frequency distribution of important terms across the dataset. Term importance measured by a Term-Frequency Inverse Document Frequency (TF-IDF) score $W_{t,d}$ was

$$W_{t,d} = F_{t,d}\left[\log\left(\frac{N}{D_t}\right) + 1\right]. \tag{1}$$

The variable $F_{t,d}$ is the term frequency (TF) for term t in document d, D_t is the number of patent summaries that contained term t, and N was the total number of patent summaries that were relevant [31]. The 'bag-of-words' approach is a standard method of deriving the TF-IDF distribution for a corpus of documents. The 'word cloud' method is a popular means of visualizing the TF-IDF distribution for an empirical understanding of the topics covered.

However, the results can be noisy or incoherent due to the presence of 'stop' words that are common in the English language. Standard natural language processing (NLP) libraries define stop words such as "and", "the", and "it" that occur frequently among individual documents. Although removing standard stop words increased the 'signal' in the word cloud, the remaining noise arose from words commonly used in patent description lingo. Therefore, expertise guided the selection of patent-related stop words, such as "invention", "prior art", "claim", "disclosure", "method", "apparatus", "patent", "application", and "embodiment" for removal. It was also important to remove the keywords "connected", "vehicle", and "vehicles" because they were common across all documents.

3.3. Datasets

Subject matter expertise guided the selection of publicly available datasets to enable the datamining and patent mining workflows defined above. Table 1 summarizes the datasets. The road and rail length data originated from the U.S. Bureau of Transportation Statistics (BTS). The agency maintained annual road mileage data from the U.S. Department of Transportation, Federal Highway Administration (FHWA), and the Association of American Railroads (AAR). Federal legislation and policy required the States to report such data to assess the health of the highway system for the U.S. Congress and other stakeholders [32]. The BTS also maintains multimodal freight production data by regularly collecting information from various agencies, including the FHWA, the AAR, and the U.S. Army Corps of Engineers [33].

Table 1. Summary of datasets analyzed in the data mining workflow.

Data	Description	Source
Road Infrastructure	All public road and street mileage in the 50 states and the District of Columbia from 1985 to 2021.	U.S. Bureau of Transportation Statics [34]
Rail Infrastructure	Miles of railroad tracks owned, excluding yard tracks, sidings, and parallel lines.	U.S. Bureau of Transportation Statics [34]
Multimodal Production	Multimodal freight ton-miles by truck, rail, pipeline, airways, and waterways from 2000 to 2020.	U.S. Bureau of Transportation Statics [33]
Production Forecasts	Multimodal freight movements by mode, origin, and destination regional zones in 2017 with forecasts to 2050.	U.S. Bureau of Transportation Statics [35]
eCommerce	Trends in eCommerce values from 2017 to 2027.	Statista [36]
Economic Growth	U.S. real gross domestic product (GDP)	U.S. Bureau of Economic Analysis [37]
Patents Issued	Summary of U.S. patents issued from 1776 to the present year.	U.S. Patent and Trademark Office [38]

Data on multimodal production forecast are from the BTS based on initial data from the FHWA [35]. The dataset includes 2.4 million records of origin-destination (OD) estimates of multimodal freight movement. Each record provided an origin zone, a destination zone, the commodity category, the weight in kilotons, the production in millions of ton-miles, the mode of transportation used, and the value in millions of dollars based on a 2017 valuation. The data incorporated forecasts of multimodal freight movements from 2017 through 2050.

The U.S. Bureau of Economic Analysis (BEA) provides a monthly update of economic statistics influencing decisions by government officials, businesses, and individuals [37]. Key metrics of economic growth reported include personal income, personal savings, and the growth in U.S. real gross domestic products (GDP). With respect to trends in e-commerce, Statista estimated that the compound annual growth rate (CAGR) for the U.S. e-commerce market will be 11.2% from 2022 to 2027 [36]. A series of quality checks with multiple alternative sources of the data assured its accuracy.

The U.S. Patent and Trademark Office (USPTO) maintains a comprehensive dataset of patents issued since 1776. The USPTO also maintains a large database containing summaries of the issued patents [38]. The structure of the patent summary data was simply one column containing the patent number and another column containing the summary text. One drawback of using patent summary data lies in its exclusion of information such as the patent title, inventors, companies, and specific claims. However, utilizing patent summaries offered a lower requirement for computational capacity to process the text.

4. Results

The subsections that follow discuss the results of analyzing multimodal freight movement trends, the capacity gap analysis, forecasts for freight weight moved, and the patent analysis.

4.1. Multimodal Trends

Figure 4 plots the results from data mining the multimodal freight movement data. Trucks and rail dominated with ton-mile share of 45% and 27%, respectively. The truck and rail trends have diverged since 2010. The proportion of freight moved by pipelines and waterways declined gradually after the early 1990s. The proportion of freight weight moved by air was consistently less than 1%, so the chart excludes it. These findings suggested that trucks will continue to dominate freight movements in the future.

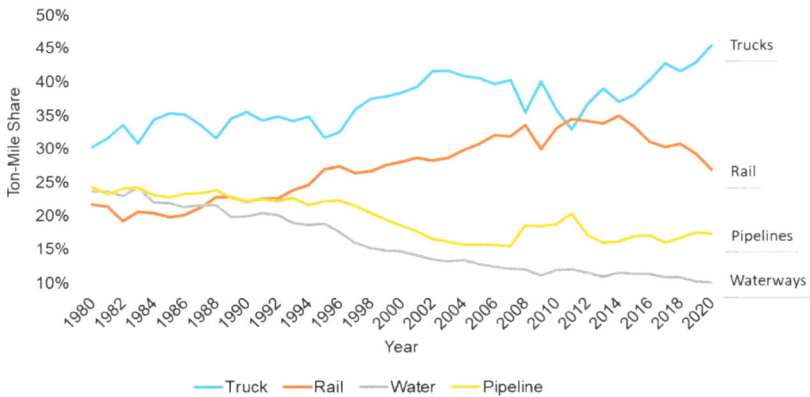

Figure 4. U.S. transportation mode share.

The analysis of co-dependency between U.S. real GDP and total ton-miles yielded a correlation coefficient of 0.88. This result indicated that there was a strong positive correlation. The compound annual growth rate (CAGR) of the U.S. GDP between 1980 and 2020 was 5.1%.

One trend in Figure 4 suggested that there had been competing mode shifts between rails and trucks since 2008 because the respective ton-mile share trends inverted. Trucks experienced a sharp increase in ton-mile share since 2011 (12.4%), whereas railroads experienced a sharp decrease (−7.6%). The trend since 2011 corresponds to an average of 2.3% increase in the U.S. GDP for nine consecutive years before declining after COVID-19. Although the trend indicates that railroads lost mode share to trucks since 2011, railroads are likely to continue leading all modes in moving the heaviest and bulkiest freight most efficiently across long distances [39].

4.2. Capacity Gap Analysis

Figure 5 plots the results from combining the freight movement and infrastructure span data. The length of highways has increased over the 30-year-span, but only by a modest 8%. This contrasts with a 46% increase in the movement of freight by trucks over the same period. Railroads, on the other hand, decreased the length of their infrastructure by

23% over the same period, even though their production increased by 39%, peaking at 79% in 2014. This suggested that railroads have been producing more with less infrastructure.

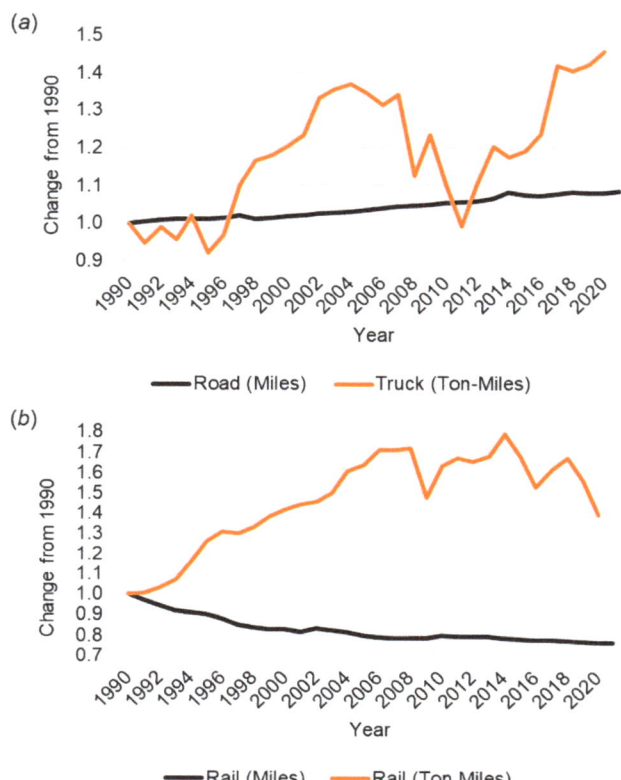

Figure 5. Traffic demand outstrips supply for both (a) trucks and (b) railroads.

The sharp decline for both modes around 2008 aligned with the global fiscal crisis. The sharp decline in rail production in 2020 aligns with a decline in demand during COVID-19 for bulk goods such as vehicles and construction material because people worked from home. However, trucks sustained their production during COVID-19 because of sustained demand for home deliveries and food items.

Overall, the most significant finding is that U.S. truck and rail traffic continued to outpace their infrastructure capacity. The data showed that short-term economic fluctuations do not sway trends in infrastructure development, which changes at a much slower pace. Therefore, in the absence of adequate investment to enhance capacity and maintain infrastructure, the disparity between availability and need will only continue to grow. Consequently, the growing demand to move more heavy loads will increase the extent and severity of infrastructure deterioration.

4.3. Production Forecasting

The sustained growth of e-commerce increased the demand for both truck and road capacity. Analysts forecasted the CAGR for the U.S. e-commerce market as 14.6% from 2022 to 2025 [36]. Increasing demands for capacity will further increase congestion levels across the multi-modal surface transportation network. Additionally, as infrastructure continues to deteriorate, it leads to reduced capacity, decreased performance, and heightened risks.

Therefore, maintaining road and rail infrastructure in optimal conditions becomes ever more crucial in the face of rising GDP and population growth.

Figure 6 shows that the weight of freight moved by both truck and rail will increase steadily from 2023 to 2050. The dip in 2020 corresponds to impacts from COVID-19. The figure shows a separate forecast for short haul (SH) and long haul (LH) movements. The analysis defined LH movements as those greater than 250 miles to be consistent with the segmentation of the data into distance bands below and above that threshold. Figure 6a shows that the weight moved by LH and SH trucks will increase by 72% and 47%, respectively, from 2017 to 2050. Figure 6b shows that the weight moved by LH and SH rail will increase 9% and 36%, respectively, from 2017 to 2050. Interestingly, for trucks, the demand for LH will outpace SH by 2050, whereas the opposite will be true for rail. This suggested that roadway agencies needed to prioritize monitoring highways that carry LH truck traffic over shorter routes, whereas railroads should prioritize monitoring SH over LH routes. The modest increase in LH movements by rail suggests that the industry has predicted a shift from LH rail to LH trucks, perhaps spurred by an expected increase autonomous truck adoption by 2035.

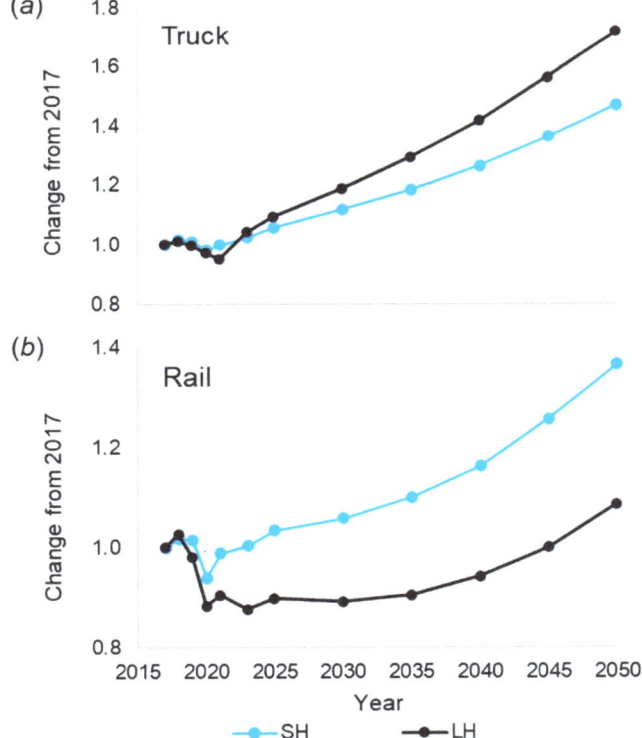

Figure 6. Forecast for (**a**) truck and (**b**) rail tons normalized to 2017.

4.4. Patent Analysis

The USPTO dataset contained 1,637,725 summaries of patents issued from 2018 to 2022. Table 2 summarizes the results from the data filtering and relevance filtering procedures of the patent mining workflow presented above. On average, patents that mentioned the term "connected vehicle" or its plural form at least once accounted for less than 0.04% of the patents awarded. The relevance filtering reduced the number of patents that contained the term "connected vehicle" or "connected vehicles" by 58.5% on average. From 2018 to 2022, inventors received an average of 45 patents annually that were relevant to connected vehicles. However, the number of those patents more than tripled from 2018 to 2022, highlighting an increasing trend of innovation in the field. A surprising finding was that none of the patents described using connected vehicles to monitor road or rail infrastructure condition.

Table 2. Summary of the data filtering and relevance filtering procedures of the patent mining.

Procedure	2022	2021	2020	2019	2018	Mean
USPTO Summaries	283,075	330,645	355,647	357,790	310,568	327,545
Isolate Patents	166	145	136	104	71	124
Duplicate Removal	164	144	135	100	64	121
Similarity Reduction	160	141	135	100	64	120
Frequency Filter	75	60	54	42	23	51
Percent Reduction	53.1%	57.4%	60.0%	58.0%	64.1%	58.5%
Term < Position	67	51	50	37	19	45
Outlier Threshold	698	488	690	811	332	604

Figure 7 shows the result of the word cloud to elicit some insights into the type of problems that companies are addressing to advance the CV technology. A subsequent empirical analysis suggested that a sizable portion of those patents discussed information networks, wireless communications, computing, vehicle micro cloud development, data storage, interactions with traffic light, and sensor data about the location of moving objects like other vehicles and pedestrian. These are all topics that one would expect a patent to cover in the field of connected vehicles. The word cloud, therefore, validated the utility and accuracy of the patent data mining workflow developed.

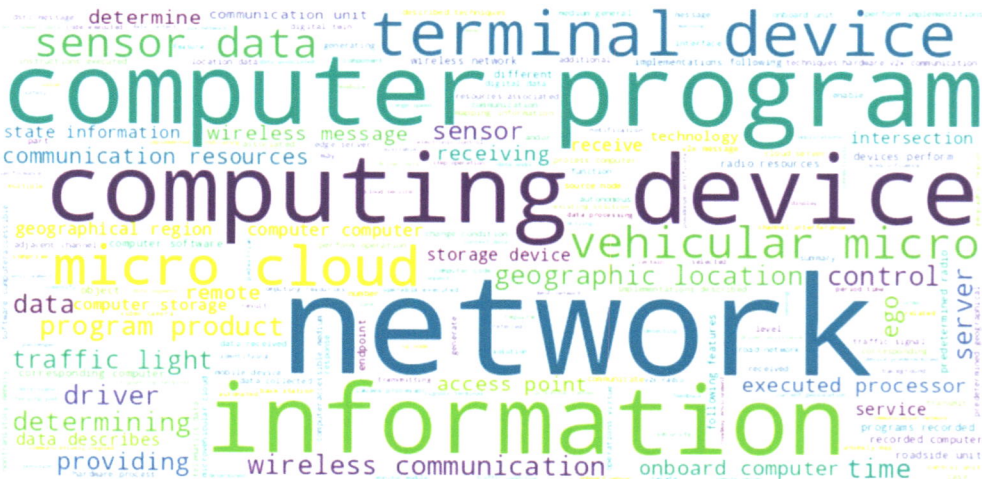

Figure 7. Word cloud from the patent analysis.

5. Discussion

Utilizing connected vehicles to gather and transmit data on ride quality and road condition offers a more stable, long-term measure of the rate at which infrastructure deteriorates [40]. Systems based on cloud computing can process data from integrated mobile sensors, providing precise evaluations of infrastructure conditions and their performance. On railroads, sensors fitted on trains operating in revenue service can pinpoint specific areas of track and train defects for subsequent remediation [15]. Models analyzing the decay of infrastructure can leverage these data to determine the best times for maintenance, thus allowing both railroad and highway authorities to efficiently schedule repairs and allocate funds. This method of ongoing monitoring will naturally adapt and expand alongside increases in traffic and infrastructural growth.

The industry gained experience in using smartphones to log and upload geospatial coordinates, roughness measurements, and roadway imagery. Hence, planners can use insights from these experiences and data analysis to promote adoption. Agencies can combine road-condition monitoring data from other sources such as environmental sensor stations, wayside sensors, weigh-in-motion (WIM) systems, and remote sensing via drones and satellites [41]. Early successes will lead more agencies to realize the benefits of using a CV approach for data acquisition. Over time, the extensive deployment of sensors and the automation of data collection will enhance the visibility and understanding of the condition of infrastructure significantly.

In review, the findings included that with a sustained GDP growth, both trucks and rail will continue to dominate freight movements and outpace the capacity of multimodal infrastructure. One interesting finding was that the demand for LH trucks will outpace SH by 2050, whereas the opposite will be true for rail. This suggested that roadway agencies can leverage the efficiency of CV technology to monitor the longest span of highways that support the heaviest load movements. These results suggest that agencies can use the data from CVs to optimize maintenance schedules and strategically allocate resources for infrastructure repair and upkeep, potentially transforming how these tasks are currently managed. Another finding was that the existing literature and practices lack a standardized methodology in this domain. A key finding was the absence of patented solutions in the field of infrastructure monitoring using connected vehicles.

One limitation of this study is that patent analysis does not necessarily indicate a lack of innovation. Several reasons exist. First, companies do not necessarily patent all inventions due to cost reasons or to protect trade secrets. Second, this study considered only U.S. issued patents, potentially missing relevant inventions filed abroad. Third, prior academic publications can discourage the filing of related patents due to fears of finding "prior art". Lastly, patents serve as lagging indicators of technology development. Nevertheless, the lack of U.S. patents for using connected vehicles to monitor road and rail infrastructure condition reduces the barrier to standardization because there will be no need to seek a licensee to implement products based on the standard. Even so, other complications may arise that include unpublished intellectual property and proprietary technology concerns. This study paves the way for future work to expand the scope to include practical implementation strategies that could include technical, financial, and legislative challenges, including the need to address privacy, cybersecurity, cloud capacity, and network reliability issues. Even with a ratified standard, there will be variations in technology adoption across different nations, which could significantly affect the broad applicability of the CV monitoring solution.

6. Conclusions

The multimodal surface transportation infrastructure is critical to the economic competitiveness of any nation. This research motivated the use of connected vehicles (CVs) to achieve a critical advancement in road infrastructure monitoring. The study highlights the growing gap in road infrastructure capacity and the increasing demand due to economic and population growth. By leveraging CVs, road agencies can benefit from real-time data collection, enabling more efficient monitoring and maintenance of road networks. This approach is not only cost-effective but also enhances the safety and longevity of roads. The absence of patented solutions in this domain could simplify the adoption and standardization process, making it a viable solution for widespread implementation. The findings will encourage policymakers and industry stakeholders to consider this proposed approach for road infrastructure monitoring, which has the potential to revolutionize current practices.

In the realm of rail infrastructure, trains with the appropriate onboard sensors can enable continuous monitoring of the entire network. This research proposes that adopting the CV approach to rail infrastructure monitoring offers a novel solution to the traditional, costly methods. This approach can significantly enhance the capability of railroads to conduct frequent and comprehensive assessments of rail health, leading to timely maintenance and improved safety. The findings highlight the lack of existing patents in this area, indicating an open field for innovation and standardization in rail infrastructure monitoring. The implications for rail infrastructure are profound, proposing a shift towards more technologically driven, efficient, and standardized monitoring methods.

Future work will examine broader trends in connected vehicle innovations by extending the data and patent mining techniques developed in this work.

Funding: This research was funded by the United States Department of Transportation under a grant to the Mountain Plains Consortium.

Data Availability Statement: The data presented in this study are openly available as cited.

Conflicts of Interest: The author declares no conflict of interest.

References

1. USDOT. *Pocket Guide to Transportation*; Bureau of Transportation Statistics (BTS): Washington, DC, USA, 2023.
2. Bridgelall, R. *Pavement Performance Evaluation Using Connected Vehicles*; North Dakota State University: Fargo, ND, USA, 2015.
3. Jeong, J.-H.; Jo, H. Toward Real-world Implementation of Deep Learning for Smartphone-Crowdsourced Pavement Condition Assessment. *IEEE Internet Things J.* **2023**. [CrossRef]
4. Zhao, J.; Lee, J.Y. Effect of Connected and Autonomous Vehicles on Supply Chain Performance. *Transp. Res. Rec. J. Transp. Res. Board* **2022**, *2677*, 402–424. [CrossRef]
5. Huang, K.; Cai, S.; Birgisson, B. Performance impact of autonomous trucks on flexible pavements: An evaluation framework and case studies. *Int. J. Pavement Eng.* **2023**, *24*, 2229481. [CrossRef]
6. FHWA. *Demonstrating the Application of Life Cycle Planning (LCP) on a Pavement Network*; Federal Highway Administration (FHWA): Washington, DC, USA, 2022.
7. NCHRP. *Optimal Timing of Pavement Preventive Maintenance Treatment Applications*; National Academies of Sciences, Engineering, and Medicine; National Cooperative Highway Research Program (NCHRP): Washington, DC, USA, 2004.
8. Bridgelall, R.; Chia, L.A.; Bhardwaj, B.; Lu, P.; Tolliver, D.D.; Dhingra, N. Enhancement of signals from connected vehicles to detect roadway and railway anomalies. *Meas. Sci. Technol.* **2019**, *31*, 035105. [CrossRef]
9. ASCE. *Report Card History*; American Society of Civil Engineers (ASCE): Reston, VA, USA, 2022. Available online: https://www.infrastructurereportcard.org/making-the-grade/report-card-history/ (accessed on 9 July 2022).
10. TRIP. *Funding America's Transportation System*; TRIP—A National Transportation Research Nonprofit: Washington, DC, USA, 2022.
11. Hu, J.; Gao, X.; Wang, R.; Sun, S. Research on Comfort and Safety Threshold of Pavement Roughness. *Transp. Res. Rec. J. Transp. Res. Board* **2017**, *2641*, 149–155. [CrossRef]
12. Li, J.; Zhou, W.; Gong, W.; Lu, Z.; Yan, H.; Wei, W.; Wang, Z.; Shen, C.; Pang, J. LiDAR-Assisted UAV Stereo Vision Detection in Railway Freight Transport Measurement. *Drones* **2022**, *6*, 367. [CrossRef]
13. Patra, A.P.; Bidhar, S.; Kumar, U. Failure prediction of rail considering rolling contact fatigue. *Int. J. Reliab. Qual. Saf. Eng.* **2010**, *17*, 167–177. [CrossRef]
14. BLS. Employment in Rail Transportation Heads Downhill between November 2018 and December 2020. October 2021. Available online: https://www.bls.gov/opub/mlr/2021/article/employment-in-rail-transportation-heads-downhill-between-november-2018-and-december-2020.htm (accessed on 9 July 2022).
15. Bridgelall, R.; Tolliver, D.D. Budgeting for the Adoption of Sensors on Connected Trains. *Transp. Plan. Technol.* **2021**, *45*, 100–116. [CrossRef]
16. Ranyal, E.; Sadhu, A.; Jain, K. Road condition monitoring using smart sensing and artificial intelligence: A Review. *Sensors* **2022**, *22*, 3044. [CrossRef]
17. Dennis, E.P.; Hong, Q.; Wallace, R.; Tansil, W.; Smith, M. Pavement condition monitoring with crowdsourced connected vehicle data. *Transp. Res. Rec. J. Transp. Res. Board* **2014**, *2460*, 31–38. [CrossRef]
18. Nguyen, T.; Lechner, B.; Wong, Y.D. Response-based methods to measure road surface irregularity: A state-of-the-art review. *Eur. Transp. Res. Rev.* **2019**, *11*, 1–18. [CrossRef]
19. Yu, Q.; Fang, Y.; Wix, R. Pavement roughness index estimation and anomaly detection using smartphones. *Autom. Constr.* **2022**, *141*, 104409. [CrossRef]
20. Yang, X.; Hu, L.; Ahmed, H.U.; Bridgelall, R.; Huang, Y. Calibration of smartphone sensors to evaluate the ride quality of paved and unpaved roads. *Int. J. Pavement Eng.* **2020**, *23*, 1529–1539. [CrossRef]

21. Janani, L.; Doley, R.; Sunitha, V.; Mathew, S. Precision enhancement of smartphone sensor-based pavement roughness estimation by standardizing host vehicle speed. *Can. J. Civ. Eng.* **2022**, *49*, 716–730. [CrossRef]
22. Mahlberg, J.A.; Li, H.; Zachrisson, B.; Leslie, D.K.; Bullock, D.M. Pavement Quality Evaluation Using Connected Vehicle Data. *Sensors* **2022**, *22*, 9109. [CrossRef] [PubMed]
23. Samie, M.; Golroo, A.; Tavakoli, D.; Fahmani, M. Potential applications of connected vehicles in pavement condition evaluation: A brief review. *Road Mater. Pavement Des.* **2023**, 1–25. [CrossRef]
24. Ruseruka, C.; Mwakalonge, J.; Comert, G.; Siuhi, S.; Perkins, J. Road Condition Monitoring Using Vehicle Built-in Cameras and GPS Sensors: A Deep Learning Approach. *Vehicles* **2023**, *5*, 931–948. [CrossRef]
25. Hijji, M.; Iqbal, R.; Pandey, A.K.; Doctor, F.; Karyotis, C.; Rajeh, W.; Alshehri, A.; Aradah, F. 6G Connected vehicle framework to support intelligent road maintenance using deep learning data fusion. *IEEE Trans. Intell. Transp. Syst.* **2023**, *24*, 7726–7735. [CrossRef]
26. Hu, J.; Huang, M.-C.; Yu, X.B. Deep learning based on connected vehicles for icing pavement detection. *AI Civ. Eng.* **2023**, *2*, 1. [CrossRef]
27. Askarzadeh, T.; Bridgelall, R.; Tolliver, D.D. Systematic Literature Review of Drone Utility in Railway Condition Monitoring. *J. Transp. Eng. Part A Syst.* **2023**, *149*, 04023041. [CrossRef]
28. Afsharnia, H.; Ghavami, S.M. Comparison of Smartphone-Based and Drone-Based Approaches for Assessing Road Roughness. *Transp. Res. Rec.* **2023**. [CrossRef]
29. Vieira, J.; Poças Martins, J.; Marques de Almeida, N.; Patrício, H.; Gomes Morgado, J. Towards resilient and sustainable rail and road networks: A systematic literature review on digital twins. *Sustainability* **2022**, *14*, 7060. [CrossRef]
30. Boslaugh, S. *Statistics in a Nutshell: A Desktop Quick Reference*; O'Reilly Media, Inc.: Sebastopol, CA, USA, 2013.
31. Aggarwal, C.C. *Data Mining*; Springer International Publishing: New York, NY, USA, 2015; p. 734.
32. FHWA. *Highway Statistics Series*; Federal Highway Administration (FHWA): Washington, DC, USA, 17 November 2022. Available online: https://www.fhwa.dot.gov/policyinformation/statistics.cfm (accessed on 2 October 2023).
33. BTS. U.S. Ton-Miles of Freight. 24 June 2023. Available online: https://www.bts.gov/content/us-ton-miles-freight (accessed on 30 September 2023).
34. BTS. *System Mileage within the United States*; United States Department of Transportation: Washington, DC, USA, 2023. Available online: https://www.bts.gov/content/system-mileage-within-united-states (accessed on 1 July 2023).
35. BTS. *Bureau of Transportation Statistics (BTS)*; United States Department of Transportation, Bureau of Transportation Statistics (BTS): Washington, DC, USA, 2023. Available online: https://www.bts.gov/faf (accessed on 1 July 2023).
36. Statista. Digital Markets eCommerce United States. August 2023. Available online: https://www.statista.com/outlook/dmo/ecommerce/united-states (accessed on 2 October 2023).
37. BEA. *U.S. Economy at a Glance*; Bureau of Economic Analysis (BEA): Washington, DC, USA, 2023. Available online: https://www.bea.gov/news/glance (accessed on 2 October 2023).
38. USPTO. *Data Download Tables*; U.S. Patent and Trademark Office (USPTO): Washington, DC, USA, 2023. Available online: https://patentsview.org/download/brf_sum_text (accessed on 2 October 2023).
39. AAR. *10 Freight Rail Fast Facts*; Association of American Railroads (AAR): Washington, DC, USA, 2022.
40. Bridgelall, R. Characterizing Ride Quality with a Composite Roughness Index. *IEEE Trans. Intell. Transp. Syst.* **2022**, *23*, 15288–15297. [CrossRef]
41. Liu, P.; Wang, Q.; Yang, G.; Li, L.; Zhang, H. Survey of Road Extraction Methods in Remote Sensing Images Based on Deep Learning. *PFG–J. Photogramm. Remote Sens. Geoinformation Sci.* **2022**, *90*, 135–159. [CrossRef]

Disclaimer/Publisher's Note: The statements, opinions and data contained in all publications are solely those of the individual author(s) and contributor(s) and not of MDPI and/or the editor(s). MDPI and/or the editor(s) disclaim responsibility for any injury to people or property resulting from any ideas, methods, instructions or products referred to in the content.

Article

Enhancing Safety Assessment of Automated Driving Systems with Key Enabling Technology Assessment Templates

Martin Skoglund [1,*], Fredrik Warg [1,†], Anders Thorsén [1,†] and Mats Bergman [2]

1 RISE—Research Institutes of Sweden, 504 62 Borås, Sweden; fredrik.warg@ri.se (F.W.); anders.thorsen@ri.se (A.T.)
2 Telia Company, 169 94 Solna, Sweden; mats.bergman@teliacompany.com
* Correspondence: martin.skoglund@ri.se; Tel.: +46-705-14-5949
† These authors contributed equally to this work.

Abstract: The emergence of Automated Driving Systems (ADSs) has transformed the landscape of safety assessment. ADSs, capable of controlling a vehicle without human intervention, represent a significant shift from traditional driver-centric approaches to vehicle safety. While traditional safety assessments rely on the assumption of a human driver in control, ADSs require a different approach that acknowledges the machine as the primary driver. Before market introduction, it is necessary to confirm the vehicle safety claimed by the manufacturer. The complexity of the systems necessitates a new comprehensive safety assessment that examines and validates the hazard identification and safety-by-design concepts and ensures that the ADS meets the relevant safety requirements throughout the vehicle lifecycle. The presented work aims to enhance the effectiveness of the assessment performed by a homologation service provider by using assessment templates based on refined requirement attributes that link to the operational design domain (ODD) and the use of Key Enabling Technologies (KETs), such as communication, positioning, and cybersecurity, in the implementation of ADSs. The refined requirement attributes can serve as safety-performance indicators to assist the evaluation of the design soundness of the ODD. The contributions of this paper are: (1) outlining a method for deriving assessment templates for use in future ADS assessments; (2) demonstrating the method by analysing three KETs with respect to such assessment templates; and (3) demonstrating the use of assessment templates on a use case, an unmanned (remotely assisted) truck in a limited ODD. By employing assessment templates tailored to the technology reliance of the identified use case, the evaluation process gained clarity through assessable attributes, assessment criteria, and functional scenarios linked to the ODD and KETs.

Keywords: safety assessment; operational design domain; automated driving; communication; connectivity; positioning; cybersecurity

Citation: Skoglund, M.; Warg, F.; Thorsén, A.; Bergman, M. Enhancing Safety Assessment of Automated Driving Systems with Key Enabling Technology Assessment Templates. *Vehicles* 2023, 5, 1818–1843. https://doi.org/10.3390/vehicles5040098

Academic Editors: Deogratias Eustace, Bhaven Naik, Heng Wei, Parth Bhavsar and Chen Lyu

Received: 30 August 2023
Revised: 10 November 2023
Accepted: 6 December 2023
Published: 13 December 2023

Copyright: © 2023 by the authors. Licensee MDPI, Basel, Switzerland. This article is an open access article distributed under the terms and conditions of the Creative Commons Attribution (CC BY) license (https://creativecommons.org/licenses/by/4.0/).

1. Introduction

The introduction of Automated Driving Systems (ADSs) has created a shift in the approach to safety assurance in the automotive industry. Contrasting with an advanced driver-assistance system (ADAS), an ADS can completely take over the driving task from the human driver for a portion of the trip [1]. Examples of ADS features include Traffic Jam Chauffeur, Highway Autopilot, Valet Parking, and Automated Truck Platooning.

Safety standards and regulation conformance form a basis for what needs to be satisfied by a vehicle before it can be commercially available. A successful fulfilment assessment, called a type approval, must be made before the market introduction of any vehicle to ensure that it is safe for use on public roads while using the new feature, e.g., Automated Lane-Keeping Systems [2].

Introducing an ADS represents a significant change in the scope of the road-vehicle approval procedures. Safety-assurance claims made by original equipment manufacturers

(OEM) must demonstrate that the ADS can operate safely in all traffic situations, including in rare circumstances such as sensor failures, cyberattacks, or environmental changes. Type approval becomes particularly important to ensure that these systems are safe and reliable to build trust and acceptance in the eyes of the public for this emerging technology. Key entities in the new type-approval process include the OEM, Homologation Authority, and Homologation Technical Service Provider, as seen in Figure 1.

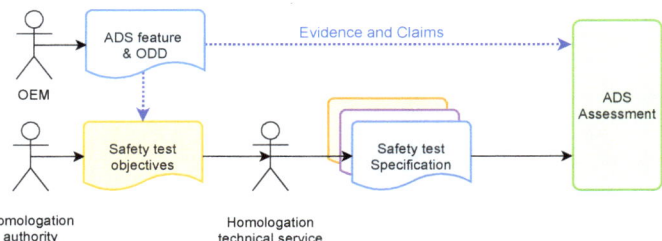

Figure 1. In the type-approval process, key entities include the OEM, Homologation Authority, and Homologation Technical Service Provider.

The OEM is responsible for designing, developing, and producing the vehicle or automotive component, seeking type approval. They ensure compliance with regulations and standards, providing necessary documentation, test reports, and technical information. The Homologation Authority is the regulatory body granting type approval. They verify compliance with regulations, assessing the safety, environmental impact, and legal requirements. They review documentation, conduct tests, and issue type-approval certificates. The Homologation Technical Service Provider is an independent organisation authorised by the Homologation Authority. They perform testing, evaluation, and certification services. Following standardised procedures, they assess the product performance, safety, and environmental characteristics that support the type-approval process.

An ADS assessment scheme must consider complex sensors, algorithms, and the vehicle's decision-making process to operate in automated mode. To meet the challenge of assessing an ADS, the United Nations Economic Commission for Europe (UNECE) World Forum for the Harmonisation of Vehicle Regulations (WP.29) drafted a "New Assessment and Test Method" (NATM) that may become part of the future type approval for ADSs.

The procedural goal of NATM is to conduct an empirical, objective, practical, and repeatable independent safety assessment of any ADS while maintaining technology neutrality. The assessment is based on high-level safety requirements [3] aiming to determine whether the vehicle can operate safely within its operational design domain (ODD) by examining scenarios linked to road users' behaviour, environmental conditions, and driver behaviour. A consensus exists that to evaluate an ADS implementation reliably, there is a need to employ a combination of methods to validate the capabilities; hence, NATM's multimethodologies (pillars) approach includes a scenarios catalogue that combines accelerated (simulation) testing, the test track, real-world testing, audit/assessment procedures, and in-service monitoring and reporting.

This work focuses on the challenges related to an independent assessment of the safety of automated vehicles and the importance of robust safety-assessment frameworks. Such a testing framework must bridge the gap between the marketing portrayal and the actual performance of such systems in real operating conditions. It requires industry, government, and academia collaboration to develop a framework that ensures this technology's safe and responsible development and deployment.

Despite the availability of safety-assessment frameworks, standards, and guidelines, there remains a need for detailed practical guidance in conducting safety assessments for ADSs. This necessity arises from the current work's general nature, which often lacks the specificity required to address the challenges posed by the complex operational contexts of ADSs. This is especially true for the assessment tasks envisioned by a technical service provider, which are complex and require expertise in multiple domains, including technology, human factors, risk management, and safety regulations. Moreover, ADS technology is rapidly evolving, and new safety and performance requirements are emerging as the technology advances. However, a significant challenge arises due to the limited availability of information before the evaluation process begins, necessitating the need for proactive guidance. By providing technical service providers with anticipatory practical guidance, they can better prepare and navigate the assessment process, identify relevant tests, and address the challenges of establishing confidence in ADSs' safety and user awareness. An assessment template can be crucial in conducting comprehensive evaluations of ADSs by capturing all assessable attributes. Yet, given the complexity and evolving nature of ADSs, achieving a fully comprehensive evaluation using a single template is currently unattainable.

To address this challenge, our contribution is threefold. First, we introduce a novel method for constructing specialised subsets of assessment templates tailored to ADSs and their specific reliance on KETs. We employ an approach that involves gathering requirements through stakeholder data collection and use cases. From these requirement groups, we derive relevant attributes that serve as the foundation for our assessment templates. In this context, requirement attributes are precisely defined as properties of a requirement that capture essential information that is well-suited for evaluation. Secondly, we put our proposed method into action by exploring requirements associated with two ubiquitous enabling technologies in ADSs: positioning and communication. Furthermore, we address the quality attribute of cybersecurity in the context of its intersection with safety considerations. This analysis results in creating specialised templates that offer a more focused and targeted approach. These templates provide forward-thinking, practical guidance tailored to assessing ADSs, ensuring an effective and thorough evaluation process. Third, we demonstrate the effectiveness of the assessment templates through a use case involving a remotely assisted truck. This practical application showcases the template content of attributes and assessable performance indicators in test scenarios.

Our threefold contribution introduces a method for developing specialised assessment templates tailored for ADSs. To the best of our knowledge, no existing approach investigates technology-aware assessment criteria to enhance safety assessments in this manner. Finally, we demonstrate the practical utility of these templates through a real-world use case, collectively advancing the field of ADS safety assessments. The KET-specific assessment templates significantly facilitate structured, technology-aware evaluations of ADS safety and performance. They establish a knowledge-driven, consistent, and repeatable assessment framework. However, it is important to note that the assessment template approach has limitations, primarily relying on predefined scenarios. As such, it is designed to complement data-driven methodologies that incorporate real-world data for a more comprehensive assessment. Additionally, these templates should be subject to continuous updates and refinements to align with ongoing technology developments.

This paper is organised as follows: the problem is introduced in Section 1, the background and related works are presented in Section 2, the method to produce templates is introduced in Section 3, the creation of fit-for-purpose templates for the considered KETs is elaborated upon in Section 4, the templates are utilised and evaluated in Section 5, and the results and future work are discussed in Section 6.

2. Background and Related Work

Automated driving technology, also known as autonomous or self-driving vehicle technology, uses a combination of complex sensors and advanced algorithms to navigate and interact with their surroundings without human intervention.

As with any new technology, the development and deployment of automated vehicles come with potential risks and challenges that must be addressed. These risks and challenges are related to the safety and reliability of the technology, the ethical and legal implications of its use, and the overall impact on society and the environment [4].

SAE J3016 is widely recognised as a taxonomy and definition reference for Automated Driving Systems (ADSs) [1]. ADS features are categorised under SAE automation levels three to five. These systems are designed to take over the driving task for a portion of a trip, performing operational functions such as vehicle motion control (lateral and longitudinal) and tactical functions like route planning, following, and object and event detection and response (OEDR). Similar to a human driver, ADSs must be able to perceive their location and surroundings, which requires various functionalities. These functional, nonfunctional, and technical requirements are crucial considerations throughout the development, type approval, and consumer testing of ADSs. The assessment of ADS features is significantly influenced by the concept of the operational design domain (ODD) [5,6]. The ODD refers to the specific operating conditions in which an ADS is designed to function and must be integrated into safety-related functions. The dynamic driving task (DDT) encompasses the real-time operational and tactical functions necessary to operate a vehicle within the ADS's ODD. Several efforts have been made and are ongoing to define and describe an ODD, including standards such as those set by the British Standards Institution (BSI) [7], the International Organization for Standardization (ISO) [8], and the Association for the Standardization of Automation and Measuring Systems (ASAM) OpenODD [9].

Safety-assessment approaches for autonomous systems encompass a range of methodologies and techniques, but many are at least relatable to scenario-based testing and the SAE taxonomy. Another important aspect is the use of scenario-based testing [10–12]. Scenario-based testing aims to identify and test scenarios that are safety critical for the ADS feature in scope to ensure automated vehicles' safe operation [13].

The approach complements real-world testing and allows for a more comprehensive evaluation of the system's capabilities and limitations. By systematically designing and evaluating scenarios representing realistic and critical situations, developers can gain valuable insights into the system's performance and identify potential failure modes. Other safety-assessment approaches include real-world testing, distance-based evaluation, staged introduction, function-based testing, shadow-mode evaluation, formal verification, and traffic-simulation-based testing [14]. These approaches all enable the assessment of the system's safety and performance in various contexts. However, ensuring that autonomous systems meet the requirements and can operate safely in diverse environments requires a holistic approach.

Several efforts are made to develop standardised testing methodologies for ADSs, and some focus on assessments [15]. Examples of standardised testing are the National Highway Traffic Safety Administration (NHTSA) Framework for Automated Driving System Testable Cases and Scenarios (ref. [16]) and the New Assessment/Test Method for Automated Driving (NATM) [17] proposed by the United Nations Economic Commission for Europe (UNECE). We primarily concentrate on NATM due to its significance in the European context.

Within the NATM certification process, accelerated testing is combined with validity documentation supplied by the manufacturer in the audit and assessment procedure to cover system-related aspects. However, it is important to note that this is meant to complement, rather than replace, classical test track certification. Combining multiple methods, as depicted in Figure 2, represents a prevalent practice within numerous assessment initiatives [10,15,17], with a sequential flow of activities in scenario-based evaluations of ADS, starting with a scenario catalogue. The efficacy and efficiency of the assessment process are heavily contingent upon the data in the scenario catalogue. Safety-performance indicators (SPIs) are tools for monitoring the validity of safety claims throughout the design, simulation, testing, and deployment stages [15]. The effective use of SPIs lies in their ability to prompt timely improvements by linking specific metrics to safety claims, ensuring a direct connection between the observed data and overall safety objectives.

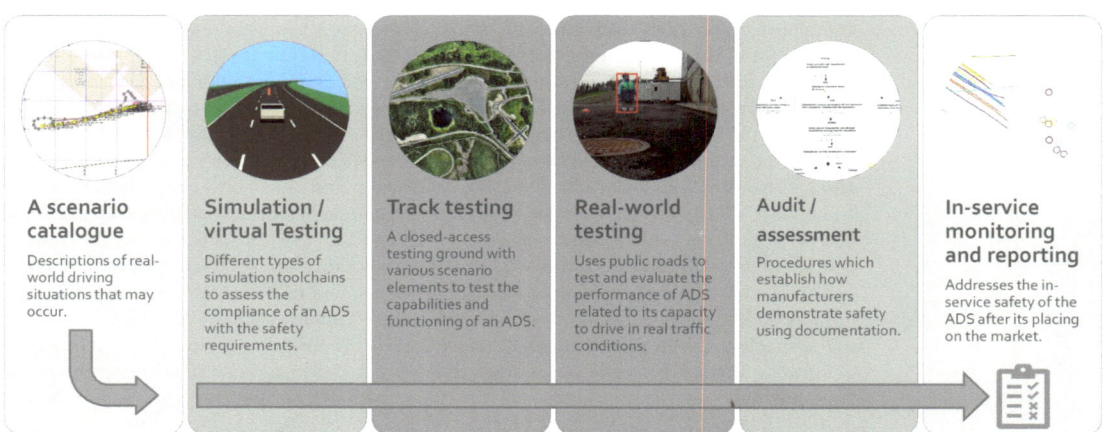

Figure 2. The envisioned procedural instance of the assessment framework. Functional scenarios related to KETs can be added to the scenario catalogue.

One limitation is that NATM is still in the proposal stage and has not been widely adopted or implemented. As a result, limited data are available to assess its effectiveness and suitability [18] for different Automated Driving Systems. Since NATM is technology-neutral, it may be difficult for assessors to apply the framework consistently and effectively across different ADS applications. Another difficulty is the dynamic nature of automated systems and the rapid pace of technological advancements. Safety assessments must keep up with the evolving technology, requiring continuous updates and adaptations to assessment frameworks and standards. The emergence of new sensor technologies, AI algorithms, and connectivity features further complicates the assessment process. The authors argue that the method of using the assessment templates proposed in this paper can help mitigate these limitations. An assessment template can add general scenarios to the scenario catalogue that cover conditions in the ODD by examining scenarios linked to road users' behaviour, environmental conditions, driver behaviour, and technology reliance, and provides some consistency of evaluation across applications.

3. Method to Derive Assessment Templates

Safety-performance indicators are important in a safety-assessment process. Striking a balance between test representativeness and reliable performance indicators is essential. These indicators encompass many factors that require evaluation, which should strongly reflect the overall vehicle's safety performance. Our thesis asserts that analysing KETs is fundamental to develop practical guidance to evaluate the soundness and comprehensiveness of the ODD and functional scenarios to test automated vehicles. This guidance, in the form of requirement attributes, serves as safety-performance indicators that enable the examination and evaluation of automated vehicle systems.

Safety-performance indicators cover various critical aspects and can be categorised as follows:

1. Indicators of system reliability include assessing the system's failure rate, response time, redundancy, and more.
2. Indicators of safety by design, evaluating hazard identification, safety-critical scenarios, cybersecurity, and safety-feature activation.
3. Indicators of design soundness and coverage of the ODD.
4. Indicators of human–machine interaction, assessing driver engagement, monitoring, and user interface design.
5. Indicators of verification and validation, analysing test coverage, scenario replication, and validation.

6. Indicators of regulatory compliance, confirming adherence to legal and ethical standards.
7. Indicators of user perception, evaluating user feedback to understand perceived safety and acceptance.

It is important to note that each vehicle-level performance indicator must be further broken down into indicators relevant to the KET under assessment. For instance, if we are assessing the communication technology of the autonomous system, specific indicators related to communication reliability, latency, and data security should be considered. These KET-specific indicators must then be aggregated into a system-level indicator to provide an overarching assessment of the autonomous vehicle's safety performance. This comprehensive approach ensures that the KETs, integral to the vehicle's functionality, are thoroughly evaluated within the broader safety framework.

Analysing all major KETs is essential in providing complete guidance to evaluate any use case of automated vehicle systems. We believe this approach should be prioritised regardless of the technologies being analysed. The process of deriving assessment templates can be summarised as follows:

1. Collect ADS use-case requirements: Engage with stakeholders, including manufacturers, researchers, regulators, and industry experts, to gather their requirements and perspectives. Identify and analyse various use cases to understand technology reliance and testing needs. Assess the reliance of each requirement on KETs.
2. Allocate requirements based on technology reliance: Determine which requirements directly or indirectly depend on specific KETs. Allocate and associate the requirements with the corresponding KET.
3. Derive attributes for the KET category: Derive attributes that capture the essential characteristics of each category. These attributes should primarily reflect safety considerations, but functionality, reliability, and other relevant technological group aspects can also be considered.
4. Establish safety-performance indicators: Based on the derived attributes and safety objectives, establish KET-specific contributions to safety-performance indicators that can be used to assess and measure the safety performance of the automated system. These indicators should provide quantifiable and meaningful measures to evaluate the system's compliance with safety requirements. Create functional scenarios that cover diverse KET-related operational conditions and situations that will be added to the scenario database. The derived test scenarios, when executed, should exercise the system's capabilities and evaluate its performance against the criteria that set the standards and indicators that provide the means and methods for measurement.

A panel of experts was used, refining the collected requirements into attributes for each KET, as illustrated in Figure 3. However, this approach comes with certain limitations. Subjectivity is a limitation, as attribute selection relies on expert opinions, potentially leading to definitions and perceived importance variations. Furthermore, a limited representation of diverse stakeholders may result in the inadvertent oversight of requirements. Additionally, the absence of standardisation can give rise to inconsistent attribute definitions, complicating meaningful comparisons.

Nevertheless, it is worth noting that building a requirement landscape for each KET is additive, with each contribution enriching the overall understanding. These limitations become less significant when a substantial input volume is collected and aggregated to a limited number of attributes, on which consensus can be reached.

When evaluating requirements against multiple quality attributes, it is imperative to acknowledge that these attributes can inherently conflict. Consequently, addressing these conflicts requires careful consideration to establish definitive assessment criteria. One approach to managing such tradeoff conflicts is the Analytic Hierarchy Process (AHP) [19], a valuable decision-making technique. In our specific case, no real conflict was detected, and the requirement-selection process sufficiently facilitated the identification of suitable requirements.

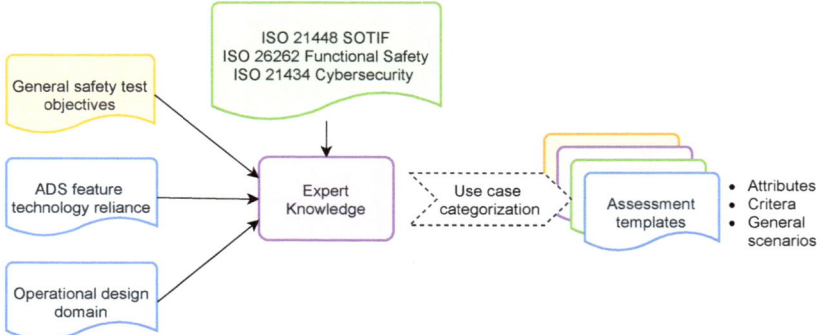

Figure 3. Schematic of assessment-templates-creation process.

Ultimately, the practical value of this approach is demonstrated in Section 5, where the viability and benefits of the method are exemplified.

Following this process, stakeholders can systematically collect requirements, identify technology dependencies, and derive requirement attributes per KET and safety-performance indicators. This structured approach systematically addresses safety considerations, leading to a more thorough and uniform evaluation of an automated system's safety performance.

4. Derive Assessment Templates

The method delineated in Section 3 serves as a blueprint for crafting assessment templates. This section provides a condensed overview of the template-creation steps for the KETs: communication, positioning, and cybersecurity. These KET categories were integral to the HEADSTART [20] project. Our main focus lies in elaborating on the attributes and assessment templates, which represent an extension of this work. At the same time, we touch upon the rudimentary aspects of requirement collection and allocating categories. Subsequent sections and Figure 4 delve into these steps, underscoring their significance. Our analysis zeroes in on these three KETs, illustrating how they were employed to validate our hypothesis concerning the role of technology-aware guidance in ADS assessments. This approach underscores the importance of encompassing a relevant array of KETs when evaluating automated-vehicle ODDs and in scenario-based testing.

4.1. Collection Requirements

The initial phase, marked as 1 in Figure 4, involves the comprehensive collection of requirements. In our previous study [21], conducted within the HEADSTART project, a rigorous effort was made to amass functional and technical requirements pertaining to the three Key Enabling Technologies (KETs). These technologies are key for ensuring automated vehicles' proper functioning and safety, hence the term. The requirements for the KETs were identified through a three-step process. Firstly, we delved into ongoing activities within standardisation organisations and other relevant interest groups. This provided valuable insights into evolving standards and industry expectations. Subsequently, we conducted surveys, questionnaires, and interviews involving stakeholders like OEMs, Tier 1 suppliers, and regulatory bodies. This direct engagement was instrumental in understanding their distinct needs and perspectives. Lastly, we integrated requirements and insights from other pertinent research projects to enrich our analysis. This comprehensive approach ensured the collection of various requirements and needs related to the KETs.

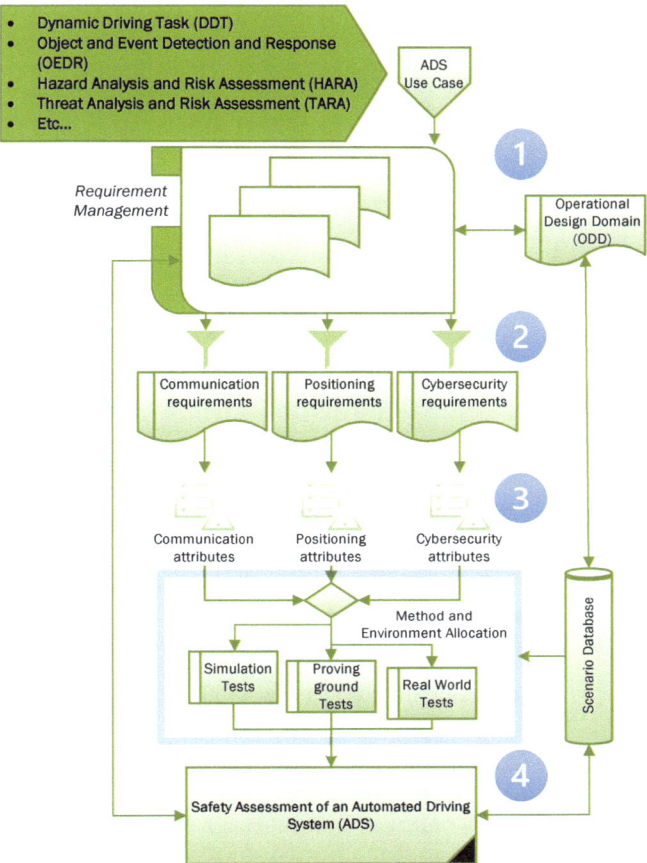

Figure 4. Method to derive assessment attributes for KETs.

The data-collection efforts were conducted closely with stakeholders, including participants and affiliates of the HEADSTART project, spanning industry, research institutes, and policymakers. The data collection of stakeholder considerations revealed diverse requirements, ranging from high-level strategic needs to intricate technical specifications. The high-level strategic needs, such as the functional requirements, illuminate the specific functions that KETs are tasked with within automated systems. These functions encompass tasks like sensor-data processing, real-time data communication, and implementing cybersecurity measures. Technical requirements delve into intricate technical details, encompassing communication protocols, data-transmission rates, encryption methods, and network architecture. These details provide insights into the technical aspects that underpin KETs. Furthermore, performance attributes closely tied to these technical requirements support the evaluation of the effectiveness of KETs. These attributes cover essential factors like latency, data throughput, reliability, and redundancy, collectively contributing to the assessment of KETs' performance and their significance in ensuring the safety of automated vehicles.

Moreover, we considered the requirements for testing and validation procedures. It covers identifying relevant test scenarios simulating real-world conditions and requirements for the testing environment to ensure accurate assessments, categorising it as development, consumer-oriented, or type-approval testing. Here, the only interest is the latter and how the two prior categories can support type approval. An insight gained is that future requirement collection efforts would benefit from the regulatory-compliance perspective. This analysis revealed numerous requirements tailored to the KETs, often intricately linked to particular use cases. One notable challenge was that some of these requirements were based on desired functionalities and needs that might not be readily available with today's technology. Given the ongoing development in all three KETs, adaptation may be needed to align the requirements with the capabilities of contemporary technology. These identified requirements and constraints pertinent to the KETs have been documented and disseminated in various publications [21–23]. These publications provide a comprehensive guideline framework for developing harmonised testing and validation procedures, a key component of the HEADSTART method's overarching objectives [24].

The use cases analysed in the project, e.g., highway pilot and highway truck platooning, are used to explore various aspects of critical enabling technologies. Understanding the variation in the reliance on these underlying technologies in developing a practical assessment procedure is important. By understanding the specific requirements and challenges associated with each use case, an assessment procedure can be developed to ensure the safety and performance of automated vehicles. The derived attributes presented in Section 4.3 are based on these collected requirements.

4.2. Allocation Requirements Based on Technology Reliance

As indicated in Step Two in Figure 4, the method integrates the gathered and categorised requirements, aiming to include all pertinent technology-specific parameters within the ODD and scenario specifications. The framework includes a separate analysis of the KETs to address their requirements comprehensively. Doing so ensures that the framework considers each technology's specific attributes and considerations. The effects of these technology-specific requirements are continuously monitored as they propagate and permeate the framework and give rise to attributes, performance indicators, and test scenarios.

4.2.1. V2X Communication

Communication and associated requirements can be crucial in ADSs. Vehicle-to-everything (V2X) communication technologies enable vehicles to wirelessly communicate with various entities that can impact their operation, including vehicle-to-infrastructure (V2I), vehicle-to-network (V2N), vehicle-to-vehicle (V2V), vehicle-to-pedestrian (V2P), vehicle-to-device (V2D), vehicle-to-grid (V2G), and Tele-operated Driving (ToD). This communication capability facilitates cooperative driving, optimising collective behaviour regarding throughput, fuel consumption, emissions, and safety [25]. In the automotive industry, there are two main types of V2X communication technologies: WLAN-based, which utilises IEEE 802.11p and is used in standards such as ETSI ITS G5 and DSRC, and cellular-based, which is defined by 3GPP and includes short-distance communication using PC5 sidelink and traditional cellular interfaces through 3G/4G/LTE/5G networks. The testing of V2X communication involves various organisations such as 3GPP, 5GAA, ETSI, GCF, IEEE, OmniAir, SAE, C-ITS, C-SAE, and NTCATS. Test-equipment vendors are actively developing instruments designed explicitly for V2X testing, with many of them also incorporating Global Navigation Satellite System (GNSS) testing capabilities. A 5G-based predictability system was evaluated to forecast the service quality [26], proposing an enhanced scheme for vehicle-to-network communication optimisation. This method effectively alleviates base station congestion while preserving key quality attributes like delay time and throughput, showcasing efficient network management even in high-demand scenarios.

4.2.2. Positioning

Positioning is a capability required for high levels of automation. It involves determining the position of the ego vehicle (the vehicle under test) and estimating and tracking the position of objects in its vicinity within the traffic system. Different applications within the scope of connected ADSs have varying positioning needs, with the main aspects being absolute and relative positioning. The accuracy, precision, refresh rate, and integrity are subattributes associated with these aspects.

Global Navigation Satellite System (GNSS)-based positioning and High-Definition (HD) maps can be utilised for absolute positioning. HD maps provide relevant information, such as traffic signs, beams, or poles, which can be trust anchors to determine the vehicle's position without active connections. V2X communications can also improve positioning by transferring information, provided a mechanism exists to establish sufficient trust in the received data. In the realm of GNSS technologies, ongoing standardisation efforts are spearheaded by organisations like ETSI, with test-equipment vendors actively enhancing GNSS testing capabilities. Furthermore, the interrelation of cybersecurity with GNSS positioning in Intelligent Transportation Systems (ITSs) is detailed in EN 16803-1 [27]. The European COST Action SaPPART [28] established standardised methodologies, enhancing integrity metrics, PVT error models, and positioning solutions for future ITSs. This involved validating positioning terminals and analysing GNSS receivers under the guidance of EN 16803-1.

4.2.3. Cybersecurity

In the realm of defining cybersecurity requirements, it becomes paramount to factor in potential threats. Notably, the NIST FIPS 199 [29] delineates three fundamental facets of cybersecurity, often referred to as CIA:

- Confidentiality: this dimension revolves around safeguarding authorised information access and disclosure restrictions, an endeavour encompassing the protection of personal privacy and proprietary data.
- Integrity: integrity focuses on thwarting improper information modification or destruction, thus ensuring information nonrepudiation and authenticity.
- Availability: the cornerstone of this facet is ensuring the timely and reliable access to and use of information.

Crucially, the technical and functional requirements identified emphasise that the latter two aspects are intrinsically linked to safety considerations. This underscores the importance of adhering to cybersecurity best practices throughout the product-development journey. Cybersecurity is a vital quality attribute, wielding substantial influence over the safety of ADS applications. Diverging from safety considerations, cybersecurity maintains a constant state of evolution, marked by the continuous development of new techniques and capabilities by potential attackers. Consequently, addressing cybersecurity concerns becomes an enduring requirement throughout the lifecycle of an ADS.

It is noteworthy that cybersecurity requirements deviate from those of communication and positioning. Cybersecurity is a vital quality attribute that permeates both domains. In vehicle-to-everything (V2X) communication, establishing a chain of trust through verified signatures and certificates proves indispensable. Rigorous state-of-the-art cybersecurity testing should be diligently executed across all aspects. Many best practices and design principles for cybersecurity in vehicular systems exist. These are outlined in standards such as SAE J3061 [30], NIST FIPS 2004 [29], and ISO/SAE 21434. Additionally, various studies and discussions delve into security and privacy in Connected vehicle-to-everything (C-V2X) communications [25,31,32]. Both the SAE and ISO [33] collaborate extensively in standardisation activities relevant to vehicle cybersecurity. Cybersecurity's intricacies stem from the perpetual evolution of techniques and the ever-present threats that can imperil safety.

Shifting our focus to the responsibilities of type-approval assessors concerning cybersecurity, their purview primarily centres on the system level. However, this scope intentionally remains narrower than the expansive realm of cybersecurity. The primary objective revolves around conducting a comprehensive evaluation of documentation, ensuring the coverage of critical cybersecurity dimensions. These encompass cybersecurity-management systems, software-update management, cybersecurity measures, risk identification, risk-mitigation strategies, and measures enacted to ensure unwavering compliance with legislative requirements. This evaluation carries significance as it aligns with the overarching goal of uncovering potential cyberattack vulnerabilities, which could compromise the vehicle's safety. The evaluation encompasses the scrutiny of the physical testing environment, spanning proving grounds and public roads. It also entails examining the manufacturer's documentation on the virtual toolchain. Lastly, it is at the assessor's discretion to determine whether comprehensive tests of the integrated toolchain are warranted. Such tests aim to affirm the credibility of the toolchain's cybersecurity safeguards, fortifying the commitment to a robust cybersecurity framework.

4.3. Derive Attributes for KETs

The general safety objectives include potential hazards during a generalised ADS operation, including internal system and external environmental hazards. The process denoted three in Figure 4 deals with assessing the risks associated with identified hazards [33–35], relevant to the reliance of KETs by analysing the likelihood and severity of potential incidents or accidents. Furthermore, strategies and measures, such as safety implementation, are devised to alleviate these identified risks.

To evaluate the influence exerted by KETs on the ODD of an ADS, the ISO 34503 "Test scenarios for Automated Driving Systems—Specification for operational design domain" is used as a baseline [8]. ISO 34503 applies to ADS levels 3–4 and provides requirements for a hierarchical taxonomy that identifies the ODD, considering static and dynamic attributes.

ISO 34503 proposes dividing the operating conditions into three primary attributes: scenery, environmental conditions, and dynamic elements. Scenery refers to nonmoving elements; dynamic elements represent moving elements in the operating environment; and environmental conditions encompass factors between geographical and temporal attributes, including meteorological weather parameters relevant to the ODD. The hierarchy in ISO 34503 provides a base set of attributes that can be expanded based on stakeholder needs. To better incorporate KETs into the ODD taxonomy, the connectivity category in ISO 34503 can be refined to include communication, positioning, and cybersecurity. Communication requirements can include coverage, latency, throughput, and predictability, as listed in Table 1. Positioning requirements encompass absolute and relative positioning with subattributes like accuracy, precision, and refresh rate integrity, as shown in Table 2. Cybersecurity requirements can be derived based on the categorisation proposed by Firesmith [36], as presented in Table 3.

Table 1. When assessing a V2X communication solution, the following attributes should be considered.

Attributes	Description
Coverage	The geographic area or range within a carrier's defined service. Indicates the solution's ability to establish and maintain connectivity.
Latency	Time delay between a message being sent by a sender and being received by the intended recipient. Indicates the responsiveness of the communication solution.
Throughput	Number of data packets that can be transferred within a specific time. Indicates the solution's capacity to handle data traffic.
Predictability	Consistency and reliability of solution performance. Indicates the ability to pre-empt and plan for degraded coverage, latency, and throughput.

Table 2. When assessing a positioning solution, the following attributes should be considered.

Attributes	Description
Position priority	Absolute, relative. Possible refinements: lateral, longitudinal, or elevation position.
Accuracy	How close measurements are to the true position. Indicates the solution's capability to determine an object's location accurately.
Precision	How close measurements are to each other. Indicates the consistency of the solution in providing consistent position measurements.
Refresh rate	How close measurements are to each other in time. Indicates the solution's responsiveness.
Confidence	Confidence reflects the ability to quantify the uncertainty in measurements. Indicates the ability to handle and pre-empt degraded services. Confidence and integrity are closely related indicators.
Integrity	Integrity refers to the reliability and availability of the solution. Indicates the solution's ability to function correctly and consistently, providing accurate and trustworthy position information.

Table 3. Additional quality attributes to assess when considering cybersecurity.

Type [1]	Description
Prevention	Measures that reduce the security risks. It is preferable to stop risks from being realised than to repair the damage after an incident.
Detection	Mechanisms to discern malicious activity from normal use.
Reaction	Strategies to employ after detecting malicious activity to minimise the harm.
Adaptation	Modification to improve prevention, detection, or reaction.

[1] Inspired by Firesmith's defensibility solution types [36].

Numerous vital questions still need to be addressed and recognised, including supporting cooperative functions and allocating responsibilities to ensure a safe implementation across multiple brands. Additionally, considerations of interdependence within the ODD must be examined, including the specification and testing of supported vehicle velocities and establishing a trusted chain of external data sources. These external data sources should have a seamless chain of trust and consistent uncertainty measurements and also assessments of common time-base solutions for synchronised cooperative ADS.

While the attributes presented in Tables 1–3 may not cover all the gathered requirements, and in all likelihood, not all relevant concerns are addressed, they provide useful patterns and attribute families to analyse the performance of KETs and map them to the ODD. Additional research is required to delve into coverage and comprehensiveness when mapping an ODD to specific isolated technology elements, encompassing specification and testing. However, it is crucial to initiate the process of providing proactive and practical guidance to technical service providers to enhance their preparedness and streamline the assessment process.

4.4. Establish Safety-Performance Indicators and Functional Scenarios

Much effort has been spent on the development of performance indicators [37,38] and scenario databases [39,40], focusing on data-driven aspects like longitudinal control (acceleration, braking, and road speed), lateral control (lane discipline), and environment monitoring (headway, side, and rear) as single aspects and when moving into more-complex scenarios in combination. This combination poses a challenge to proving ground capabilities due to the high level of coordination needed to realise the scenarios. As it is virtually impossible to evaluate an automated vehicle against all possible scenarios it will face in real-world traffic, balancing the representativeness of the tests and the reliable safety-performance indicators is necessary.

Conversely, we talk about the assessment criteria subset that can be created for the attributes derived previously for the enabling technologies, positioning, communication (V2X), and cybersecurity. Criteria set the evaluative standards, and safety-performance indicators provide the means for measuring compliance with those standards. Both are integral to the assessment process but serve different purposes: criteria guide what to measure, and indicators define how to measure it. Knowledge-driven indicators can be assigned to elementary behavioural aspects of the automated function that must be assessed with scenarios linked to the ODD and its monitoring, e.g.,

- Conditions for activation:
  - External and internal human–machine interfaces;
- Triggering conditions for minimum risk manoeuvres:
  - External and internal human–machine interfaces;
- Conditions for deactivation:
  - External and internal human–machine interfaces.

The assessment criteria are partly based on the existing automotive safety-assessment methods (see Figure 3), as also discussed in Section 2. In the assessment framework, we describe activities as denoted (4); see Figure 4, i.e., new assessable criteria related to KETs.

In scenario-based testing, two primary criteria come into play: pass/fail and metric criteria. Both of these criteria rely on objective observations of the executed scenario. Context-specific safety-performance indicators are used to establish success criteria and metrics. These indicators serve as data gatherers, facilitating evaluating and comparing the automated vehicle's expected and executed behaviour. Each KET introduces specific attributes that must be met during operation. Failure to meet these conditions often triggers a minimal risk manoeuvre (MRM) activation to return the system to a minimal risk state. Various failures, encompassing scenarios such as attacks on vehicle control, environmental monitoring, and interactions within the human–machine interface (HMI), both internally and externally, may trigger MRMs. It is imperative to assess the appropriateness of these MRM and HMI interactions [41].

Coverage pertains to the extent of the communication system's reach. It is important to determine the acceptable coverage values based on the specific safety requirements of the ADS and its ODD. Establishing and improving these values requires an iterative process and an understanding of real-world operational conditions. Latency, another critical metric, gauges the time it takes for data to travel between the sender and receiver. Ensuring low latency is essential, especially for safety-critical applications. This assessment should align with the safety requirements outlined for the ADS within its ODD. The throughput assesses the system's data-transfer capacity, often measured in terms of bandwidth. Defining acceptable throughput values necessitates thoroughly examining the ADS's safety prerequisites and ODD. Maintaining adequate throughput levels is essential, even in challenging operational conditions. Predictability evaluates the system's ability to deliver expected results consistently. Predictable communication is vital for the safe functioning of an ADS. Establishing criteria for predictability should align with safety requirements and ODD specifications. These metrics are the foundation for creating an assessment template, as depicted in Table 4. Stakeholders can systematically evaluate the communication system's performance and alignment with safety objectives by defining acceptable values for these metrics tailored to the specific ADS and its operational context.

Regarding activation scenarios, these tests ensure that all KET ODD conditions are met before activation occurs. Conversely, deactivation scenarios assess the appropriateness of both internal and external HMI responses when deactivation is required. This deactivation can be initiated gracefully through control-transition demands or via minimal-risk manoeuvres, ensuring safety is maintained.

Table 4. Sample template for communication with attributes and basic related HMI aspects.

KET	Test Scenario	Attribute [1]	Criteria Description	Evaluation
Communication	Activation-condition scenarios	Coverage	Activation criteria for coverage	☐ Pass ☐ Fail
Communication	Activation-condition scenarios	Latency	Activation criteria for latency	☐ Pass ☐ Fail
Communication	Activation-condition scenarios	Throughput	Activation criteria for throughput	☐ Pass ☐ Fail
Communication	Activation-condition scenarios	Predictability	Activation criteria for predictability	☐ Pass ☐ Fail
Communication	Internal HMI activation scenarios	:	Criteria for internal HMI evaluation	☐ Pass ☐ Fail
Communication	External HMI activation scenarios	:	Criteria for external HMI evaluation	☐ Pass ☐ Fail
Communication	Internal HMI control-transition scenarios	:	Criteria for control-transition evaluation	☐ Pass ☐ Fail
Communication	MRC triggering-condition scenarios	Coverage	Criteria for MRC evaluation	☐ Pass ☐ Fail
Communication	MRC triggering-condition scenarios	Latency	Criteria for MRC evaluation	☐ Pass ☐ Fail
Communication	MRC triggering-condition scenarios	Throughput	Criteria for MRC evaluation	☐ Pass ☐ Fail
Communication	MRC triggering-condition scenarios	Predictability	Criteria for MRC evaluation	☐ Pass ☐ Fail
Communication	Internal HMI of MRC triggering scenarios	:	Criteria for HMI MRC evaluation	☐ Pass ☐ Fail
Communication	External HMI of MRC triggering scenarios	:	Criteria for HMI MRC evaluation	☐ Pass ☐ Fail
Communication	Deactivation-condition scenario	Coverage	Criteria for deactivation evaluation	☐ Pass ☐ Fail
Communication	Deactivation-condition scenario	Latency	Criteria for deactivation evaluation	☐ Pass ☐ Fail
Communication	Deactivation-condition scenario	Throughput	Criteria for deactivation evaluation	☐ Pass ☐ Fail
Communication	Deactivation-condition scenarios	Predictability	Criteria for deactivation evaluation	☐ Pass ☐ Fail
Communication	Internal HMI deactivation scenarios	:	Criteria for deactivation HMI evaluation	☐ Pass ☐ Fail
Communication	External HMI deactivation scenarios	:	Criteria for deactivation HMI evaluation	☐ Pass ☐ Fail

[1] Attributes related to human–machine interfaces (HMI) are beyond the current scope.

Similarly, for positioning, evaluating the ADS's capability to determine its position and track objects in its environment may require metrics such as accuracy, precision, refresh rate, and integrity. The acceptable values for these metrics will hinge on the specific use case and the safety-critical requirements governing it.

When delving into cybersecurity aspects, metrics can encompass factors such as robustness against cyberattacks, resistance to unauthorised access, and the integrity of data transmission. Also, determining acceptable metric values is contingent upon industry best practices, relevant standards, and the criticality of the ADS's functions. In summation, as we address the requirements posed by the various KETs, the scenario catalogue expands with an array of assessment criteria for minimum risk manoeuvres, hand-over transitions, HMI (both internal and external), and driver monitoring. It is vital to review this expanding catalogue to ensure that its representation and completeness align rigorously with the demands and expectations of the system under evaluation.

5. Evaluation of the Use of Assessment Templates

Analysing how each use case relies on support technology building blocks, which are implementing the KETs, helps identify the specific requirements and dependencies of different technological components. Understanding these dependencies allows for determining which assessment templates are relevant and how they should be applied (Figure 5). It also becomes possible to tailor the assessment process to the specific needs and requirements of the system in terms of functional requirements on the ODD and the functional scenarios to test or assess. Furthermore, it is essential to consider the interdependencies between different technology building blocks and how they collectively contribute to the overall functionality and safety of the Automated Driving System. In contrast, some assessment templates may address multiple technology components simultaneously.

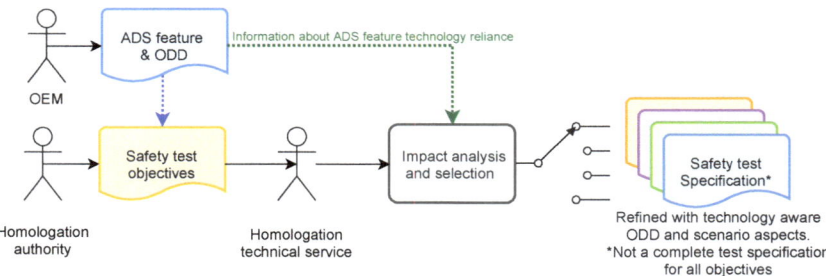

Figure 5. Schematic of selection process of assessment templates.

5.1. ADS Feature and ODD Under Evaluation

The evaluation centres on a highly automated freight vehicle in a dedicated urban area (Figure 6). The vehicle aims for SAE level 4 automation [1], indicating it can perform all driving tasks under certain conditions without human intervention. Additionally, the vehicle is equipped with remote-assistance functionality, allowing for human oversight and intervention remotely, which is essential for addressing scenarios beyond the capabilities of the automated system or during transitions between automated and manual control. It involves automated freight transport within a controlled environment, specifically for potentially uncrewed vehicles. Design options include vehicles with or without a driver's cab, focusing on lower speeds for fuel efficiency.

As depicted in Figure 5, the input is the safety objectives, function description, and intended ODD. The ADS features describe the system utilised, including the functions of remote-assistance automated-vehicle features and the infrastructure deployed within the trial environment.

The safety objectives align with the guidelines outlined by the Swedish Transport Agency (TSFS 2022:82 [42]), emphasising including a traffic safety analysis and an independent risk assessment in all exemption applications. These safety objectives ensure that the evaluation process addresses and fulfils the requirements for risk assessment, guaranteeing the safety and reliability of testing the automated freight transport system on public roads. They serve as representative surrogates for the envisioned safety objectives of future type approval.

A potential site for conducting the ADS feature trials has been identified in the urban traffic environment in Lindholmen, Gothenburg, Sweden. The intended route can be seen in Figure 6. The ODD is relevant to this specific ADS feature and can be generally described as a route encompassing parking lots and streets with parked cars on either or both sides. Traffic in the area generally operates at low speeds, with few vulnerable road users (VRUs) except during lunch and rush hour. VRUs are expected to walk and cycle throughout the area.

- Road conditions: public urban roads going straight, at intersections, and at turns.
- Geographical area: Lindholmen, Sweden. Exact geographic site determined with Geofence.
- Environmental conditions: daylight, good visibility, no or light rain, and little or no water on the road surface.
- Velocities: speed restricted to lower ranges <15 km/h.
- Other constraints: conditions must be fulfilled for the safe operation.

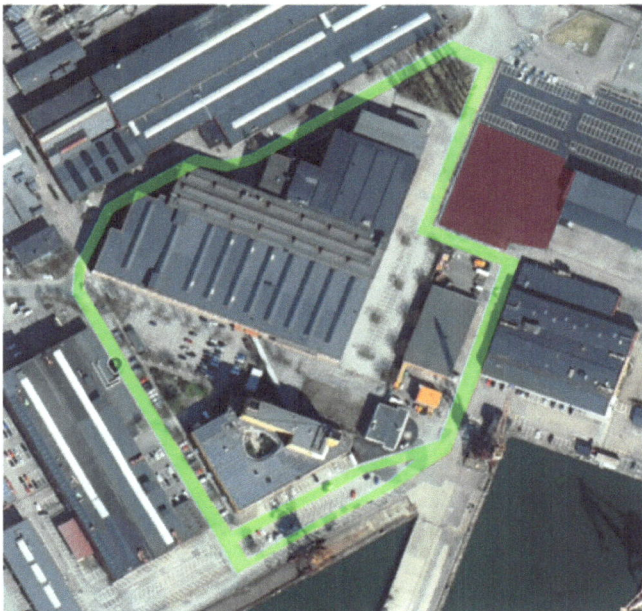

Figure 6. Potential ODD at Lindholmen. The geofenced route is denoted by green.

To ensure that the trial operation of the vehicle maintains a traffic-safe environment, the assessment plan considers multiple aspects. These include adhering to regulatory requirements within the ODD, establishing safety and security objectives for remotely assisted automated functions, and ensuring seamless control transitions during operation.

A geofence solution utilising a GNSS is a safety and cybersecurity mechanism used to mitigate vehicle-operating risks beyond the defined ODD. While geofencing is partially rooted in threat analysis, additional cybersecurity assessments currently fall beyond the scope of this study. Maintaining precise positioning within the ODD often supports the fulfilment of critical system safety and security requirements. This investigation primarily centres on KET's assessment guidance.

Hence, the relevant assessment templates encompass V2X communication; its interdependencies with cybersecurity in the context of 5G connectivity; and its position within the broader assessment plan, particularly regarding geofencing.

5.2. Guided Assessment Plan

Here, the assessment primarily focuses on positioning, V2X communication, and their interplay with cybersecurity. It leaves significant portions of object detection and event response without specific guidance.

Integrating 5G communication into the ODD expands the evaluation of operational conditions. The ODD's boundaries are extended by incorporating 5G communication attributes to encompass connectivity considerations. This evaluation covers system-performance and safety scenarios like network congestion or communication disruptions.

Including 5G communication attributes in the assessment process aids in identifying potential risks and challenges. It evaluates the system's capability to handle situations involving degraded connectivity, assesses the impact of communication delays on decision-making processes, and tests the system's resilience against potential cybersecurity threats targeting the 5G infrastructure.

Therefore, compared to existing standards like ISO 34503, which includes attributes such as vehicle-to-infrastructure (V2I) and 5G, we propose a refinement of operating conditions to focus on attributes like network coverage, latency, throughput, and predictability.

These refined attributes are designed to serve as performance indicators. The assessment metrics and use-case-specific conditions were derived from the Safety Case for Autonomous Trucks (SCAT) project [43].

The control loop for a remotely assisted automated vehicle operates as a continuous process where the vehicle's sensors gather environmental and status data, which are transmitted to a remote operator. The operator issues control commands back to the vehicle. The vehicle then executes these commands, completing the loop. The real-time demands within this loop necessitate precise latency requirements. Ensuring comprehensive coverage using minimum throughput or bandwidth is vital for the safe control of remote operations and for enabling actionable minimal-risk manoeuvres. The guarantee of this minimum throughput holds utmost importance throughout the entire ODD. Maintaining high service availability is critical to pre-empting potential service congestion and counteracting inadequate coverage, especially in adverse weather conditions, emphasising the need for predictability. This comprehensive coverage requirement must be consistently met within the ODD in alignment with the communication-assessment template specified in Table 4.

Furthermore, the Quality of Service (QoS) for bandwidth reservation involves allocating specific portions of the network capacity to certain applications or services to ensure consistent performance levels, especially for data throughput and latency. This ensures that essential services receive the bandwidth unaffected by network congestion, regardless of whether it results from natural factors or intentional actions. Predictability can be further achieved by implementing multiple redundant 5G carrier networks and real-time performance monitoring.

The assessment of GNSS-based geofence considerations follows a similar approach. It employs the prototype template in Table 2. This assessment emphasises the need for accurate absolute positioning, which GNSS systems are tasked to provide. The evaluation criteria stipulate that the geofence system should maintain accuracy within a meter, a benchmark achievable by implementing real-time kinematic (RTK) solutions. The assessment process also involves evaluating the confidence level in the positioning measurements, including analysing how the system quantifies and manages the inherent uncertainty. In the context of geofencing, the precision of positioning is paramount. The assessment, therefore, involves verifying that RTK-based positioning consistently meets the system's accuracy requirements. Additionally, the assessment includes examining the refresh rate at which the system updates its position measurements, with a rate exceeding 1 Hz being the general target for the geofencing application to ensure timeliness. Maintaining confidence in the measurements is vital. The system should be capable of quantifying the uncertainty associated with position measurements. This quantification aids in assessing the reliability and robustness of the geofence solution. Incorporating a second observer for plausibility checks and using dual-frequency receivers are evaluated to determine how they contribute to the overall integrity of the geofence system. The accuracy and reliability of GNSS-based geofence solutions can be assessed by evaluating these criteria.

In applying the cybersecurity attributes delineated in Section 4.3, the emphasis is placed on preventative measures. Specifically, the objective is to fortify the vehicle against unauthorised control by employing authenticated and encrypted communication protocols.

Detection mechanisms are essential to identify malicious activity during remote assistance, and in the case of failure or an attack, a fail-safe reaction strategy should be employed. This transition ensures the safe operation with reduced functionality and may involve over-the-air updates for security patches.

To evaluate the presence and appropriateness of cybersecurity measures, although not a direct component of the ODD, including cybersecurity criteria improves the evaluation process. The enhancement entails a direct association between threat agents, their underlying motives, and potential attack surfaces within the ODD. This refined linkage provides more precise guidance for implementing measures against prospective attacks. By integrating cybersecurity measures, the assessment plan comprehensively evaluates the system's safety and resilience, aligning with future type-approval requirements.

The attributes derived in Section 5 have enhanced assessment planning and analyses of use cases. They emphasise the importance of maintaining 5G communication coverage with a QoS bandwidth priority to ensure a consistent bandwidth, whether due to natural factors or malicious actions. Further assessments of these attributes' suitability for proving ground testing are pending. Utilising the prototype-assessment templates in collaboration with specific functional scenarios—such as minimal risk manoeuvres, activation, and deactivation—facilitates the evaluation of 19 distinct test scenarios.

These indicators are especially relevant to 5G communication and geofencing conditions. They have been categorised into different domains, including activation conditions, minimal risk manoeuvres, and external and internal HMI considerations. The distribution of these conditions within each category is not uniform; for instance, at least four conditions are explicitly pertinent to 5G communication, while six conditions are geared towards geofence considerations.

5.3. Test Scenario Execution for 5G Communication

An evaluation of the cellular coverage at the test site is conducted to ensure dependable communication and data exchange between the vehicle and infrastructure. This is vital for the seamless operation of the monitored ADS feature, encompassing functionalities like assistance and monitoring links. Maintaining a bidirectional stream with a balanced symmetric bandwidth and low latency for the control channel requires consistent capacity to attain robust connectivity. Comprehending the capabilities and limitations of the test site is instrumental for effective planning and preparation for operational deployment. This understanding is achieved by identifying areas necessitating enhancement or optimisation and verifying that essential infrastructure and connectivity prerequisites are satisfied to successfully demonstrate the ADS feature.

Remote assistance and monitoring, especially video streaming, necessitates low latency and high uplink bandwidth. The adaptive video codec should accommodate varying bitrates based on availability. Additionally, the uplink is typically more constrained than the downlink, making it a critical consideration. The site assessment has concentrated on the available uplink capacity, which is likely to be the limiting factor in a remote-assistance and monitoring scenario.

The assessment predominantly focused on measuring the Reference Signal Received Power (RSRP). The RSRP is a reliable indicator for predicting the radio uplink capacity since it gauges the cell's proximity from a radio standpoint. Uplink radio interference is mainly due to other handsets moving within the cell, making it more dynamic and more challenging to predict downlink interference.

The site assessment employed a low-adaptive-latency User Datagram Protocol (UDP) stream to validate video performance. This helped estimate the traffic that could be sent on the uplink without causing delays or overloading the network. Unlike network speed test tools prioritising high bandwidth, this approach considers the absolute latency and latency variation (jitter).

In the experimental procedure, the tester held a measurement terminal during the initial lap of the site under assessment, as shown in Figure 6; the corresponding speed profile is illustrated in Figure 7. For laps 2 to 5, the handheld device was positioned between the front seats of a car circulating the track. The first two laps were executed with a target bitrate of 20 Mbit/s, while the bitrate for the subsequent laps was elevated to 50 Mbit/s.

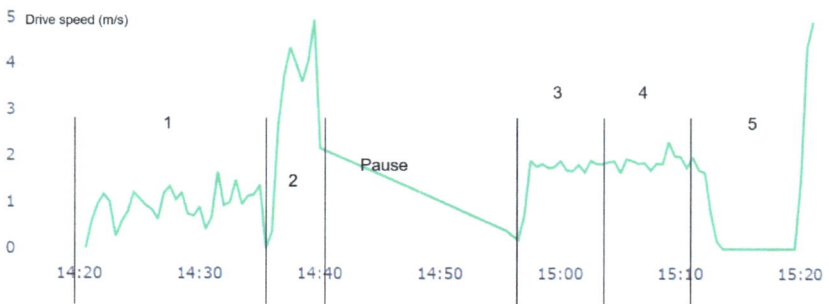

Figure 7. Data were collected over a total of 5 laps at the route at Lindholmen.

The test utilised a handheld terminal with a specialised carrier company application (Telia). This application collected and reported essential radio measurements, including the Reference Signal Received Power (RSRP), Reference Signal Received Quality (RSRQ), Signal-to-noise ratio (SNR), frequency, cell information, and absolute position using the GNSS (Global Navigation Satellite System). Figure 8 showcases the RSRP as a performance indicator for coverage while Figure 9 illustrates the related functional-handover scenario.

An adaptive UDP stream, emulating adaptive video, was used to measure the real-time bandwidth (RT BW) up to a target level. Laps 1 and 2 employed a 20 Mbit/s target bitrate, with later laps using 50 Mbit/s. The RT BW serves as a performance indicator for the throughput, as depicted in Figure 10, and the related scenarios are portrayed in Figure 11.

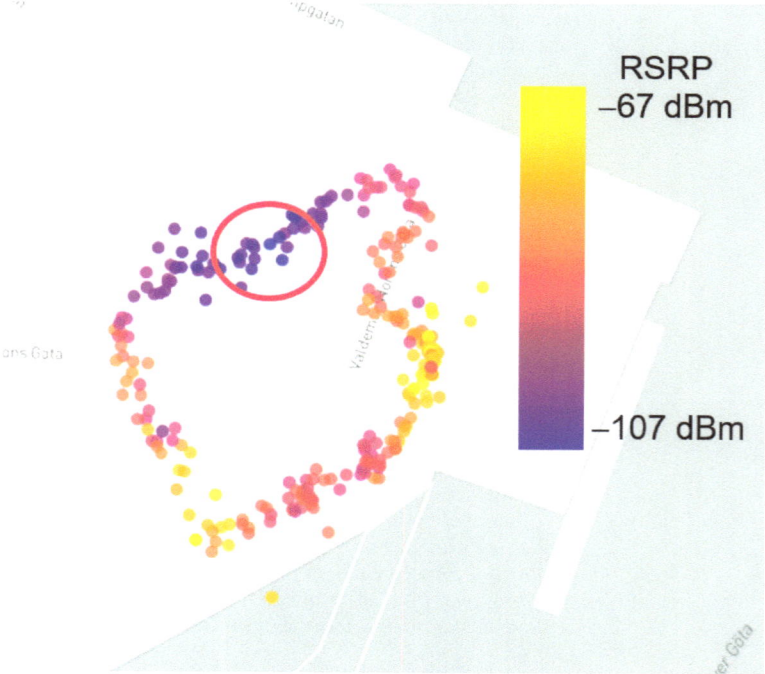

Figure 8. Reference Signal Received Power (RSRP) primary cells. Points of interest are circled in red.

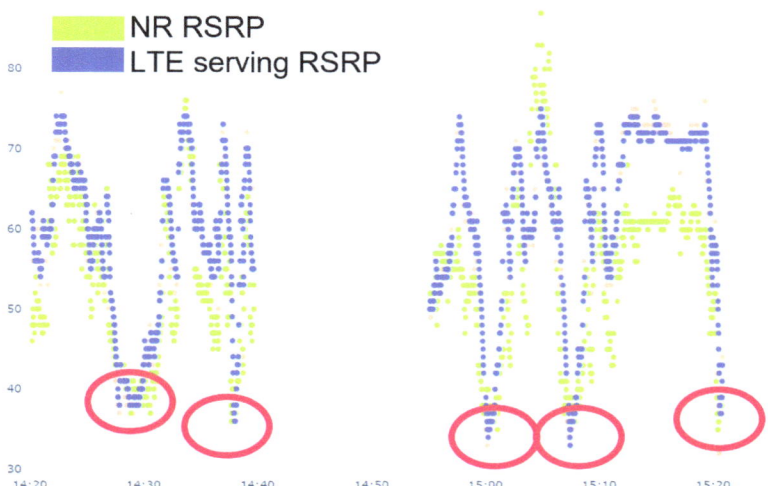

Figure 9. NR and LTE Reference Signal Received Power (RSRP) over time. Points of interest are circled in red.

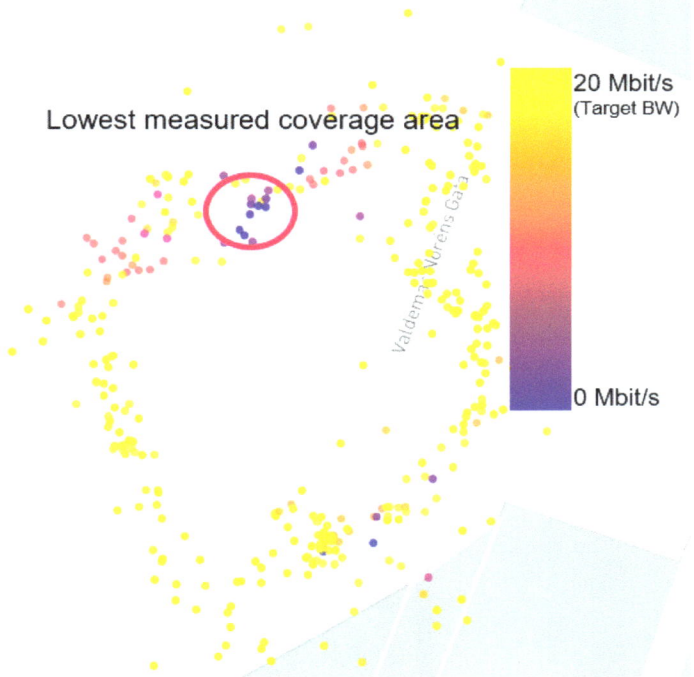

Figure 10. Measured bandwidth in the demonstration area. Points of interest are circled in red.

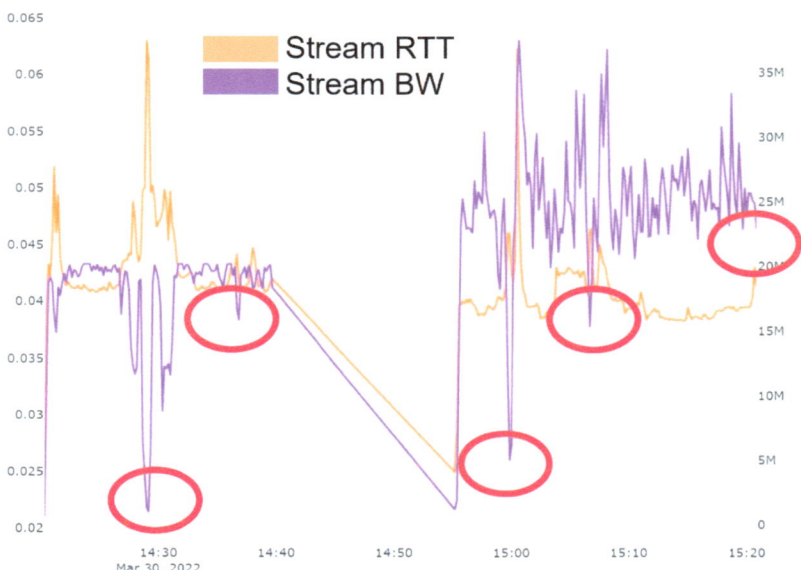

Figure 11. Stream round trip time (RTT) and stream bandwidth (BW) over time. Points of interest are circled in red.

In an unloaded network, the latency remains consistent at a specific location. The latency measured by the tool reflects the delay in the transmitted data stream. A significant relationship exists between traffic load and latency, as an increased load results in network queues. The concept of real-time bandwidth aims to maximise the bandwidth while preserving low latency.

The measurement tool employs Ericsson's SCReAM algorithm [44], a mobile-optimised congestion-control algorithm. SCReAM dynamically adjusts the bandwidth based on various metrics, including the round trip time (RTT). As depicted in Figure 11, SCReAM responds by reducing the bandwidth when the RTT increases, effectively minimising the latency. Therefore, the RT BW refers to the data delivered within a reasonably bounded RTT delay. Both the bandwidth and throughput serve as indicators of the network performance. While the bandwidth indicates the available or predicted network capacity, the throughput represents the transmitted data. Given the susceptibility of the intended networks to congestion, mainly as they are not private, the throughput is a more pertinent measurement in this context.

Accurately predicting handover issues between cellular network cells is vital for coverage testing. Assessment criteria such as the signal strength, quality, and latency are used to identify potential challenges during the handover procedure, as shown between cell one and cell two in Figure 12. Through scenario simulations and a network-performance analysis, operators can improve handover algorithms and configurations to maintain connectivity. Conducting a specialised ODD assessment is necessary to validate and assess the results.

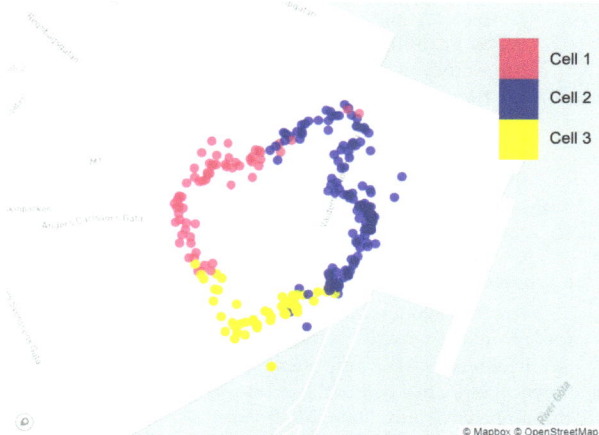

Figure 12. Three cells are involved in the coverage.

Using an assessment template, as shown in Table 4, improves the efficiency of the site assessments for ADSs. This template, which links predefined performance indicators with test scenarios, provides a structured framework that reduces the effort required. It establishes a starting point for developing a more-detailed and customised assessment strategy, as Table 5 exemplifies. The advantage of this approach lies in its focus on test scenarios closely connected to KET and ODD dependencies, especially concerning connectivity and positioning. These scenarios form a pertinent baseline suite to test whether the conditions for activating and maintaining the ADS features throughout the ODD are met. Compared to approaches without such templates, this method offers a more organised and comprehensive way to conduct tests. It ensures that pertinent scenarios and performance indicators are considered, which is crucial to accurately assess an ADS's capabilities and limitations.

Table 5. Excerpt of application of assessment template for 5G communication at Lindholmen.

KET	Test Scenario	Attribute	Criteria Description	Evaluation
⋮	⋮	⋮	⋮	⋮
5G communication	Activation-condition scenario	Coverage	Coverage is present in the whole Lindholmen ODD. Coverage is achieved by several cells. Handover must not affect throughput.	☐ Pass ☑ Fail
5G communication	Activation-condition scenario	Latency	Here, latency is assessed to be subsumed by 5G coverage and validated by video-performance tests.	☐ Pass ☑ Fail
5G communication	Activation-condition scenario	Throughput	Target bandwidth: 20 Mbit/s. Unsafe below 1 MBit/s or 15 frames per second.	☐ Pass ☑ Fail
5G communication	Activation-condition scenario	Predictability	Deployment-site test measurements and Quality of Service (QoS).	☑ Pass ☐ Fail
⋮	⋮	⋮	⋮	⋮

Using KET-assessment templates makes the assessment process a systematic exercise in evaluating the ADS's functionality and performance within the defined ODD. The perceived efficiency gain can be attributed to:

1. Structured testing: ensures the comprehensive coverage of essential test scenarios and indicators, reducing the risk of missing critical evaluation aspects.

2. Consistency and comparability: provides a uniform framework for assessing different ADSs, enabling consistent and fair comparisons.
3. Time efficiency: saves time by offering KET-relevant indicators with a predefined set of criteria and scenarios, speeding up the assessment process.
4. Customise framework: allows adjustments to fit specific ADS features or testing environments, maintaining relevance across various assessments.

In summary, the KET-assessment templates aid in a more-efficient and complete evaluation process, aligning the assessment with the specific requirements of the tested ADS. This contributes to better-informed decision making and safer Automated Driving Systems.

6. Conclusions

In conclusion, while notable strides have been taken in safety-assessment strategies for automated vehicles, certain limitations linked to practical assessment endeavours still require attention. The proposed approach underscores the significance of technology-aware practical guidance within the assessment process, which should seamlessly integrate into a comprehensive and adaptable framework.

The primary contribution of this study lies in proposing the augmentation of existing scenario-based testing frameworks with a detailed examination of the underlying supporting technologies. The approach enriches the test suite employed in scenario-based testing by factoring in the specific attributes of test scenarios linked to the Key Enabling Technologies (KETs). By blending bottom-up analysis with top-down scrutiny focused on potentially hazardous traffic scenarios at the vehicle level, a more comprehensive understanding of the system's performance can be achieved.

While the method outlined in this study demonstrates practicality and efficacy, certain areas warrant further exploration. Subsequent research should investigate the extent of coverage and completeness when mapping the ODD to precise technological elements in specification and testing and address the limitation posed by relying on predefined scenarios. One notable weakness of KET-related predefined test scenarios is their reliance on historical information, which may not adequately account for the expected novelties in KETs. As such, our method is designed to complement data-driven approaches that incorporate real-world data to create a more-comprehensive assessment. To overcome the potential limitations imposed by rigid, predefined templates, it is necessary to continuously develop, update, and refine the templates to remain aligned with ongoing technology developments—a challenge encountered in all checklist-based approaches. We highly recommend combining knowledge-driven and data-driven approaches in future safety-assurance-framework endeavours. This harmonious blend can enrich the assessment framework by capitalising on existing knowledge and real-world data. Its relevance is especially pronounced in situations where substantial real-world data are scarce. To facilitate the seamless integration of these approaches, we propose adopting a policy that underscores the importance of integrating prior knowledge into the assessment processes and any scenario databases. Such a policy can be a stepping stone for accommodating evolving challenges and fostering a comprehensive safety-assurance approach.

Therefore, developing technology-aware assessment criteria for attributes derived from enabling technologies is important. These criteria should complement the overarching high-level requirements and encompass the fundamental behavioural facets of the automated function within the defined ODD. This involves appraising the functionality of sensors and communication devices, adherence to protocols and standards, and the effective mitigation of potential cybersecurity threats. By assimilating technology-aware assessment criteria, a more-comprehensive evaluation of the automated function's performance can be achieved.

Author Contributions: Conceptualisation and methodology, M.S.; validation measurements, M.B.; investigation, M.S., A.T. and F.W.; writing—original draft preparation, M.S., A.T. and F.W.; writing—review and editing, M.S., A.T. and F.W.; supervision, F.W. All authors have read and agreed to the published version of the manuscript.

Funding: The SUNRISE project is funded by the European Union's Horizon Europe Research and Innovation Actions under grant agreement no.101069573. The views and opinions expressed are, however, those of the author(s) only and do not necessarily reflect those of the European Union or European Union's Horizon Europe Research and Innovation Actions. The SCAT project (2020-04205) has received funding from Vinnova, Sweden's innovation agency. The content reflects only the authors' views. Sweden's innovation agency is not responsible for any use that may be made of the information it contains.

Conflicts of Interest: Author Mats Bergman was employed by the company Telia Company. The remaining authors declare that the research was conducted in the absence of any commercial or financial relationships that could be construed as a potential conflict of interest.

References

1. J3016 APRIL2021; SAE J3016-Taxonomy and Definitions for Terms Related to Driving Automation Systems for On-Road Motor Vehicles. Surface Vehicle Recommended Practice J3016 APRIL2021; SAE International: Warrendale, PA, USA, 2021.
2. ECE/TRANS/WP.29/2020/81. UN Regulation No 157—Uniform Provisions Concerning the Approval of Vehicles with Regards to Automated Lane Keeping Systems [2021/389]. 2021. Available online: https://op.europa.eu/s/y6kz (accessed on 1 December 2023).
3. WP.29/GRVA. Current Draft of the Guidelines and Recommendations Concerning Safety Requirements for ADS (FRAV). 2022. Available online: https://unece.org/transport/documents/2022/05/informal-documents/frav-current-draft-guidelines-and-recommendations (accessed on 1 December 2023).
4. Chan, C.Y. Advancements, Prospects, and Impacts of Automated Driving Systems. *Int. J. Transp. Sci. Technol.* **2017**, *6*, 208–216. [CrossRef]
5. Chen, C.; Zhao, Q.; Zheng, T.; Zhai, Y.; Zhu, X. The Research on Current Automated Driving ODD Regulations, Standards and Applications. In Proceedings of the 2022 IEEE International Conference on Real-Time Computing and Robotics (RCAR), Guiyang, China, 17–22 July 2022; pp. 744–747. [CrossRef]
6. Gyllenhammar, M.; Johansson, R.; Warg, F.; Chen, D.; Heyn, H.M.; Sanfridson, M.; Söderberg, J.; Thorsén, A.; Ursing, S. Towards an Operational Design Domain That Supports the Safety Argumentation of an Automated Driving System. In Proceedings of the 10th European Congress on Embedded Real Time Systems (ERTS 2020), Toulouse, France, 29–31 January 2020.
7. PAS 1883:2020; Operational Design Domain (ODD) Taxonomy for an Automated Driving System (ADS)—Specification. Technical Report; BSIGROUP: London, UK, 2020.
8. ISO 34503:2023; Road Vehicles—Test Scenarios for Automated Driving Systems—Specification for Operational Design Domain. International Organization for Standardization: Geneva, Switzerland, 2023. Available online: https://www.iso.org/standard/78952.html (accessed on 1 December 2023).
9. Association for Standardization of Automation and Measuring Systems. ASAM OpenODD. Available online: https://www.asam.net/standards/detail/openodd/ (accessed on 1 December 2023).
10. Sunrise Project. *D3.1 Report on Baseline Analysis of Existing Methodology*; Technical Report; Sunrise Project: 2023. Available online: https://ccam-sunrise-project.eu/deliverable/d3-1-report-on-baseline-analysis-of-existing-methodology/ (accessed on 1 December 2023).
11. Ulbrich, S.; Menzel, T.; Reschka, A.; Schuldt, F.; Maurer, M. Defining and Substantiating the Terms Scene, Situation, and Scenario for Automated Driving. In Proceedings of the 2015 IEEE 18th International Conference on Intelligent Transportation Systems, Gran Canaria, Spain, 15–18 September 2015; pp. 982–988. [CrossRef]
12. Menzel, T.; Bagschik, G.; Maurer, M. Scenarios for Development, Test and Validation of Automated Vehicles. In Proceedings of the 2018 IEEE Intelligent Vehicles Symposium (IV), Changshu, China, 26–30 June 2018; pp. 1821–1827.
13. Koopman, P.; Wagner, M. Autonomous Vehicle Safety: An Interdisciplinary Challenge. *IEEE Intell. Transp. Syst. Mag.* **2017**, *9*, 90–96. [CrossRef]
14. Riedmaier, S.; Ponn, T.; Ludwig, D.; Schick, B.; Diermeyer, F. Survey on Scenario-Based Safety Assessment of Automated Vehicles. *IEEE Access* **2020**, *8*, 87456–87477. [CrossRef]
15. Underwriters Laboratories. *UL 4600: Standard for Evaluation of Autonomous Products*; Technical Report; Underwriters Laboratories: Northbrook, IL, USA, 2020.
16. Thorn, E.; Kimmel, S.C.; Chaka, M.; Virginia Tech Transportation Institute; Southwest Research Institute; Booz Allen Hamilton, Inc. *A Framework for Automated Driving System Testable Cases and Scenarios*; Technical Report DOT HS 812 623; National Highway Traffic Safety Administration: Washington, DC, USA, 2018.

17. 183rd WP.29. New Assessment/Test Method for Automated Driving (NATM) (Proposal). Informal Document WP.29-183-05, 183rd WP.29, 9–11 March 2021. Agenda items 2.3 and 3.5.5. 2021. Available online: https://unece.org/sites/default/files/2021-01/GRVA-09-07e.pdf (accessed on 1 December 2023).
18. Cieslik, I.; Expósito Jiménez, V.J.; Martin, H.; Scharke, H.; Schneider, H. State of the Art Study of the Safety Argumentation Frameworks for Automated Driving System. In *Computer Safety, Reliability, and Security. SAFECOMP 2022 Workshops, Proceedings of the International Conference on Computer Safety, Reliability, and Security, Munich, Germany, 6–9 June 2022*; Lecture Notes in Computer Science; Trapp, M., Schoitsch, E., Guiochet, J., Bitsch, F., Eds.; Springer International Publishing: Cham, Switzerland, 2022; pp. 178–191. [CrossRef]
19. Zhu, L.; Aurum, A.; Gorton, I.; Jeffery, R. Tradeoff and sensitivity analysis in software architecture evaluation using analytic hierarchy process. *Softw. Qual. J.* **2005**, *13*, 357–375. [CrossRef]
20. HEADSTART Project. Available online: https://www.headstart-project.eu/ (accessed on 1 December 2023).
21. Skoglund, M.; Thorsén, A.; Arrue, A.; Coget, J.B.; Plestan, C. Technical and Functional Requirements for V2X Communication, Positioning and Cyber-Security in the HEADSTART Project. In Proceedings of the ITS World Congress 2021, Hamburg, Germany, 11–15 October 2021.
22. Thorsén, A.; Skoglund, M.; Warg, F.; Jacobson, J.; Hult, R.; Wagener, N.; Ballis, A.; van de Sluis, J.; Perez, J.J.; Steccanella, A. HEADSTART D 1.3 Technical and Functional Requirements for KETs. HEADSTART Deliverable D 1.3 v2.0, HEADSTART Project. 2021. Available online: https://ec.europa.eu/research/participants/documents/downloadPublic?documentIds=080166e5ddc96d9f&appId=PPGMS (accessed on 1 December 2023).
23. Skoglund, M.; Hult, R.; Jacobson, J.; Jonasson, M.; Ballis, A.; Weissensteiner, P.; Coget, J.-B.; Otaegui, O.; Wiggerich, A.; Wagener, N.; et al. HEADSTART D 1.4 Functional Requirements of Selected Use Cases. HEADSTART Deliverable D 1.4, HEADSTART Project. 2019. Available online: https://ec.europa.eu/research/participants/documents/downloadPublic?documentIds=080166e5c833933d&appId=PPGMS (accessed on 1 December 2023).
24. Wagener, N. Common Methodology for Data-Driven Scenario-Based Safety Assurance in the HEADSTART Project. In Proceedings of the Virtual ITS European Congress, Virtual event, 9–10 November 2020. [CrossRef]
25. NGMN Alliance. *V2X Task Force: V2X White Paper*; Technical Report; NGMN Alliance: 2018. Available online: https://www.ngmn.org/publications/v2x-task-force-white-paper-v1-0.html (accessed on 1 December 2023).
26. Hasegawa, R.; Okamoto, E. Adaptive Transmission Suspension of V2N Uplink Communication Based on In-Advanced Quality of Service Notification. *Vehicles* **2023**, *5*, 203–222. [CrossRef]
27. *EN 16803-1:2020*; Space-Use of GNSS-Based Positioning for Road Intelligent Transport Systems (ITS)—Part 1: Definitions and System Engineering Procedures for the Establishment and Assessment of Performances. CEN/CENELEC: Brussels, Belgium, 2016.
28. Štern, A.; Kos, A. Positioning Performance Assessment of Geodetic, Automotive, and Smartphone GNSS Receivers in Standardized Road Scenarios. *IEEE Access* **2018**, *6*, 41410–41428. [CrossRef]
29. Radack, S. *Federal Information Processing Standard (Fips) 199, Standards for Security Categorization of Federal Information and Information Systems*; Technical Report; National Institute of Standards and Technology: Gaithersburg, MD, USA, 2004.
30. *SAE J3061*; Cybersecurity Guidebook for Cyber-Physical Automotive Systems. Technical Report; SAE International: Warrendale, PA, USA, 2016.
31. Lonc, B.; Cincilla, P. Cooperative ITS Security Framework: Standards and Implementations Progress in Europe. In Proceedings of the 2016 IEEE 17th International Symposium on a World of Wireless, Mobile and Multimedia Networks (WoWMoM), Coimbra, Portugal, 21–24 June 2016; pp. 1–6. [CrossRef]
32. Marojevic, V. C-V2X Security Requirements and Procedures: Survey and Research Directions. *arXiv* **2018**, arXiv:1807.09338. [CrossRef]
33. *ISO/SAE 21434:2021*; Road Vehicles—Cybersecurity Engineering. Technical Report; ISO/SAE: Geneva, Switzerland, 2021.
34. *ISO 26262:2018*; Road Vehicles: Functional Safety. Technical Report; International Organization for Standardization: Geneva, Switzerland, 2018.
35. *ISO/PAS 21448:2019*; Road Vehicles—Safety of the Intended Functionality. Technical Report; International Organization for Standardization: Geneva, Switzerland, 2019.
36. Firesmith, D.G. A Taxonomy of Security-Related Requirements. In Proceedings of the Fourth International Workshop on Requirements Engineering for High-Availability Systems (RHAS'05), Kyoto, Japan, 6 September 2005; p. 11.
37. de Gelder, E.; Paardekooper, J.P. Assessment of Automated Driving Systems Using Real-Life Scenarios. In Proceedings of the 2017 IEEE Intelligent Vehicles Symposium (IV), Los Angeles, CA, USA, 11–14 June 2017; pp. 589–594.
38. Roesener, C.; Sauerbier, J.; Zlocki, A.; Fahrenkrog, F.; Wang, L.; Várhelyi, A.; de Gelder, E.; Dufils, J.; Breunig, S.; Mejuto, P.; et al. A Comprehensive Evaluation Approach for Highly Automated Driving. In Proceedings of the 25th International Technical Conference on the Enhanced Safety of Vehicles (ESV) National Highway Traffic Safety Administration, Detroit, MI, USA, 5–8 June 2017.
39. Nalic, D.; Mihalj, T.; Bäumler, M.; Lehmann, M.; Eichberger, A.; Bernsteiner, S. Scenario Based Testing of Automated Driving Systems: A Literature Survey. In Proceedings of the FISITA Web Congress, Virtual event, 24 November 2020; Volume 10.
40. Düser, T.; Abdellatif, H.; Gutenkunst, C.; Gnandt, C. Approaches for the Homologation of Automated Driving. *ATZelectron. Worldw.* **2019**, *14*, 48–53. [CrossRef]

41. Warg, F.; Skoglund, M.; Sassman, M. Human Interaction Safety Analysis Method for Agreements with Connected Automated Vehicles. In Proceedings of the 2021 IEEE 94th Vehicular Technology Conference (VTC2021-Fall), Virtual event, 27–30 September 2021; pp. 1–7. [CrossRef]
42. Transportstyrelsen. *Transportstyrelsens Föreskrifter Och Allmänna råd om Tillstånd att Bedriva Försök Med Automatiserade Fordon*; Transportstyrelsen: Norrköping, Sweden, 2022.
43. Sobiech, C.; Berglund, P.; Bergman, M.; Johansson, V.; Lundahl, J.; Nylander, T.; Skoglund, M.; Strandberg, T. *Safety Case for Autonomous Trucks (SCAT)*; Technical Report; Research Institutes of Sweden: Göteborg, Sweden, 2023.
44. SCReAM. 2023. Available online: https://github.com/EricssonResearch/scream/ (accessed on 1 December 2023).

Disclaimer/Publisher's Note: The statements, opinions and data contained in all publications are solely those of the individual author(s) and contributor(s) and not of MDPI and/or the editor(s). MDPI and/or the editor(s) disclaim responsibility for any injury to people or property resulting from any ideas, methods, instructions or products referred to in the content.

MDPI AG
Grosspeteranlage 5
4052 Basel
Switzerland
Tel.: +41 61 683 77 34

Vehicles Editorial Office
E-mail: vehicles@mdpi.com
www.mdpi.com/journal/vehicles

Disclaimer/Publisher's Note: The title and front matter of this reprint are at the discretion of the Guest Editors. The publisher is not responsible for their content or any associated concerns. The statements, opinions and data contained in all individual articles are solely those of the individual Editors and contributors and not of MDPI. MDPI disclaims responsibility for any injury to people or property resulting from any ideas, methods, instructions or products referred to in the content.

www.ingramcontent.com/pod-product-compliance
Lightning Source LLC
LaVergne TN
LVHW072343090526
838202LV00019B/2468